A Course in Game Theory

A Course in Game Theory

Martin J. Osborne
Ariel Rubinstein

The MIT Press
Cambridge, Massachusetts
London, England

This book was typeset by the authors, who are greatly indebted to Donald Knuth (the creator of TeX), Leslie Lamport (the creator of LaTeX), and Eberhard Mattes (the creator of emTeX) for generously putting superlative software in the public domain. Camera-ready copy was produced by Type 2000, Mill Valley, California, and the book was printed and bound by The Maple-Vail Book Manufacturing Group, Binghamton, New York.

Osborne, Martin J.
    A course in game theory/Martin J. Osborne, Ariel Rubinstein.
        p.   cm.
    Includes bibliographical references and index.
    ISBN 0-262-15041-7.—ISBN 0-262-65040-1 (pbk.)
    1. Game Theory.   I. Rubinstein, Ariel.   II. Title.
HB144.O733 1994
658.4'0353–dc20   94-8308
                                                            CIP

Eighth printing, 2002

Contents

III   Extensive Games with Imperfect Information   197

IV   Coalitional Games   255

Preface

This book presents some of the main ideas of game theory. It is designed to serve as a textbook for a one-semester graduate course consisting of about 28 meetings each of 90 minutes.

The topics that we cover are those that we personally would include in such a one-semester course. We do not pretend to provide a complete reference book on game theory and do not necessarily regard the topics that we exclude as unimportant. Our selection inevitably reflects our own preferences and interests. (Were we to start writing the book now we would probably add two chapters, one on experimental game theory and one on learning and evolution.)

We emphasize the foundations of the theory and the interpretation of the main concepts. Our style is to give precise definitions and full proofs of results, sacrificing generality and limiting the scope of the material when necessary to most easily achieve these goals.

We have made a serious effort to give credit for all the concepts, results, examples, and exercises (see the "Notes" at the end of each chapter). We regret any errors and encourage you to draw our attention to them.

Structure of the Book

The book consists of four parts; in each part we study a group of related models. The chart on the next page summarizes the interactions among the chapters. A basic course could consist of Chapters 2, 3, 6, 11, 12, and 13.

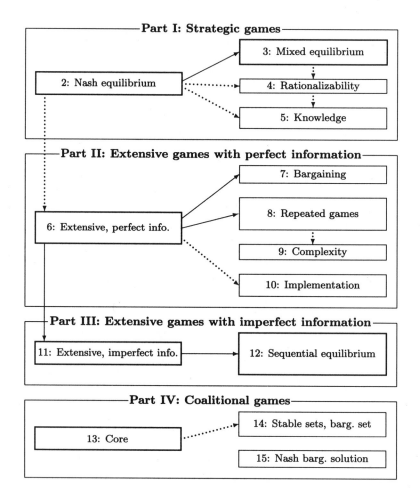

The main interactions between the chapters. The areas of the boxes in which the names of the chapters appear are proportional to the lengths of the chapters. A solid arrow connecting two boxes indicates that one chapter depends on the other; a dotted arrow indicates that only the main ideas of one chapter are used in the other. A basic course could consist of the six chapters in heavy boxes.

Exercises

Many of the exercises are challenging; we often use exercises to state subsidiary results. Instructors will probably want to assign additional straightforward problems and adjust (by giving hints) the level of our exercises to make it appropriate for their students. Solutions are available to instructors on the web site for the book (see page xv).

Disagreements Between the Authors

We see no reason why a jointly authored book should reflect a uniform view. At several points, as in the following note, we briefly discuss issues about which we disagree.

A Note on Personal Pronouns

We disagree about how to handle English third-person singular pronouns.

AR argues that we should use a "neutral" pronoun and agrees to the use of "he", with the understanding that this refers to both men and women. Continuous reminders of the he/she issue simply divert the reader's attention from the main issues. Language is extremely important in shaping our thinking, but in academic material it is not useful to wave it as a flag, as is common in some circles.

MJO argues that no language is "neutral". In particular, there is a wealth of evidence, both from experiments and from analyses of language use, that "he" is not generally perceived to encompass both females and males. To quote the *American Heritage Dictionary* (third edition, page 831), "Thus *he* is not really a gender-neutral pronoun; rather it refers to a male who is to be taken as the representative member of the group referred to by its antecedent. The traditional usage, then, is not simply a grammatical convention; it also suggests a particular pattern of thought." Further, the use of "he" to refer to an individual of unspecified sex did not even arise naturally, but was imposed as a rule by (male) prescriptive grammarians in the eighteenth and nineteenth centuries who were upset by the widespread use of "they" as a singular pronoun and decided that, since in their opinion men were more important than women, "he" should be used. The use of "he" to refer to a generic individual thus both has its origins in sexist attitudes and promotes such attitudes. There is no neat solution to the problem, especially in a book such as this in which there are so many references to generic individuals. "They" has many merits as a singular pronoun, although its use can lead to ambiguities (and complaints from editors). My preference is to use "she" for all individuals. Obviously this usage is not gender-neutral, but its use for a few decades, after a couple of centuries in which "he" has dominated, seems likely only to help to eliminate sexist ways of thought. If such usage diverts some readers' attentions from the subjects discussed in this book and leads them to contemplate sexism in the use of language, which is surely an issue at least as significant as the minutiae of sequential equilibrium, then an increase in social welfare will have been achieved. (Whether or not this book qualifies as "academic material", I see no reason why its readers should be treated differently from those of any other material.)

To conclude, we both feel strongly on this issue; we both regard the compromise that we have reached as highly unsatisfactory. When referring to specific individuals, we sometimes use "he" and sometimes "she". For example, in two-player games we treat player 1 as female and player 2 as male. We use "he" for generic individuals.

Acknowledgements

This book is an outgrowth of courses we have taught and discussions we have had with many friends and colleagues. Some of the material in Chapters 5, 8, and 9 is based on parts of a draft of a book on models of bounded rationality by AR.

MJO I had the privilege of being educated in game theory by Robert Aumann, Sergiu Hart, Mordecai Kurz, and Robert Wilson at Stanford University. It is a great pleasure to acknowledge my debt to them. Discussions over the years with Jean-Pierre Benoît, Haruo Imai, Vijay Krishna, and Carolyn Pitchik have improved my understanding of many topics. I completed my work on the book during a visit to the Department of Economics at the University of Canterbury, New Zealand; I am grateful to the members of the department for their generous hospitality. I am grateful also to the Social Science and Humanities Research Council of Canada and the Natural Sciences and Engineering Research Council of Canada for financially supporting my research in game theory over the last six years.

AR I have used parts of this book in courses at the London School of Economics (1987 and 1988), The Hebrew University (1989), Tel Aviv University (1990), and Princeton University (1992). The hospitality and collegiality of the London School of Economics, Princeton University, and Tel Aviv University are gratefully appreciated. Special thanks are due to my friend Asher Wolinsky for endless illuminating conversations. Part of my work on the book was supported by the United States–Israel Binational Science Foundation (grant number 1011-341).

We are grateful to Pierpaolo Battigalli, Takako Fujiwara, Wulong Gu, Abhinay Muthoo, Michele Piccione, and Doron Sonsino for making very detailed comments on drafts of the book, comments that led us to substantially improve the accuracy and readability of the text. We are grateful also to Dilip Abreu, Jean-Pierre Benoît, Larry Blume, In-Koo Cho, Eddie Dekel, Faruk Gul, Vijay Krishna, Bart Lipman, Bentley MacLeod, Sylvain Sorin, Ran Spiegler, and Arthur Sweetman for giving us advice and pointing out improvements. Finally, we thank Wulong Gu and Arthur Sweetman, who provided us with outstanding assistance in completing the book: Wulong worked on the exercises, correcting our solutions and providing many of his own, and Arthur constructed the index.

On the technical side we thank Ed Sznyter for cajoling the ever recalcitrant TEX to execute our numbering scheme.

It was a great pleasure to deal with Terry Vaughn of The MIT Press; his encouragement in the early stages of the project was important in motivating us to write the book.

MARTIN J. OSBORNE
martin.osborne@utoronto.ca
http://www.economics.utoronto.ca/osborne/
Department of Economics, University of Toronto
150 St. George Street, Toronto, Canada, M5S 3G7

ARIEL RUBINSTEIN
rariel@ccsg.tau.ac.il
http://www.princeton.edu/~ariel/
Department of Economics, Tel Aviv University
Tel Aviv, Israel, 69978
Department of Economics, Princeton University
Princeton, NJ 08540, USA

We maintain a web site for the book. A link to this site is provided on The MIT Press page for the book,

   http://mitpress.mit.edu/book-home.tcl?isbn=0262650401

The URL of our site is currently

        http://www.economics.utoronto.ca/osborne/cgt

---

1 Introduction

1.1 Game Theory

Game theory is a bag of analytical tools designed to help us understand the phenomena that we observe when decision-makers interact. The basic assumptions that underlie the theory are that decision-makers pursue well-defined exogenous objectives (they are *rational*) and take into account their knowledge or expectations of *other* decision-makers' behavior (they *reason strategically*).

The models of game theory are highly abstract representations of classes of real-life situations. Their abstractness allows them to be used to study a wide range of phenomena. For example, the theory of Nash equilibrium (Chapter 2) has been used to study oligopolistic and political competition. The theory of mixed strategy equilibrium (Chapter 3) has been used to explain the distributions of tongue length in bees and tube length in flowers. The theory of repeated games (Chapter 8) has been used to illuminate social phenomena like threats and promises. The theory of the core (Chapter 13) reveals a sense in which the outcome of trading under a price system is stable in an economy that contains many agents.

The boundary between pure and applied game theory is vague; some developments in the pure theory were motivated by issues that arose in applications. Nevertheless we believe that such a line can be drawn. Though we hope that this book appeals to those who are interested in applications, we stay almost entirely in the territory of "pure" theory. The art of applying an abstract model to a real-life situation should be the subject of another tome.

Game theory uses mathematics to express its ideas formally. However, the game theoretical ideas that we discuss are not *inherently* mathematical; in principle a book could be written that had essentially the same content as this one and was devoid of mathematics. A mathematical formulation makes it easy to define concepts precisely, to verify the consistency of ideas, and to explore the implications of assumptions. Consequently our style is formal: we state definitions and results precisely, interspersing them with motivations and interpretations of the concepts.

The use of mathematical models creates independent mathematical interest. In this book, however, we treat game theory not as a branch of mathematics but as a social science whose aim is to understand the behavior of interacting decision-makers; we do not elaborate on points of mathematical interest. From our point of view the mathematical results are interesting only if they are confirmed by intuition.

1.2   Games and Solutions

A game is a description of strategic interaction that includes the constraints on the actions that the players *can* take and the players' interests, but does not specify the actions that the players *do* take. A *solution* is a systematic description of the outcomes that may emerge in a family of games. Game theory suggests reasonable solutions for classes of games and examines their properties.

We study four groups of game theoretic models, indicated by the titles of the four parts of the book: strategic games (Part I), extensive games with and without perfect information (Parts II and III), and coalitional games (Part IV). We now explain some of the dimensions on which this division is based.

Noncooperative and Cooperative Games

In all game theoretic models the basic entity is a *player*. A player may be interpreted as an individual or as a group of individuals making a decision. Once we define the set of players, we may distinguish between two types of models: those in which the sets of possible actions of *individual* players are primitives (Parts I, II, and III) and those in which the sets of possible joint actions of *groups* of players are primitives (Part IV). Sometimes models of the first type are referred to as "noncooperative", while those of the second type are referred to as "cooperative" (though these terms do not express well the differences between the models).

The numbers of pages that we devote to each of these branches of the theory reflect the fact that in recent years most research has been devoted to noncooperative games; it does not express our evaluation of the relative importance of the two branches. In particular, we do not share the view of some authors that noncooperative models are more "basic" than cooperative ones; in our opinion, neither group of models is more "basic" than the other.

Strategic Games and Extensive Games

In Part I we discuss the concept of a strategic game and in Parts II and III the concept of an extensive game. A strategic game is a model of a situation in which each player chooses his plan of action once and for all, and all players' decisions are made simultaneously (that is, when choosing a plan of action each player is not informed of the plan of action chosen by any other player). By contrast, the model of an extensive game specifies the possible orders of events; each player can consider his plan of action not only at the beginning of the game but also whenever he has to make a decision.

Games with Perfect and Imperfect Information

The third distinction that we make is between the models in Parts II and III. In the models in Part II the participants are fully informed about each others' moves, while in the models in Part III they may be imperfectly informed. The former models have firmer foundations. The latter were developed intensively only in the 1980s; we put less emphasis on them not because they are less realistic or important but because they are less mature.

1.3 Game Theory and the Theory of Competitive Equilibrium

To clarify further the nature of game theory, we now contrast it with the theory of competitive equilibrium that is used in economics. Game theoretic reasoning takes into account the attempts by each decision-maker to obtain, prior to making his decision, information about the other players' behavior, while competitive reasoning assumes that each agent is interested only in some environmental parameters (such as prices), even though these parameters are determined by the actions of all agents.

To illustrate the difference between the theories, consider an environment in which the level of some activity (like fishing) of each agent depends on the level of pollution, which in turn depends on the levels of

the agents' activities. In a competitive analysis of this situation we look
for a level of pollution consistent with the actions that the agents take
when each of them regards this level as given. By contrast, in a game
theoretic analysis of the situation we require that each agent's action
be optimal given the agent's expectation of the pollution created by the
combination of his action and all the other agents' actions.

1.4   Rational Behavior

The models we study assume that each decision-maker is "rational" in
the sense that he is aware of his alternatives, forms expectations about
any unknowns, has clear preferences, and chooses his action deliberately
after some process of optimization. In the absence of uncertainty the
following elements constitute a model of rational choice.

- A set A of *actions* from which the decision-maker makes a choice.
- A set C of possible *consequences* of these actions.
- A *consequence function* $g: A \to C$ that associates a consequence
  with each action.
- A *preference relation* (a complete transitive reflexive binary rela-
  tion) $\succsim$ on the set C.

Sometimes the decision-maker's preferences are specified by giving a
utility function $U: C \to \mathbb{R}$, which defines a preference relation $\succsim$ by the
condition $x \succsim y$ if and only if $U(x) \geq U(y)$.

Given any set $B \subseteq A$ of actions that are feasible in some particular
case, a *rational decision-maker* chooses an action a^* that is feasible
(belongs to B) and optimal in the sense that $g(a^*) \succsim g(a)$ for all $a \in B$;
alternatively he solves the problem $\max_{a \in B} U(g(a))$. An assumption
upon which the usefulness of this model of decision-making depends is
that the individual uses the same preference relation when choosing from
different sets B.

In the models we study, individuals often have to make decisions under
conditions of uncertainty. The players may be

- uncertain about the objective parameters of the environment
- imperfectly informed about events that happen in the game
- uncertain about actions of the other players that are not deterministic
- uncertain about the reasoning of the other players.

To model decision-making under uncertainty, almost all game theory uses the theories of von Neumann and Morgenstern (1944) and of Savage (1972). That is, if the consequence function is stochastic and known to the decision-maker (i.e. for each $a \in A$ the consequence $g(a)$ is a lottery (probability distribution) on C) then the decision-maker is assumed to behave as if he maximizes the expected value of a (*von Neumann–Morgenstern utility*) function that attaches a number to each consequence. If the stochastic connection between actions and consequences is not given, the decision-maker is assumed to behave as if he has in mind a (subjective) probability distribution that determines the consequence of any action. In this case the decision-maker is assumed to behave as if he has in mind a "state space" Ω, a probability measure over Ω, a function $g: A \times \Omega \to C$, and a utility function $u: C \to \mathbb{R}$; he is assumed to choose an action a that maximizes the expected value of $u(g(a, \omega))$ with respect to the probability measure.

We do not discuss the assumptions that underlie the theory of a rational decision-maker. However, we do point out that these assumptions are under perpetual attack by experimental psychologists, who constantly point out severe limits to its application.

1.5 The Steady State and Deductive Interpretations

There are two conflicting interpretations of solutions for strategic and extensive games. The *steady state* (or, as Binmore (1987/88) calls it, evolutive) interpretation is closely related to that which is standard in economics. Game theory, like other sciences, deals with regularities. As Carnap (1966, p. 3) writes, "The observations we make in everyday life as well as the more systematic observations of science reveal certain repetitions or regularities in the world. ... The laws of science are nothing more than statements expressing these regularities as precisely as possible." The steady state interpretation treats a game as a model designed to explain some regularity observed in a family of similar situations. Each participant "knows" the equilibrium and tests the optimality of his behavior given this knowledge, which he has acquired from his long experience. The *deductive* (or, as Binmore calls it, eductive) interpretation, by contrast, treats a game in isolation, as a "one-shot" event, and attempts to infer the restrictions that rationality imposes on the outcome; it assumes that each player deduces how the other players will behave simply from principles of rationality. We try to avoid the confusion between the two interpretations that frequently arises in game theory.

1.6   Bounded Rationality

When we talk in real life about games we often focus on the asymmetry between individuals in their abilities. For example, some players may have a clearer perception of a situation or have a greater ability to analyze it. These differences, which are so critical in life, are missing from game theory in its current form.

To illustrate the consequences of this fact, consider the game of chess. In an actual play of chess the players may differ in their knowledge of the legal moves and in their analytical abilities. In contrast, when chess is modeled using current game theory it is assumed that the players' knowledge of the rules of the game is perfect and their ability to analyze it is ideal. Results we prove in Chapters 2 and 6 (Propositions 22.2 and 99.2) imply that chess is a trivial game for "rational" players: an algorithm exists that can be used to "solve" the game. This algorithm defines a pair of strategies, one for each player, that leads to an "equilibrium" outcome with the property that a player who follows his strategy can be sure that the outcome will be at least as good as the equilibrium outcome no matter what strategy the other player uses. The existence of such strategies suggests that chess is uninteresting because it has only one possible outcome. Nevertheless, chess remains a very popular and interesting game. Its equilibrium outcome is yet to be calculated; currently it is impossible to do so using the algorithm. Even if White, for example, is shown one day to have a winning strategy, it may not be possible for a human being to implement that strategy. Thus while the abstract model of chess allows us to deduce a significant fact about the game, at the same time it omits the most important determinant of the outcome of an actual play of chess: the players' "abilities".

Modeling asymmetries in abilities and in perceptions of a situation by different players is a fascinating challenge for future research, which models of "bounded rationality" have begun to tackle.

1.7   Terminology and Notation

We presume little familiarity with mathematical results, but throughout use deductive reasoning. Our notation and mathematical definitions are standard, but to avoid ambiguities we list some of them here.

We denote the set of real numbers by $\mathbb{R}$, the set of nonnegative real numbers by $\mathbb{R}_+$, the set of vectors of n real numbers by $\mathbb{R}^n$, and the set of vectors of n nonnegative real numbers by $\mathbb{R}^n_+$. For $x \in \mathbb{R}^n$ and $y \in \mathbb{R}^n$ we use $x \geq y$ to mean $x_i \geq y_i$ for $i = 1, \ldots, n$ and $x > y$ to mean $x_i > y_i$ for $i = 1, \ldots, n$. We say that a function $f : \mathbb{R} \to \mathbb{R}$ is *increasing* if $f(x) > f(y)$ whenever $x > y$ and is *nondecreasing* if $f(x) \geq f(y)$ whenever $x > y$. A function $f : \mathbb{R} \to \mathbb{R}$ is *concave* if $f(\alpha x + (1 - \alpha)x') \geq \alpha f(x) + (1 - \alpha)f(x')$ for all $x \in \mathbb{R}$, all $x' \in \mathbb{R}$, and all $\alpha \in [0, 1]$. Given a function $f : X \to \mathbb{R}$ we denote by $\arg\max_{x \in X} f(x)$ the set of maximizers of f; for any $Y \subseteq X$ we denote by $f(Y)$ the set $\{f(x) : x \in Y\}$.

Throughout we use N to denote the set of players. We refer to a collection of values of some variable, one for each player, as a *profile*; we denote such a profile by $(x_i)_{i \in N}$, or, if the qualifier "$i \in N$" is clear, simply (x_i). For any profile $x = (x_j)_{j \in N}$ and any $i \in N$ we let x_{-i} be the list $(x_j)_{j \in N \setminus \{i\}}$ of elements of the profile x for all players except i. Given a list $x_{-i} = (x_j)_{j \in N \setminus \{i\}}$ and an element x_i we denote by (x_{-i}, x_i) the profile $(x_i)_{i \in N}$. If X_i is a set for each $i \in N$ then we denote by X_{-i} the set $\times_{j \in N \setminus \{i\}} X_j$.

A binary relation $\succsim$ on a set A is complete if $a \succsim b$ or $b \succsim a$ for every $a \in A$ and $b \in A$, reflexive if $a \succsim a$ for every $a \in A$, and transitive if $a \succsim c$ whenever $a \succsim b$ and $b \succsim c$. A *preference relation* is a complete reflexive transitive binary relation. If $a \succsim b$ but not $b \succsim a$ then we write $a \succ b$; if $a \succsim b$ and $b \succsim a$ then we write $a \sim b$. A preference relation $\succsim$ on A is *continuous* if $a \succsim b$ whenever there are sequences $(a^k)_k$ and $(b^k)_k$ in A that converge to a and b respectively for which $a^k \succsim b^k$ for all k. A preference relation $\succsim$ on $\mathbb{R}^n$ is *quasi-concave* if for every $b \in \mathbb{R}^n$ the set $\{a \in \mathbb{R}^n : a \succsim b\}$ is convex; it is *strictly quasi-concave* if every such set is strictly convex.

Let X be a set. We denote by $|X|$ the number of members of X. A *partition* of X is a collection of disjoint subsets of X whose union is X. Let N be a finite set and let $X \subseteq \mathbb{R}^N$ be a set. Then $x \in X$ is *Pareto efficient* if there is no $y \in X$ for which $y_i > x_i$ for all $i \in N$; $x \in X$ is *strongly Pareto efficient* if there is no $y \in X$ for which $y_i \geq x_i$ for all $i \in N$ and $y_i > x_i$ for some $i \in N$.

A *probability measure* μ on a finite (or countable) set X is an additive function that associates a nonnegative real number with every subset of X (that is, $\mu(B \cup C) = \mu(B) + \mu(C)$ whenever B and C are disjoint) and satisfies $\mu(X) = 1$. In some cases we work with probability measures over spaces that are not necessarily finite. If you are unfamiliar with such measures, little is lost by restricting attention to the finite case; for a definition of more general measures see, for example, Chung (1974, Ch. 2).

Notes

Von Neumann and Morgenstern (1944) is the classic work in game theory. Luce and Raiffa (1957) is an early textbook; although now out-of-date, it contains superb discussions of the basic concepts in the theory. Schelling (1960) provides a verbal discussion of some of the main ideas of the theory.

A number of recent books cover much of the material in this book, at approximately the same level: Shubik (1982), Moulin (1986), Friedman (1990), Kreps (1990a, Part III), Fudenberg and Tirole (1991a), Myerson (1991), van Damme (1991), and Binmore (1992). Gibbons (1992) is a more elementary introduction to the subject.

Aumann (1985b) contains a discussion of the aims and achievements of game theory, and Aumann (1987b) is an account of game theory from a historical perspective. Binmore (1987/88) is a critical discussion of game theory that makes the distinction between the steady state and deductive interpretations. Kreps (1990b) is a reflective discussion of many issues in game theory.

For an exposition of the theory of rational choice see Kreps (1988).

I   Strategic Games

In this part we study a model of strategic interaction known as a *strategic game*, or, in the terminology of von Neumann and Morgenstern (1944), a "game in normal form". This model specifies for each player a set of possible actions and a preference ordering over the set of possible action profiles.

In Chapter 2 we discuss Nash equilibrium, the most widely used solution concept for strategic games. In Chapter 3 we consider the closely related solutions of mixed strategy equilibrium and correlated equilibrium, in which the players' actions are not necessarily deterministic. Nash equilibrium is a steady state solution concept in which each player's decision depends on knowledge of the equilibrium. In Chapter 4 we study the deductive solution concepts of rationalizability and iterated elimination of dominated actions, in which the players are not assumed to know the equilibrium. Chapter 5 describes a model of knowledge that allows us to examine formally the assumptions that underlie the solutions that we have defined.

2 Nash Equilibrium

Nash equilibrium is one of the most basic concepts in game theory. In this chapter we describe it in the context of a strategic game and in the related context of a Bayesian game.

2.1 Strategic Games

2.1.1 Definition

A strategic game is a model of interactive decision-making in which each decision-maker chooses his plan of action once and for all, and these choices are made simultaneously. The model consists of a finite set N of *players* and, for each player i, a set A_i of *actions* and a *preference relation* on the set of action profiles. We refer to an action profile $a = (a_j)_{j \in N}$ as an *outcome*, and denote the set $\times_{j \in N} A_j$ of outcomes by A. The requirement that the preferences of each player i be defined over A, rather than A_i, is the feature that distinguishes a strategic game from a decision problem: each player may care not only about his own action but also about the actions taken by the other players. To summarize, our definition is the following.

▶ DEFINITION 11.1 A **strategic game** consists of

- a finite set N (the set of **players**)
- for each player $i \in N$ a nonempty set A_i (the set of **actions** available to player i)
- for each player $i \in N$ a preference relation $\succsim_i$ on $A = \times_{j \in N} A_j$ (the **preference relation** of player i).

If the set A_i of actions of every player i is finite then the game is *finite*.

The high level of abstraction of this model allows it to be applied to a
wide variety of situations. A player may be an individual human being
or any other decision-making entity like a government, a board of direc-
tors, the leadership of a revolutionary movement, or even a flower or an
animal. The model places no restrictions on the set of actions available
to a player, which may, for example, contain just a few elements or be a
huge set containing complicated plans that cover a variety of contingencies. However, the range of application of the model is limited by the
requirement that we associate with each player a preference relation. A
player's preference relation may simply reflect the player's feelings about
the possible outcomes or, in the case of an organism that does not act
consciously, the chances of its reproductive success.

The fact that the model is so abstract is a merit to the extent that
it allows applications in a wide range of situations, but is a drawback
to the extent that the implications of the model cannot depend on any
specific features of a situation. Indeed, very few conclusions can be
reached about the outcome of a game at this level of abstraction; one
needs to be much more specific to derive interesting results.

In some situations the players' preferences are most naturally defined
not over action profiles but over their consequences. When modeling an
oligopoly, for example, we may take the set of players to be a set of firms
and the set of actions of each firm to be the set of prices; but we may
wish to model the assumption that each firm cares only about its profit,
not about the profile of prices that generates that profit. To do so we
introduce a set C of *consequences*, a function $g: A \to C$ that associates
consequences with action profiles, and a profile $(\succsim_i^*)$ of preference relations over C. Then the preference relation $\succsim_i$ of each player i in the
strategic game is defined as follows: $a \succsim_i b$ if and only if $g(a) \succsim_i^* g(b)$.

Sometimes we wish to model a situation in which the consequence
of an action profile is affected by an exogenous random variable whose
realization is not known to the players before they take their actions.
We can model such a situation as a strategic game by introducing a set
C of consequences, a probability space Ω, and a function $g: A \times \Omega \to C$
with the interpretation that $g(a, \omega)$ is the consequence when the action
profile is $a \in A$ and the realization of the random variable is $\omega \in \Omega$. A
profile of actions induces a lottery on C; for each player i a preference
relation $\succsim_i^*$ must be specified over the set of all such lotteries. Player i's
preference relation in the strategic game is defined as follows: $a \succsim_i b$
if and only if the lottery over C induced by $g(a, \cdot)$ is at least as good
according to $\succsim_i^*$ as the lottery induced by $g(b, \cdot)$.

Figure 13.1 A convenient representation of a two-player strategic game in which each player has two actions.

Under a wide range of circumstances the preference relation $\succsim_i$ of player i in a strategic game can be represented by a **payoff function** $u_i \colon A \to \mathbb{R}$ (also called a *utility function*), in the sense that $u_i(a) \geq u_i(b)$ whenever $a \succsim_i b$. We refer to values of such a function as **payoffs** (or utilities). Frequently we specify a player's preference relation by giving a payoff function that represents it. In such a case we denote the game by $\langle N, (A_i), (u_i) \rangle$ rather than $\langle N, (A_i), (\succsim_i) \rangle$.

A finite strategic game in which there are two players can be described conveniently in a table like that in Figure 13.1. One player's actions are identified with the rows and the other player's with the columns. The two numbers in the box formed by row r and column c are the players' payoffs when the row player chooses r and the column player chooses c, the first component being the payoff of the row player. Thus in the game in Figure 13.1 the set of actions of the row player is $\{T, B\}$ and that of the column player is $\{L, R\}$, and for example the row player's payoff from the outcome (T, L) is w_1 and the column player's payoff is w_2. If the players' names are "1" and "2" then the convention is that the row player is player 1 and the column player is player 2.

2.1.2 Comments on Interpretation

A common interpretation of a strategic game is that it is a model of an event that occurs only once; each player knows the details of the game and the fact that all the players are "rational" (see Section 1.4), and the players choose their actions simultaneously and independently. Under this interpretation each player is unaware, when choosing his action, of the choices being made by the other players; there is no information (except the primitives of the model) on which a player can base his expectation of the other players' behavior.

Another interpretation, which we adopt through most of this book, is that a player can form his expectation of the other players' behavior on the basis of information about the way that the game or a similar game was played in the past (see Section 1.5). A sequence of plays of the game can be modeled by a strategic game only if there are no strategic links between the plays. That is, an individual who plays the game many times must be concerned only with his instantaneous payoff and ignore the effects of his current action on the other players' future behavior. In this interpretation it is thus appropriate to model a situation as a strategic game only in the absence of an intertemporal strategic link between occurrences of the interaction. (The model of a repeated game discussed in Chapter 8 deals with series of strategic interactions in which such intertemporal links *do* exist.)

When referring to the actions of the players in a strategic game as "simultaneous" we do not necessarily mean that these actions are taken at the same point in time. One situation that can be modeled as a strategic game is the following. The players are at different locations, in front of terminals. First the players' possible actions and payoffs are described publicly (so that they are common knowledge among the players). Then each player chooses an action by sending a message to a central computer; the players are informed of their payoffs when all the messages have been received. However, the model of a strategic game is much more widely applicable than this example suggests. For a situation to be modeled as a strategic game it is important only that the players make decisions independently, no player being informed of the choice of any other player prior to making his own decision.

2.2   Nash Equilibrium

The most commonly used solution concept in game theory is that of Nash equilibrium. This notion captures a *steady state* of the play of a strategic game in which each player holds the correct expectation about the other players' behavior and acts rationally. It does not attempt to examine the process by which a steady state is reached.

▶ DEFINITION 14.1 A **Nash equilibrium of a strategic game** $\langle N, (A_i), (\succsim_i) \rangle$ is a profile $a^* \in A$ of actions with the property that for every player $i \in N$ we have

$$(a^*_{-i}, a^*_i) \succsim_i (a^*_{-i}, a_i) \text{ for all } a_i \in A_i.$$

Thus for a^* to be a Nash equilibrium it must be that no player i has an action yielding an outcome that he prefers to that generated when he chooses a_i^*, given that every other player j chooses his equilibrium action a_j^*. Briefly, no player can profitably deviate, given the actions of the other players.

The following restatement of the definition is sometimes useful. For any $a_{-i} \in A_{-i}$ define $B_i(a_{-i})$ to be the set of player i's best actions given a_{-i}:

$$B_i(a_{-i}) = \{a_i \in A_i : (a_{-i}, a_i) \succsim_i (a_{-i}, a_i') \text{ for all } a_i' \in A_i\}. \quad (15.1)$$

We call the set-valued function B_i the **best-response function** of player i. A Nash equilibrium is a profile a^* of actions for which

$$a_i^* \in B_i(a_{-i}^*) \text{ for all } i \in N. \quad (15.2)$$

This alternative formulation of the definition points us to a (not necessarily efficient) method of finding Nash equilibria: first calculate the best response function of each player, then find a profile a^* of actions for which $a_i^* \in B_i(a_{-i}^*)$ for all $i \in N$. If the functions B_i are singleton-valued then the second step entails solving $|N|$ equations in the $|N|$ unknowns $(a_i^*)_{i \in N}$.

2.3 Examples

The following classical games represent a variety of strategic situations. The games are very simple: in each game there are just two players and each player has only two possible actions. Nevertheless, each game captures the essence of a type of strategic interaction that is frequently present in more complex situations.

◇ EXAMPLE 15.3 (*Bach or Stravinsky? (BoS)*) Two people wish to go out together to a concert of music by either Bach or Stravinsky. Their main concern is to go out together, but one person prefers Bach and the other person prefers Stravinsky. Representing the individuals' preferences by payoff functions, we have the game in Figure 16.1.

This game is often referred to as the "Battle of the Sexes"; for the standard story behind it see Luce and Raiffa (1957, pp. 90–91). For consistency with this nomenclature we call the game "BoS".

BoS models a situation in which players wish to coordinate their behavior, but have conflicting interests. The game has two Nash equilibria: (*Bach, Bach*) and (*Stravinsky, Stravinsky*). That is, there are two steady states: one in which both players always choose *Bach* and one in which they always choose *Stravinsky*.

	Bach	Stravinsky
Bach	2, 1	0, 0
Stravinsky	0, 0	1, 2

Figure 16.1 *Bach or Stravinsky?* (BoS) (Example 15.3).

	Mozart	Mahler
Mozart	2, 2	0, 0
Mahler	0, 0	1, 1

Figure 16.2 A coordination game (Example 16.1).

◇ EXAMPLE 16.1 (*A coordination game*) As in BoS, two people wish to go
out together, but in this case they agree on the more desirable concert.
A game that captures this situation is given in Figure 16.2.

Like BoS, the game has two Nash equilibria: (*Mozart*, *Mozart*) and
(*Mahler*, *Mahler*). In contrast to BoS, the players have a mutual interest
in reaching one of these equilibria, namely (*Mozart*, *Mozart*); however,
the notion of Nash equilibrium does not rule out a steady state in which
the outcome is the inferior equilibrium (*Mahler*, *Mahler*).

◇ EXAMPLE 16.2 (The *Prisoner's Dilemma*) Two suspects in a crime are
put into separate cells. If they both confess, each will be sentenced to
three years in prison. If only one of them confesses, he will be freed
and used as a witness against the other, who will receive a sentence of
four years. If neither confesses, they will both be convicted of a minor
offense and spend one year in prison. Choosing a convenient payoff
representation for the preferences, we have the game in Figure 17.1.

This is a game in which there are gains from cooperation—the best
outcome for the players is that neither confesses—but each player has
an incentive to be a "free rider". Whatever one player does, the other
prefers *Confess* to *Don't Confess*, so that the game has a unique Nash
equilibrium (*Confess*, *Confess*).

◇ EXAMPLE 16.3 (*Hawk–Dove*) Two animals are fighting over some prey.
Each can behave like a dove or like a hawk. The best outcome for

	Don't Confess	Confess
Don't confess	3, 3	0, 4
Confess	4, 0	1, 1

Figure 17.1 The *Prisoner's Dilemma* (Example 16.2).

	Dove	Hawk
Dove	3, 3	1, 4
Hawk	4, 1	0, 0

Figure 17.2 *Hawk–Dove* (Example 16.3).

each animal is that in which it acts like a hawk while the other acts like a dove; the worst outcome is that in which both animals act like hawks. Each animal prefers to be hawkish if its opponent is dovish and dovish if its opponent is hawkish. A game that captures this situation is shown in Figure 17.2. The game has two Nash equilibria, (*Dove, Hawk*) and (*Hawk, Dove*), corresponding to two different conventions about the player who yields.

◇ EXAMPLE 17.1 (*Matching Pennies*) Each of two people chooses either Head or Tail. If the choices differ, person 1 pays person 2 a dollar; if they are the same, person 2 pays person 1 a dollar. Each person cares only about the amount of money that he receives. A game that models this situation is shown in Figure 17.3. Such a game, in which the interests of the players are diametrically opposed, is called "strictly competitive". The game *Matching Pennies* has no Nash equilibrium.

	Head	Tail
Head	1, −1	−1, 1
Tail	−1, 1	1, −1

Figure 17.3 *Matching Pennies* (Example 17.1).

The notion of a strategic game encompasses situations much more complex than those described in the last five examples. The following are representatives of three families of games that have been studied extensively: auctions, games of timing, and location games.

◇ EXAMPLE 18.1 (*An auction*) An object is to be assigned to a player in the set $\{1, \ldots, n\}$ in exchange for a payment. Player i's valuation of the object is v_i, and $v_1 > v_2 > \cdots > v_n > 0$. The mechanism used to assign the object is a (sealed-bid) auction: the players simultaneously submit bids (nonnegative numbers), and the object is given to the player with the lowest index among those who submit the highest bid, in exchange for a payment.

In a *first price* auction the payment that the winner makes is the price that he bids.

? EXERCISE 18.2 Formulate a first price auction as a strategic game and analyze its Nash equilibria. In particular, show that in all equilibria player 1 obtains the object.

In a *second price auction* the payment that the winner makes is the highest bid among those submitted by the players who do not win (so that if only one player submits the highest bid then the price paid is the *second* highest bid).

? EXERCISE 18.3 Show that in a second price auction the bid v_i of any player i is a *weakly dominant* action: player i's payoff when he bids v_i is at least as high as his payoff when he submits any other bid, regardless of the actions of the other players. Show that nevertheless there are ("inefficient") equilibria in which the winner is not player 1.

◇ EXAMPLE 18.4 (*A war of attrition*) Two players are involved in a dispute over an object. The value of the object to player i is $v_i > 0$. Time is modeled as a continuous variable that starts at 0 and runs indefinitely. Each player chooses when to concede the object to the other player; if the first player to concede does so at time t, the other player obtains the object at that time. If both players concede simultaneously, the object is split equally between them, player i receiving a payoff of $v_i/2$. Time is valuable: until the first concession each player loses one unit of payoff per unit of time.

? EXERCISE 18.5 Formulate this situation as a strategic game and show that in all Nash equilibria one of the players concedes immediately.

◇ EXAMPLE 18.6 (*A location game*) Each of n people chooses whether or not to become a political candidate, and if so which position to take.

There is a continuum of citizens, each of whom has a favorite position; the distribution of favorite positions is given by a density function f on $[0, 1]$ with $f(x) > 0$ for all $x \in [0, 1]$. A candidate attracts the votes of those citizens whose favorite positions are closer to his position than to the position of any other candidate; if k candidates choose the same position then each receives the fraction $1/k$ of the votes that the position attracts. The winner of the competition is the candidate who receives the most votes. Each person prefers to be the unique winning candidate than to tie for first place, prefers to tie for first place than to stay out of the competition, and prefers to stay out of the competition than to enter and lose.

? EXERCISE 19.1 Formulate this situation as a strategic game, find the set of Nash equilibria when $n = 2$, and show that there is no Nash equilibrium when $n = 3$.

2.4 Existence of a Nash Equilibrium

Not every strategic game has a Nash equilibrium, as the game *Matching Pennies* (Figure 17.3) shows. The conditions under which the set of Nash equilibria of a game is nonempty have been investigated extensively. We now present an existence result that is one of the simplest of the genre. (Nevertheless its mathematical level is more advanced than most of the rest of the book, which does not depend on the details.)

An existence result has two purposes. First, if we have a game that satisfies the hypothesis of the result then we know that there is some hope that our efforts to find an equilibrium will meet with success. Second, and more important, the existence of an equilibrium shows that the game is consistent with a steady state solution. Further, the existence of equilibria for a family of games allows us to study properties of these equilibria (by using, for example, "comparative static" techniques) without finding them explicitly and without taking the risk that we are studying the empty set.

To show that a game has a Nash equilibrium it suffices to show that there is a profile a^* of actions such that $a_i^* \in B_i(a_{-i}^*)$ for all $i \in N$ (see (15.2)). Define the set-valued function $B: A \to A$ by $B(a) = \times_{i \in N} B_i(a_{-i})$. Then (15.2) can be written in vector form simply as $a^* \in B(a^*)$. *Fixed point theorems* give conditions on B under which there indeed exists a value of a^* for which $a^* \in B(a^*)$. The fixed point theorem that we use is the following (due to Kakutani (1941)).

■ LEMMA 20.1 (Kakutani's fixed point theorem) *Let X be a compact convex subset of $\mathbb{R}^n$ and let $f: X \to X$ be a set-valued function for which*

- *for all $x \in X$ the set $f(x)$ is nonempty and convex*
- *the graph of f is closed (i.e. for all sequences $\{x_n\}$ and $\{y_n\}$ such that $y_n \in f(x_n)$ for all n, $x_n \to x$, and $y_n \to y$, we have $y \in f(x)$).*

Then there exists $x^* \in X$ such that $x^* \in f(x^*)$.

? EXERCISE 20.2 Show that each of the following four conditions is necessary for Kakutani's theorem. (i) X is compact. (ii) X is convex. (iii) $f(x)$ is convex for each $x \in X$. (iv) f has a closed graph.

Define a preference relation $\succsim_i$ over A to be *quasi-concave on A_i* if for every $a^* \in A$ the set $\{a_i \in A_i : (a^*_{-i}, a_i) \succsim_i a^*\}$ is convex.

■ PROPOSITION 20.3 *The strategic game $\langle N, (A_i), (\succsim_i) \rangle$ has a Nash equilibrium if for all $i \in N$*

- *the set A_i of actions of player i is a nonempty compact convex subset of a Euclidian space*

and the preference relation $\succsim_i$ is

- *continuous*
- *quasi-concave on A_i.*

Proof. Define $B: A \to A$ by $B(a) = \times_{i \in N} B_i(a_{-i})$ (where B_i is the best-response function of player i, defined in (15.1)). For every $i \in N$ the set $B_i(a_{-i})$ is nonempty since $\succsim_i$ is continuous and A_i is compact, and is convex since $\succsim_i$ is quasi-concave on A_i; B has a closed graph since each $\succsim_i$ is continuous. Thus by Kakutani's theorem B has a fixed point; as we have noted any fixed point is a Nash equilibrium of the game. □

Note that this result asserts that a strategic game satisfying certain conditions has *at least* one Nash equilibrium; as we have seen, a game can have more than one equilibrium. (Results that we do not discuss identify conditions under which a game has a unique Nash equilibrium.) Note also that Proposition 20.3 does not apply to any game in which some player has finitely many actions, since such a game violates the condition that the set of actions of every player be convex.

? EXERCISE 20.4 (*Symmetric games*) Consider a two-person strategic game that satisfies the conditions of Proposition 20.3. Let $N = \{1, 2\}$ and assume that the game is *symmetric*: $A_1 = A_2$ and $(a_1, a_2) \succsim_1 (b_1, b_2)$ if and only if $(a_2, a_1) \succsim_2 (b_2, b_1)$ for all $a \in A$ and $b \in A$. Use Kakutani's theorem to prove that there is an action $a_1^* \in A_1$ such that (a_1^*, a_1^*) is a Nash equilibrium of the game. (Such an equilibrium is called a *symmetric equilibrium*.) Give an example of a finite symmetric game that has only asymmetric equilibria.

2.5 Strictly Competitive Games

We can say little about the set of Nash equilibria of an arbitrary strategic game; only in limited classes of games can we say something about the qualitative character of the equilibria. One such class of games is that in which there are two players, whose preferences are diametrically opposed. We assume for convenience in this section that the names of the players are "1" and "2" (i.e. $N = \{1, 2\}$).

▶ DEFINITION 21.1 A strategic game $\langle \{1, 2\}, (A_i), (\succsim_i) \rangle$ is **strictly competitive** if for any $a \in A$ and $b \in A$ we have $a \succsim_1 b$ if and only if $b \succsim_2 a$.

A strictly competitive game is sometimes called *zerosum* because if player 1's preference relation $\succsim_1$ is represented by the payoff function u_1 then player 2's preference relation is represented by u_2 with $u_1 + u_2 = 0$.

We say that player i *maxminimizes* if he chooses an action that is best for him on the assumption that whatever he does, player j will choose her action to hurt him as much as possible. We now show that for a strictly competitive game that possesses a Nash equilibrium, a pair of actions is a Nash equilibrium if and only if the action of each player is a maxminimizer. This result is striking because it provides a link between individual decision-making and the reasoning behind the notion of Nash equilibrium. In establishing the result we also prove the strong result that for strictly competitive games that possess Nash equilibria all equilibria yield the same payoffs. This property of Nash equilibria is rarely satisfied in games that are not strictly competitive.

▶ DEFINITION 21.2 Let $\langle \{1, 2\}, (A_i), (u_i) \rangle$ be a strictly competitive strategic game. The action $x^* \in A_1$ is a **maxminimizer for player 1** if

$$\min_{y \in A_2} u_1(x^*, y) \geq \min_{y \in A_2} u_1(x, y) \text{ for all } x \in A_1.$$

Similarly, the action $y^* \in A_2$ is a **maxminimizer for player 2** if

$$\min_{x \in A_1} u_2(x, y^*) \geq \min_{x \in A_1} u_2(x, y) \text{ for all } y \in A_2.$$

In words, a maxminimizer for player i is an action that maximizes the payoff that player i can *guarantee*. A maxminimizer for player 1 solves the problem $\max_x \min_y u_1(x, y)$ and a maxminimizer for player 2 solves the problem $\max_y \min_x u_2(x, y)$.

In the sequel we assume for convenience that player 1's preference relation is represented by a payoff function u_1 and, without loss of generality, that $u_2 = -u_1$. The following result shows that the maxminimization of player 2's payoff is equivalent to the minmaximization of player 1's payoff.

■ LEMMA 22.1 *Let* $\langle \{1,2\}, (A_i), (u_i) \rangle$ *be a strictly competitive strategic game. Then* $\max_{y \in A_2} \min_{x \in A_1} u_2(x, y) = -\min_{y \in A_2} \max_{x \in A_1} u_1(x, y)$. *Further,* $y \in A_2$ *solves the problem* $\max_{y \in A_2} \min_{x \in A_1} u_2(x, y)$ *if and only if it solves the problem* $\min_{y \in A_2} \max_{x \in A_1} u_1(x, y)$.

Proof. For any function f we have $\min_z(-f(z)) = -\max_z f(z)$ and $\arg\min_z(-f(z)) = \arg\max_z f(z)$. It follows that for every $y \in A_2$ we have $-\min_{x \in A_1} u_2(x, y) = \max_{x \in A_1}(-u_2(x, y)) = \max_{x \in A_1} u_1(x, y)$. Hence $\max_{y \in A_2} \min_{x \in A_1} u_2(x, y) = -\min_{y \in A_2}[-\min_{x \in A_1} u_2(x, y)] = -\min_{y \in A_2} \max_{x \in A_1} u_1(x, y)$; in addition $y \in A_2$ is a solution of the problem $\max_{y \in A_2} \min_{x \in A_1} u_2(x, y)$ if and only if it is a solution of the problem $\min_{y \in A_2} \max_{x \in A_1} u_1(x, y)$.   □

The following result gives the connection between the Nash equilibria of a strictly competitive game and the set of pairs of maxminimizers.

■ PROPOSITION 22.2 *Let* $G = \langle \{1,2\}, (A_i), (u_i) \rangle$ *be a strictly competitive strategic game.*

a. *If* (x^*, y^*) *is a Nash equilibrium of* G *then* x^* *is a maxminimizer for player 1 and* y^* *is a maxminimizer for player 2.*

b. *If* (x^*, y^*) *is a Nash equilibrium of* G *then* $\max_x \min_y u_1(x, y) = \min_y \max_x u_1(x, y) = u_1(x^*, y^*)$, *and thus all Nash equilibria of* G *yield the same payoffs.*

c. *If* $\max_x \min_y u_1(x, y) = \min_y \max_x u_1(x, y)$ *(and thus, in particular, if* G *has a Nash equilibrium (see part b)),* x^* *is a maxminimizer for player 1, and* y^* *is a maxminimizer for player 2, then* (x^*, y^*) *is a Nash equilibrium of* G.

Proof. We first prove parts (a) and (b). Let (x^*, y^*) be a Nash equilibrium of G. Then $u_2(x^*, y^*) \geq u_2(x^*, y)$ for all $y \in A_2$ or, since $u_2 = -u_1$, $u_1(x^*, y^*) \leq u_1(x^*, y)$ for all $y \in A_2$. Hence $u_1(x^*, y^*) = \min_y u_1(x^*, y) \leq \max_x \min_y u_1(x, y)$. Similarly, $u_1(x^*, y^*) \geq u_1(x, y^*)$

for all $x \in A_1$ and hence $u_1(x^*, y^*) \geq \min_y u_1(x, y)$ for all $x \in A_1$, so that $u_1(x^*, y^*) \geq \max_x \min_y u_1(x, y)$. Thus $u_1(x^*, y^*) = \max_x \min_y u_1(x, y)$ and x^* is a maxminimizer for player 1.

An analogous argument for player 2 establishes that y^* is a maxminimizer for player 2 and $u_2(x^*, y^*) = \max_y \min_x u_2(x, y)$, so that $u_1(x^*, y^*) = \min_y \max_x u_1(x, y)$.

To prove part (c) let $v^* = \max_x \min_y u_1(x, y) = \min_y \max_x u_1(x, y)$. By Lemma 22.1 we have $\max_y \min_x u_2(x, y) = -v^*$. Since x^* is a maxminimizer for player 1 we have $u_1(x^*, y) \geq v^*$ for all $y \in A_2$; since y^* is a maxminimizer for player 2 we have $u_2(x, y^*) \geq -v^*$ for all $x \in A_1$. Letting $y = y^*$ and $x = x^*$ in these two inequalities we obtain $u_1(x^*, y^*) = v^*$ and, using the fact that $u_2 = -u_1$, we conclude that (x^*, y^*) is a Nash equilibrium of G. □

Note that by part (c) a Nash equilibrium can be found by solving the problem $\max_x \min_y u_1(x, y)$. This fact is sometimes useful when calculating the Nash equilibria of a game, especially when the players randomize (see for example Exercise 36.1).

Note also that it follows from parts (a) and (c) that Nash equilibria of a strictly competitive game are *interchangeable*: if (x, y) and (x', y') are equilibria then so are (x, y') and (x', y).

Part (b) shows that $\max_x \min_y u_1(x, y) = \min_y \max_x u_1(x, y)$ for any strictly competitive game that has a Nash equilibrium. Note that the inequality $\max_x \min_y u_1(x, y) \leq \min_y \max_x u_1(x, y)$ holds more generally: for any x' we have $u_1(x', y) \leq \max_x u_1(x, y)$ for all y, so that $\min_y u_1(x', y) \leq \min_y \max_x u_1(x, y)$. (If the maxima and minima are not well-defined then max and min should be replaced by sup and inf respectively.) Thus in *any* game (whether or not it is strictly competitive) the payoff that player 1 can guarantee herself is *at most* the amount that player 2 can hold her down to. The hypothesis that the game has a Nash equilibrium is essential in establishing the opposite inequality. To see this, consider the game *Matching Pennies* (Figure 17.3), in which $\max_x \min_y u_1(x, y) = -1 < \min_y \max_x u_1(x, y) = 1$.

If $\max_x \min_y u_1(x, y) = \min_y \max_x u_1(x, y)$ then we say that this payoff, the equilibrium payoff of player 1, is the **value** of the game. It follows from Proposition 22.2 that if v^* is the value of a strictly competitive game then any equilibrium strategy of player 1 guarantees that her payoff is at least her equilibrium payoff v^*, and any equilibrium strategy of player 2 guarantees that his payoff is at least his equilibrium payoff $-v^*$, so that any such strategy of player 2 guarantees that player 1's payoff is at most her equilibrium payoff. In a game that is not strictly competitive a player's equilibrium strategy does not in general have these properties (consider, for example, BoS (Figure 16.1)).

☐ EXERCISE 24.1 Let G be a strictly competitive game that has a Nash equilibrium.

    a. Show that if some of player 1's payoffs in G are increased in such a way that the resulting game G' is strictly competitive then G' has no equilibrium in which player 1 is worse off than she was in an equilibrium of G. (Note that G' may have no equilibrium at all.)

    b. Show that the game that results if player 1 is prohibited from using one of her actions in G does not have an equilibrium in which player 1's payoff is higher than it is in an equilibrium of G.

    c. Give examples to show that neither of the above properties necessarily holds for a game that is not strictly competitive.

2.6   Bayesian Games: Strategic Games with Imperfect Information

2.6.1 *Definitions*

We frequently wish to model situations in which some of the parties are not certain of the characteristics of some of the other parties. The model of a Bayesian game, which is closely related to that of a strategic game, is designed for this purpose.

As for a strategic game, two primitives of a Bayesian game are a set N of players and a profile (A_i) of sets of actions. We model the players' uncertainty about each other by introducing a set Ω of possible "states of nature", each of which is a description of all the players' relevant characteristics. For convenience we assume that Ω is finite. Each player i has a *prior belief* about the state of nature given by a probability measure p_i on Ω. In any given play of the game some state of nature $\omega \in \Omega$ is realized. We model the players' information about the state of nature by introducing a profile (τ_i) of *signal functions*, $\tau_i(\omega)$ being the signal that player i observes, before choosing his action, when the state of nature is ω. Let T_i be the set of all possible values of τ_i; we refer to T_i as the set of *types* of player i. We assume that $p_i(\tau_i^{-1}(t_i)) > 0$ for all $t_i \in T_i$ (player i assigns positive prior probability to every member of T_i). If player i receives the signal $t_i \in T_i$ then he deduces that the state is in the set $\tau_i^{-1}(t_i)$; his *posterior belief* about the state that has been realized assigns to each state $\omega \in \Omega$ the probability $p_i(\omega)/p_i(\tau_i^{-1}(t_i))$ if $\omega \in \tau_i^{-1}(t_i)$ and the probability zero otherwise (i.e. the probability of ω conditional on $\tau_i^{-1}(t_i)$). As an example, if $\tau_i(\omega) = \omega$ for all $\omega \in \Omega$ then player i has full information about the state of nature. Alternatively, if $\Omega = \times_{i \in N} T_i$ and for each player i the probability measure p_i is a product measure on Ω and $\tau_i(\omega) = \omega_i$ then the players' signals are independent and player i does not learn from his signal anything about the other players' information.

As in a strategic game, each player cares about the action profile; in addition he may care about the state of nature. Now, even if he knows the action taken by every other player in every state of nature, a player may be uncertain about the pair (a, ω) that will be realized given any action that he takes, since he has imperfect information about the state of nature. Therefore we include in the model a profile $(\succsim_i)$ of preference relations over *lotteries* on $A \times \Omega$ (where, as before, $A = \times_{j \in N} A_j$). To summarize, we make the following definition.

▸ DEFINITION 25.1 A **Bayesian game** consists of

- a finite set N (the set of **players**)
- a finite set Ω (the set of **states**)

and for each player $i \in N$

- a set A_i (the set of **actions** available to player i)
- a finite set T_i (the set of **signals** that may be observed by player i) and a function $\tau_i \colon \Omega \to T_i$ (the **signal function** of player i)
- a probability measure p_i on Ω (the **prior belief** of player i) for which $p_i(\tau_i^{-1}(t_i)) > 0$ for all $t_i \in T_i$
- a preference relation $\succsim_i$ on the set of probability measures over $A \times \Omega$ (the **preference relation** of player i), where $A = \times_{j \in N} A_j$.

Note that this definition allows the players to have different prior beliefs. These beliefs may be related; commonly they are identical, co-incident with an "objective" measure. Frequently the model is used in situations in which a state of nature is a profile of parameters of the players' preferences (for example, profiles of their valuations of an object). However, the model is much more general; in Section 2.6.3 we consider its use to capture situations in which each player is uncertain about what the others *know*.

Note also that sometimes a Bayesian game is described not in terms of an underlying state space Ω, but as a "reduced form" in which the

basic primitive that relates to the players' information is the profile of
the sets of possible types.

We now turn to a definition of equilibrium for a Bayesian game. In
any given play of a game each player knows his type and does not need to
plan what to do in the hypothetical event that he is of some other type.
Consequently, one might think that an equilibrium should be defined
for each state of nature in isolation. However, in any given state a
player who wishes to determine his best action may need to hold a belief
about what the other players would do in other states, since he may be
imperfectly informed about the state. Further, the formation of such a
belief may depend on the action that the player himself would choose in
other states, since the other players may also be imperfectly informed.

Thus we are led to define a Nash equilibrium of a Bayesian game $\langle N, \Omega,$
$(A_i), (T_i), (\tau_i), (p_i), (\succsim_i) \rangle$ to be a Nash equilibrium of the strategic game
G^* in which for each $i \in N$ and each possible signal $t_i \in T_i$ there is
a player, whom we refer to as (i, t_i) ("type t_i of player i"). The set of
actions of each such player (i, t_i) is A_i; thus the set of action profiles in G^*
is $\times_{j \in N}(\times_{t_j \in T_j} A_j)$. The preferences of each player (i, t_i) are defined as
follows. The posterior belief of player i, together with an action profile a^*
in G^*, generates a lottery $L_i(a^*, t_i)$ over $A \times \Omega$: the probability assigned
by $L_i(a^*, t_i)$ to $((a^*(j, \tau_j(\omega)))_{j \in N}, \omega)$ is player i's posterior belief that the
state is ω when he receives the signal t_i $(a^*(j, \tau_j(\omega))$ being the action
of player $(j, \tau_j(\omega))$ in the profile a^*). Player (i, t_i) in G^* prefers the
action profile a^* to the action profile b^* if and only if player i in the
Bayesian game prefers the lottery $L_i(a^*, t_i)$ to the lottery $L_i(b^*, t_i)$. To
summarize, we have the following.

▶ DEFINITION 26.1 **A Nash equilibrium of a Bayesian game** $\langle N,$
$\Omega, (A_i), (T_i), (\tau_i), (p_i), (\succsim_i) \rangle$ is a Nash equilibrium of the strategic game
defined as follows.

- The set of players is the set of all pairs (i, t_i) for $i \in N$ and $t_i \in T_i$.
- The set of actions of each player (i, t_i) is A_i.
- The preference ordering $\succsim^*_{(i, t_i)}$ of each player (i, t_i) is defined by

$$a^* \succsim^*_{(i, t_i)} b^* \text{ if and only if } L_i(a^*, t_i) \succsim_i L_i(b^*, t_i),$$

  where $L_i(a^*, t_i)$ is the lottery over $A \times \Omega$ that assigns probabil-
  ity $p_i(\omega)/p_i(\tau_i^{-1}(t_i))$ to $((a^*(j, \tau_j(\omega)))_{j \in N}, \omega)$ if $\omega \in \tau_i^{-1}(t_i)$, zero
  otherwise.

In brief, in a Nash equilibrium of a Bayesian game each player chooses
the best action available to him given the signal that he receives and his

belief about the state and the other players' actions that he deduces from this signal. Note that to determine whether an action profile is a Nash equilibrium of a Bayesian game we need to know only how each player in the Bayesian game compares lotteries over $A \times \Omega$ in which the distribution over Ω is the same: a player never needs to compare lotteries in which this distribution is different. Thus from the point of view of Nash equilibrium the specification of the players' preferences in a Bayesian game contains more information than is necessary. (This redundancy has an analog in a strategic game: to define a Nash equilibrium of a strategic game we need to know only how any player i compares any outcome (a_{-i}, a_i) with any other outcome (a_{-i}, b_i).)

2.6.2 Examples

◇ EXAMPLE 27.1 (*Second-price auction*) Consider a variant of the second-price sealed-bid auction described in Example 18.1 in which each player i knows his own valuation v_i but is uncertain of the other players' valuations. Specifically, suppose that the set of possible valuations is the finite set V and each player believes that every other player's valuation is drawn independently from the same distribution over V. We can model this situation as the Bayesian game in which

- the set N of players is $\{1, \ldots, n\}$
- the set Ω of states is V^n (the set of profiles of valuations)
- the set A_i of actions of each player i is $\mathbb{R}_+$
- the set T_i of signals that i can receive is V
- the signal function τ_i of i is defined by $\tau_i(v_1, \ldots, v_n) = v_i$
- the prior belief p_i of i is given by $p_i(v_1, \ldots, v_n) = \Pi_{j=1}^n \pi(v_j)$ for some probability distribution π over V
- player i's preference relation is represented by the expectation of the random variable whose value in state $(v_1, \ldots, v_n)$ is $v_i - \max_{j \in N \setminus \{i\}} a_j$ if i is the player with the lowest index for whom $a_i \geq a_j$ for all $j \in N$, and 0 otherwise.

This game has a Nash equilibrium a^* in which $a^*(i, v_i) = v_i$ for all $i \in N$ and $v_i \in V = T_i$ (each player bids his valuation). In fact (as in Exercise 18.3) it is a weakly dominant action for each type of each player to bid his valuation.

? EXERCISE 27.2 Two players wish to go out together to a concert of music by either Bach or Stravinsky. As in BoS their main concern is

to go out together; but neither player knows whether the other prefers
Bach to Stravinsky, or the reverse. Each player's preferences are represented by the expectation of his payoff, the payoffs to pure outcomes
being analogous to those given in Figure 16.1. Model this situation as
a Bayesian game and find the Nash equilibria for all possible beliefs.
Show in particular that there are equilibria in which there is a positive
probability that the players do not go to the same concert.

⟨?⟩ EXERCISE 28.1 (*An exchange game*) Each of two players receives a ticket
on which there is a number in some finite subset S of the interval $[0, 1]$.
The number on a player's ticket is the size of a prize that he may receive. The two prizes are identically and independently distributed, with
distribution function F. Each player is asked independently and simultaneously whether he wants to exchange his prize for the other player's
prize. If both players agree then the prizes are exchanged; otherwise
each player receives his own prize. Each player's objective is to maximize his expected payoff. Model this situation as a Bayesian game and
show that in any Nash equilibrium the highest prize that either player
is willing to exchange is the smallest possible prize.

⟨?⟩ EXERCISE 28.2 Show that more information may hurt a player by
constructing a two-player Bayesian game with the following features.
Player 1 is fully informed while player 2 is not; the game has a unique
Nash equilibrium, in which player 2's payoff is higher than his payoff in
the unique equilibrium of any of the related games in which he knows
player 1's type.

2.6.3 Comments on the Model of a Bayesian Game

The idea that a situation in which the players are unsure about each
other's characteristics can be modeled as a Bayesian game, in which the
players' uncertainty is captured by a probability measure over some set
of "states", is due to Harsanyi (1967/68). Harsanyi assumes that the
prior belief of every player is the same, arguing that all differences in the
players' knowledge should be derived from an objective mechanism that
assigns information to each player, not from differences in the players'
initial beliefs. In Section 5.3 we show that the assumption of a common
prior belief has strong implications for the relationship between the players' posterior beliefs. (For example, after a pair of players receive their
signals it cannot be "common knowledge" between them that player 1
believes the probability that the state of nature is in some given set to

be α and that player 2 believes this probability to be $\beta \neq \alpha$, though it *is* possible that player 1 believes the probability to be α, player 2 believes it to be β, and one of them is unsure about the other's belief.)

A Bayesian game can be used to model not only situations in which each player is uncertain about the other players' payoffs, as in Example 27.1, but also situations in which each player is uncertain about the other players' *knowledge*.

Consider, for example, a Bayesian game in which the set of players is $N = \{1, 2\}$, the set of states is $\Omega = \{\omega_1, \omega_2, \omega_3\}$, the prior belief of each player assigns probability $\frac{1}{3}$ to each state, the signal functions are defined by $\tau_1(\omega_1) = \tau_1(\omega_2) = t'_1$, $\tau_1(\omega_3) = t''_1$, and $\tau_2(\omega_1) = t'_2$, $\tau_2(\omega_2) = \tau_2(\omega_3) = t''_2$, and player 1's preferences satisfy $(b, \omega_j) \succ_1 (c, \omega_j)$ for $j = 1, 2$ and $(c, \omega_3) \succ_1 (b, \omega_3)$ for some action profiles b and c, while player 2 is indifferent between all pairs (a, ω). In state ω_1 in such a game player 2 knows that player 1 prefers b to c, while in state ω_2 he does not know whether player 1 prefers b to c or c to b. Since in state ω_1 player 1 does not know whether the state is ω_1 or ω_2, she does not know in this case whether (i) player 2 knows that she prefers b to c, or (ii) player 2 is not sure whether she prefers b to c or c to b.

Can every situation in which the players are uncertain about each other's knowledge be modeled as a Bayesian game? Assume that the players' payoffs depend only on a parameter $\theta \in \Theta$. Denote the set of possible beliefs of each player i by X_i. Then a belief of any player j is a probability distribution over $\Theta \times X_{-j}$. That is, the set of beliefs of any player has to be defined in terms of the sets of beliefs of all the other players. Thus the answer to the question we posed is not trivial and is equivalent to the question of whether we can find a collection $\{X_j\}_{j \in N}$ of sets with the property that for all $i \in N$ the set X_i is isomorphic to the set of probability distributions over $\Theta \times X_{-i}$. If so, we can let $\Omega = \Theta \times (\times_{i \in N} X_i)$ be the state space and use the model of a Bayesian game to capture any situation in which players are uncertain not only about each other's payoffs but also about each other's beliefs. A positive answer is given to the question by Mertens and Zamir (1985); we omit the argument.

Notes

The notion of an abstract strategic game has its origins in the work of Borel (1921) and von Neumann (1928). The notion of Nash equilibrium was formalized in the context of such a game by Nash (1950a); the basic idea behind it goes back at least to Cournot (1838). The idea of the proof of Proposition 20.3 originated with Nash (1950a, 1951) and Glicksberg (1952), though the results they prove are slightly different. As stated the result is similar to Theorem 3.1 of Nikaidô and Isoda (1955). The idea of maxminimization dates back at least to the early eighteenth century (see Kuhn (1968)). The main ideas of Proposition 22.2 are due to von Neumann (1928); the theory of strictly competitive games was developed by von Neumann and Morgenstern (1944). Bayesian games were defined and studied by Harsanyi (1967/68).

The *Prisoner's Dilemma* appears to have first entered the literature in unpublished papers by Raiffa (in 1951) and Flood (in 1952, reporting joint work with Dresher); the standard interpretation of the game is due to Tucker (see Raiffa (1992, p. 173)). BoS is due to Luce and Raiffa (1957). *Hawk–Dove* is known also as "Chicken". Auctions (Examples 18.1 and 27.1) were first studied formally by Vickrey (1961). The war of attrition in Example 18.4 is due to Maynard Smith (1974), the location game in Example 18.6 is due to Hotelling (1929), and the game in Exercise 28.1 is due to Brams, Kilgour, and Davis (1993).

3 Mixed, Correlated, and Evolutionary Equilibrium

In this chapter we examine two concepts of equilibrium in which the players' actions are not deterministic: mixed strategy Nash equilibrium and correlated equilibrium. We also briefly study a variant of Nash equilibrium designed to model the outcome of an evolutionary process.

3.1 Mixed Strategy Nash Equilibrium

3.1.1 Definitions

The notion of mixed strategy Nash equilibrium is designed to model a steady state of a game in which the participants' choices are not deterministic but are regulated by probabilistic rules. We begin with formal definitions, then turn to their interpretation.

In the previous chapter we define a strategic game to be a triple $\langle N, (A_i), (\succsim_i) \rangle$, where the preference relation $\succsim_i$ of each player i is defined over the set $A = \times_{i \in N} A_i$ of action profiles (Definition 11.1). In this chapter we allow the players' choices to be nondeterministic and thus need to add to the primitives of the model a specification of each player's preference relation over *lotteries* on A. Following the current convention in game theory, we assume that the preference relation of each player i satisfies the assumptions of von Neumann and Morgenstern, so that it can be represented by the expected value of some function $u_i: A \to \mathbb{R}$. Thus our basic model of strategic interaction in this chapter is a triple $\langle N, (A_i), (u_i) \rangle$ that differs from a strategic game as we previously defined it in that $u_i: A \to \mathbb{R}$ for each $i \in N$ is a function whose expected value represents player i's preferences over the set of *lotteries* on A. Nevertheless, we refer to the model simply as a **strategic game**.

Let $G = \langle N, (A_i), (u_i) \rangle$ be such a strategic game. We denote by $\Delta(A_i)$ the set of probability distributions over A_i and refer to a member of $\Delta(A_i)$ as a **mixed strategy** of player i; we assume that the players' mixed strategies are independent randomizations. For clarity, we sometimes refer to a member of A_i as a **pure strategy**. For any finite set X and $\delta \in \Delta(X)$ we denote by $\delta(x)$ the probability that δ assigns to $x \in X$ and define the *support* of δ to be the set of elements $x \in X$ for which $\delta(x) > 0$. A profile $(\alpha_j)_{j \in N}$ of mixed strategies induces a probability distribution over the set A; if, for example, each A_j is finite then given the independence of the randomizations the probability of the action profile $a = (a_j)_{j \in N}$ is $\Pi_{j \in N} \alpha_j(a_j)$, so that player i's evaluation of $(\alpha_j)_{j \in N}$ is $\sum_{a \in A} (\Pi_{j \in N} \alpha_j(a_j)) u_i(a)$.

We now derive from G another strategic game, called the "mixed extension" of G, in which the set of actions of each player i is the set $\Delta(A_i)$ of his mixed strategies in G.

▶ DEFINITION 32.1 The **mixed extension** of the strategic game $\langle N, (A_i), (u_i) \rangle$ is the strategic game $\langle N, (\Delta(A_i)), (U_i) \rangle$ in which $\Delta(A_i)$ is the set of probability distributions over A_i, and $U_i \colon \times_{j \in N} \Delta(A_j) \to \mathbb{R}$ assigns to each $\alpha \in \times_{j \in N} \Delta(A_j)$ the expected value under u_i of the lottery over A that is induced by α (so that $U_i(\alpha) = \sum_{a \in A} (\Pi_{j \in N} \alpha_j(a_j)) u_i(a)$ if A is finite).

Note that each function U_i is multilinear. That is, for any mixed strategy profile α, any mixed strategies β_i and γ_i of player i, and any number $\lambda \in [0, 1]$, we have $U_i(\alpha_{-i}, \lambda \beta_i + (1 - \lambda) \gamma_i) = \lambda U_i(\alpha_{-i}, \beta_i) + (1 - \lambda) U_i(\alpha_{-i}, \gamma_i)$. Note also that when each A_i is finite we have

$$U_i(\alpha) = \sum_{a_i \in A_i} \alpha_i(a_i) U_i(\alpha_{-i}, e(a_i)) \qquad (32.2)$$

for any mixed strategy profile α, where $e(a_i)$ is the degenerate mixed strategy of player i that attaches probability one to $a_i \in A_i$.

We now define the main equilibrium notion we study in this chapter.

▶ DEFINITION 32.3 A **mixed strategy Nash equilibrium of a strategic game** is a Nash equilibrium of its mixed extension.

Suppose that $\alpha^* \in \times_{j \in N} \Delta(A_j)$ is a mixed strategy Nash equilibrium of $G = \langle N, (A_i), (u_i) \rangle$ in which each player i's mixed strategy α_i^* is degenerate in the sense that it assigns probability one to a single member—say a_i^*—of A_i. Then, since A_i can be identified with a subset of $\Delta(A_i)$, the action profile a^* is a Nash equilibrium of G. Conversely, suppose that a^* is a Nash equilibrium of G. Then by the linearity of U_i in α_i no probability distribution over actions in A_i yields player i a payoff higher than that generated by $e(a_i^*)$, and thus the profile $(e(a_i^*))$ is a mixed strategy Nash equilibrium of G.

We have just argued that the set of Nash equilibria of a strategic game is a subset of its set of mixed strategy Nash equilibria. In Chapter 2 we saw that there are games for which the set of Nash equilibria is empty. There are also games for which the set of mixed strategy Nash equilibria is empty. However, every game in which each player has finitely many actions has at least one mixed strategy Nash equilibrium, as the following result shows.

▪ PROPOSITION 33.1 *Every finite strategic game has a mixed strategy Nash equilibrium.*

Proof. Let $G = \langle N, (A_i), (u_i) \rangle$ be a strategic game, and for each player i let m_i be the number of members of the set A_i. Then we can identify the set $\Delta(A_i)$ of player i's mixed strategies with the set of vectors $(p_1, \ldots, p_{m_i})$ for which $p_k \geq 0$ for all k and $\sum_{k=1}^{m_i} p_k = 1$ (p_k being the probability with which player i uses his kth pure strategy). This set is nonempty, convex, and compact. Since expected payoff is linear in the probabilities, each player's payoff function in the mixed extension of G is both quasi-concave in his own strategy and continuous. Thus the mixed extension of G satisfies all the requirements of Proposition 20.3. □

Essential to this proof is the assumption that the set of actions of each player is finite. Glicksberg (1952) shows that a game in which each action set is a convex compact subset of a Euclidian space and each payoff function is continuous has a mixed strategy Nash equilibrium. (If each player's payoff function is also quasi-concave in his own action then Proposition 20.3 shows that such a game has a pure strategy Nash equilibrium.)

The following result gives an important property of mixed strategy Nash equilibria that is useful when calculating equilibria.

▪ LEMMA 33.2 *Let $G = \langle N, (A_i), (u_i) \rangle$ be a finite strategic game. Then $\alpha^* \in \times_{i \in N} \Delta(A_i)$ is a mixed strategy Nash equilibrium of G if and only if for every player $i \in N$ every pure strategy in the support of α_i^* is a best response to α_{-i}^*.*

Proof. First suppose that there is an action a_i in the support of α_i^* that is not a best response to α_{-i}^*. Then by linearity of U_i in α_i (see (32.2)) player i can increase his payoff by transferring probability from a_i to an action that *is* a best response; hence α_i^* is not a best response to α_{-i}^*.

Second suppose that there is a mixed strategy α_i' that gives a higher expected payoff than does α_i^* in response to α_{-i}^*. Then again by the linearity of U_i at least one action in the support of α_i' must give a higher payoff than some action in the support of α_i^*, so that not all actions in the support of α_i^* are best responses to α_{-i}^*.   □

It follows that *every action in the support of any player's equilibrium mixed strategy yields that player the same payoff.*

If the set of actions of some player is not finite the result needs to be modified. In this case, α^* is a mixed strategy Nash equilibrium of G if and only if (i) for every player i no action in A_i yields, given α_{-i}^*, a payoff to player i that exceeds his equilibrium payoff, and (ii) the set of actions that yield, given α_{-i}^*, a payoff less than his equilibrium payoff has α_i^*-measure zero.

Note that the assumption that the players' preferences can be represented by expected payoff functions plays a key role in these characterizations of mixed strategy equilibrium. The results do not necessarily hold for other theories of decision-making under uncertainty.

3.1.2 Examples

The following example illustrates how one can find mixed strategy Nash equilibria of finite games.

◇ EXAMPLE 34.1 (*BoS*) Consider the game BoS, reproduced in the top of Figure 35.1. In Chapter 2 we interpreted the payoffs of player i in this table as representing player i's preferences over the set of (pure) outcomes. Here, given our interest in mixed strategy equilibria, we interpret the payoffs as von Neumann–Morgenstern utilities.

As we noted previously this game has two (pure) Nash equilibria, (B, B) and (S, S), where $B = Bach$ and $S = Stravinsky$. Suppose that (α_1, α_2) is a mixed strategy Nash equilibrium. If $\alpha_1(B)$ is zero or one, we obtain the two pure Nash equilibria. If $0 < \alpha_1(B) < 1$ then, given α_2, by Lemma 33.2 player 1's actions B and S must yield the same payoff, so that we must have $2\alpha_2(B) = \alpha_2(S)$ and thus $\alpha_2(B) = \frac{1}{3}$. Since $0 < \alpha_2(B) < 1$ it follows from the same result that player 2's actions B and S must yield the same payoff, so that $\alpha_1(B) = 2\alpha_1(S)$, or $\alpha_1(B) = \frac{2}{3}$. Thus the only nondegenerate mixed strategy Nash equilibrium of the game is $((\frac{2}{3}, \frac{1}{3}), (\frac{1}{3}, \frac{2}{3}))$.

It is illuminating to construct the players' best response functions in the mixed extension of this game. If $0 \leq \alpha_2(B) < \frac{1}{3}$ then player 1's

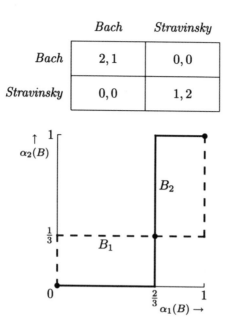

Figure 35.1 The strategic game BoS (top) and the players' best response functions in the mixed extension of this game (bottom). The best response function of player 1 is given by the dashed line; that of player 2 is given by the solid line. The small disks indicate the two pure strategy Nash equilibria and the mixed strategy Nash equilibrium.

unique best response α_1 has $\alpha_1(B) = 0$; if $\frac{1}{3} < \alpha_2(B) \leq 1$ then her unique best response has $\alpha_1(B) = 1$; and if $\alpha_2 = \frac{1}{3}$ then, as we saw above, *all* of her mixed strategies are best responses. Making a similar computation for player 2 we obtain the functions shown at the bottom of Figure 35.1.

? EXERCISE 35.1 (*Guess the average*) Each of n people announces a number in the set $\{1, \ldots, K\}$. A prize of \$1 is split equally between all the people whose number is closest to $\frac{2}{3}$ of the average number. Show that the game has a unique mixed strategy Nash equilibrium, in which each player's strategy is pure.

? EXERCISE 35.2 (*An investment race*) Two investors are involved in a competition with a prize of \$1. Each investor can spend any amount in the interval $[0, 1]$. The winner is the investor who spends the most; in the event of a tie each investor receives \$0.50. Formulate this situation as a strategic game and find its mixed strategy Nash equilibria. (Note that the players' payoff functions are discontinuous, so that Glicksberg's result does not apply; nevertheless the game has a mixed strategy Nash equilibrium.)

In Section 2.5 we define and study the class of strictly competitive games. We show (Proposition 22.2) that in any strictly competitive strategic game that has a Nash equilibrium the set of equilibria coincides with the set of pairs of maxminimizers. This fact can be used to find the set of mixed strategy Nash equilibria of games whose mixed extensions are strictly competitive. (Note that the fact that a game is strictly competitive does not imply that its mixed extension is strictly competitive. To see this, consider a game in which there are three possible outcomes a^1, a^2, and a^3. Then we may have $a^1 \succ_1 a^2 \succ_1 a^3$ and $a^3 \succ_2 a^2 \succ_2 a^1$, so that the game is strictly competitive, while *both* players may prefer a^2 to the lottery in which a^1 and a^3 occur with equal probabilities, so that its mixed extension is not strictly competitive.)

[?] EXERCISE 36.1 (*Guessing right*) Players 1 and 2 each choose a member of the set $\{1, \ldots, K\}$. If the players choose the same number then player 2 pays \$1 to player 1; otherwise no payment is made. Each player maximizes his expected monetary payoff. Find the mixed strategy Nash equilibria of this (strictly competitive) game.

[?] EXERCISE 36.2 (*Air strike*) Army A has a single plane with which it can strike one of three possible targets. Army B has one anti-aircraft gun that can be assigned to one of the targets. The value of target k is v_k, with $v_1 > v_2 > v_3 > 0$. Army A can destroy a target only if the target is undefended and A attacks it. Army A wishes to maximize the expected value of the damage and army B wishes to minimize it. Formulate the situation as a (strictly competitive) strategic game and find its mixed strategy Nash equilibria.

[?] EXERCISE 36.3 Show the following mathematical result, which we use in Exercise 64.2. For any two compact convex subsets X and Y of $\mathbb{R}^k$ there exist $x^* \in X$ and $y^* \in Y$ such that $x^* \cdot y \le x^* \cdot y^* \le x \cdot y^*$ for all $x \in X$ and $y \in Y$. (You can prove this result either by appealing to the existence of a Nash equilibrium in a strategic game (Proposition 20.3), or by the following elementary argument (which avoids the implicit use of Kakutani's fixed point theorem). Let (x^k) and (y^k) be sequences dense in X and Y respectively, and for each positive integer n consider the strictly competitive game in which each player has n actions and the payoff function of player 1 is given by $u_1(i,j) = x^i \cdot y^j$; use Propositions 33.1 and 22.2.)

3.2   Interpretations of Mixed Strategy Nash Equilibrium

In this section we discuss a number of interpretations of mixed strategy equilibrium. We disagree on some points; paragraphs that express the views of only one of us are preceded by that author's initials.

3.2.1   Mixed Strategies as Objects of Choice

Viewed naïvely, a mixed strategy entails a deliberate decision by a player to introduce randomness into his behavior: a player who chooses a mixed strategy commits himself to a random device that probabilistically selects members of his set of actions. After all the players have so committed themselves, the devices are operated and an action profile is realized. Thus each player i chooses a member of $\Delta(A_i)$ in the same way that he chooses a member of A_i in the strategic games discussed in Chapter 2.

There certainly are cases in which players introduce randomness into their behavior. For example, players randomly "bluff" in poker, governments randomly audit taxpayers, and some stores randomly offer discounts.

AR However, the notion of a mixed strategy equilibrium in a strategic game does not capture the players' motivation to introduce randomness into their behavior. Usually a player deliberately randomizes in order to influence the other players' behavior. Consider, for example, the children's version of *Matching Pennies* (Example 17.1) in which the players choose to display an odd or even number of fingers. This game is classically used to motivate the notion of mixed strategy equilibrium, but randomization is a bizarre description of a player's deliberate strategy in the game. A player's action is a response to his guess about the other player's choice; guessing is a psychological operation that is very much deliberate and not random. Alternatively, consider another example often given to motivate mixed strategy equilibrium, namely the relationship between the tax authorities and a taxpayer. The authorities' aim is to deter the taxpayer from tax evasion; considerations of cost lead them to audit only randomly. They would like the taxpayer to know their strategy and are not indifferent between a strategy in which they audit the taxpayer and one in which they do not do so, as required in a mixed strategy equilibrium. The situation should be modeled as a game in which the authorities first choose the probability of auditing, and then, being informed of this probability, the taxpayer takes an action. In such a model the set of possible randomizations is the set of pure strategies.

MJO The main problem with interpreting a player's equilibrium mixed strategy as a deliberate choice is the fact that in a mixed strategy equilibrium each player is indifferent between all mixed strategies whose supports are subsets of her equilibrium strategy: her equilibrium strategy is only one of many strategies that yield her the same expected payoff, given the other players' equilibrium behavior. However, this problem is not limited to mixed strategy equilibria. For example, it afflicts equilibria in many sequential games (including all repeated games), in which a player is indifferent between her equilibrium strategy and many non-equilibrium strategies. Further, in some games there may be other reasons to choose an equilibrium mixed strategy. In strictly competitive games, for example, we have seen that an equilibrium mixed strategy may *strictly* maximize the payoff that a player can guarantee. (This is so, for example, in *Matching Pennies*.) Finally, the ingenious argument of Harsanyi (1973) (considered below in Section 3.2.4) provides some relief from this feature of an equilibrium mixed strategy.

MJO It seems likely that the mixed strategy equilibrium of *Matching Pennies* provides a good description of the steady state behavior of players who play the game repeatedly against randomly selected opponents. In such a situation a player has no way of guessing the action of her opponent in any particular encounter, and it is reasonable for her to adopt the strategy that maximizes the payoff that she can guarantee. If two players interact repeatedly then the psychology of guessing may offer insights into their behavior, though even in this case the mixed strategy equilibrium of the game may provide a good description of their behavior. The tax auditing situation can equally well be modeled as a strategic game in which the choices of the players are simultaneous. The equilibrium audit probability chosen by the authorities is the same in this game as it is in the game in which the authorities move first; given the behavior of the taxpayer, the authorities are indifferent between auditing and not.

3.2.2 Mixed Strategy Nash Equilibrium as a Steady State

In Chapter 2 we interpreted a Nash equilibrium as a steady state of an environment in which players act repeatedly and ignore any strategic link that may exist between plays. We can interpret a mixed strategy Nash equilibrium similarly as a stochastic steady state. The players have information about the frequencies with which actions were taken in the past ("80% of the time a player in the role of player 1 in this game took the action a_1 and 20% of the time such a player took the action b_1"); each player uses these frequencies to form his belief about the future behavior of the other players, and hence formulate his action. In equilibrium these frequencies remain constant over time and are stable in the sense that any action taken with positive probability by a player is optimal given the steady state beliefs.

A mixed strategy equilibrium predicts that the outcome of a game is stochastic, so that for a single play of a game its prediction is less precise than that of a pure strategy equilibrium. But as we argued in Section 1.5, the role of the theory is to explain regularities; the notion of mixed strategy equilibrium captures stochastic regularity.

A variant of this interpretation is based on an interpretation of an n-player game as a model of the interaction of n large populations. Each occurrence of the game takes place after n players are randomly drawn, one from each population. The probabilities in player i's equilibrium mixed strategy are interpreted as the steady state frequencies with which the members of A_i are used in the ith population. In this interpretation the game is a reduced form of a model in which the populations are described explicitly.

An assumption that underlies the steady state interpretation is that no player detects any correlation among the other players' actions or between the other players' actions and his own behavior. Removing this assumption leads to the notion of correlated equilibrium, which we discuss in Section 3.3.

3.2.3 Mixed Strategies as Pure Strategies in an Extended Game

Before selecting his action a player may receive random private information, inconsequential from the point of view of the other players, on which his action may depend. The player may not consciously choose the connection between his action and the realization of his private information; it may just happen that there is a correlation between the two that causes his action to appear to be "random" from the point of view of another player or outside observer. In modeling a player's behavior as random, a mixed strategy Nash equilibrium captures the dependence of behavior on factors that the players perceive as irrelevant. Alternatively, a player may be aware that external factors determine his opponents' behavior, but may find it impossible or very costly to determine the relationship. (For the same reason we model the outcome of a coin toss as random rather than describe it as the result of the interaction of its starting position and velocity, the wind speed, and other factors.) To summarize, a mixed strategy Nash equilibrium, viewed in this way, is a description of a steady state of the system that reflects elements missing from the original description of the game.

To be more concrete, consider the game BoS (Example 34.1). As we saw, this game has a mixed strategy Nash equilibrium $((\frac{2}{3}, \frac{1}{3}), (\frac{1}{3}, \frac{2}{3}))$. Now suppose that each player has three possible "moods", determined by factors he does not understand. Each player is in each of these moods one-third of the time, independently of the other player's mood; his mood has no effect on his payoff. Assume that player 1 chooses *Bach* whenever she is in moods 1 or 2 and *Stravinsky* when she is in mood 3, and player 2 chooses *Bach* when he is in mood 1 and *Stravinsky* when he is in moods 2 or 3. Viewing the situation as a Bayesian game in which the three types of each player correspond to his possible moods, this behavior defines a *pure* strategy equilibrium corresponding exactly to the mixed strategy Nash equilibrium of the original game BoS. Note that this interpretation of the mixed strategy equilibrium does not depend on each player's having three equally likely and independent moods; we need the players' private information only to be rich enough that they can create the appropriate random variables. Nevertheless, the requirement that such an informational structure exist limits the interpretation.

AR There are three criticisms of this interpretation. First, it is hard to accept that the deliberate behavior of a player depends on factors that have no effect on his payoff. People usually give reasons for their choices; in any particular situation a modeler who wishes to apply the notion of mixed strategy equilibrium should point out the reasons that are payoff irrelevant and explain the required dependency between the player's private information and his choice.

MJO In a mixed strategy equilibrium each player is indifferent between all the actions in the support of her equilibrium strategy, so that it is not implausible that the action chosen depends upon factors regarded by the modeler as "irrelevant". When asked why they chose a certain action from a set whose members are equally attractive, people often give answers like "I don't know—I just felt like it".

AR Second, the behavior predicted by an equilibrium under this interpretation is very fragile. If a manager's behavior is determined by the type of breakfast he eats, then factors outside the model, such as a change in his diet or the price of eggs, may change the frequency with which he chooses his actions, thus inducing changes in the beliefs of the other players and causing instability.

MJO For each structure of the random events there is a pattern of behavior that leads to the *same* equilibrium. For example, if, before an increase in the price of eggs, there was an equilibrium in which a manager offered a discount on days when she ate eggs for breakfast and got up before 7:30 A.M., then after the price increase there may be an equilibrium in which she offers the discount when she eats eggs and gets up before 8 A.M. After the price change her old pattern of behavior is no longer a best response to the other players' strategies; whether or not the system will adjust in a *stable* way to the new equilibrium depends on the process of adjustment. A mixed strategy Nash equilibrium is fragile in the sense that the players have no positive incentive to adhere to their equilibrium patterns of behavior (since the equilibrium strategies are not *uniquely* optimal); beyond this, an equilibrium under this interpretation is no more fragile than under any other interpretation. (And, once again, this is a problem that is addressed by Harsanyi's model, discussed in the next section.)

AR Third, in order to interpret an equilibrium of a particular problem in this way one needs to indicate the "real life" exogenous variables on which the players base their behavior. For example, to interpret a mixed strategy Nash equilibrium in a model of price competition one should both specify the unmodeled factors that serve as the basis for the firms' pricing policies and show that the information structure is rich enough to span the set of all mixed strategy Nash equilibria. Those who apply the notion of mixed strategy equilibrium rarely do so.

MJO A player in the world has access to a multitude of random variables on which her actions may depend: the time she wakes up in the morning, the "mood" she is in, the time her newspaper is delivered, ... The structure of these random variables is so rich that it is unnecessary to spell them out in every application of the theory. To interpret mixed strategies as pure strategies in a larger game nicely captures the idea that the action chosen by a player may depend on factors outside the model.

3.2.4 Mixed Strategies as Pure Strategies in a Perturbed Game

We now present a rationale for mixed strategy equilibrium due to Harsanyi (1973). A game is viewed as a frequently occurring situation in which the players' preferences are subject to small random variations. (Thus, as in the argument of the previous section, random factors are introduced, but here they are payoff-relevant.) In each occurrence of the situation each player knows his own preferences but not those of the other players. A mixed strategy equilibrium is a summary of the frequencies with which the players choose their actions over time.

Let $G = \langle N, (A_i), (u_i) \rangle$ be a finite strategic game and let $\epsilon = (\epsilon_i(a))_{i \in N, a \in A}$ be a collection of random variables with range $[-1, 1]$, where $\epsilon_i = (\epsilon_i(a))_{a \in A}$ has a continuously differentiable density function and an absolutely continuous distribution function, and the random vectors $(\epsilon_i)_{i \in N}$ are independent. Consider a family of perturbed games in which each player i's payoff at the outcome a is subject to the small random variation $\epsilon_i(a)$, each player i knowing the realization $(\epsilon_i(a))_{a \in A}$ of ϵ_i but not the realizations of the other players' random variables. That is, consider the Bayesian game $G(\epsilon)$ in which the set of states of nature is the set of all possible values of the realizations of ϵ, the (common) prior belief of each player is the probability distribution specified by ϵ, the signal function of player i informs him only of the realizations $(\epsilon_i(a))_{a \in A}$, and the payoff of player i at the outcome a and state ϵ is $u_i(a) + \epsilon_i(a)$. (Note that each player has infinitely many types.)

Harsanyi's main result (1973, Theorems 2 and 7) is that for almost any game G and *any* collection ϵ^* of random variables satisfying the conditions above, almost any mixed strategy Nash equilibrium of G is the mixed strategy profile associated with the limit, as the size γ of the perturbation vanishes, of a sequence of pure strategy equilibria of the Bayesian games $G(\gamma \epsilon^*)$ in each of which the action chosen by each type is *strictly* optimal. Further, the limit of any such convergent sequence is associated with a mixed strategy equilibrium of G (Harsanyi (1973, Theorem 5)). That is, when the random variations in payoffs are small, almost any mixed equilibrium of the game G is close to a pure equilibrium of the associated Bayesian game and *vice versa*. We say that a mixed strategy equilibrium of G with this property is *approachable* under ϵ^*. (Because of the relative mathematical complexity of these results we do not include proofs.)

☐ EXERCISE 42.1 Consider two-player games in which each player i has two pure strategies, a_i and b_i. Let δ_i for $i = 1, 2$ be independent random variables, each uniformly distributed on $[-1, 1]$, and let the random variables $\epsilon_i(a)$ for $i = 1, 2$ and $a \in A$ have the property that $\epsilon_1(a_1, x) - \epsilon_1(b_1, x) = \delta_1$ for $x = a_2, b_2$ and $\epsilon_2(x, a_2) - \epsilon_2(x, b_2) = \delta_2$ for $x = a_1, b_1$.

　　a. Show that all the equilibria of *BoS* (Example 15.3) are approachable under ϵ.

b. For the game in which $u_i(a_1, a_2) = 1$ for $i = 1$, 2 and all other payoffs are zero, show that only the pure strategy Nash equilibrium (a_1, a_2) is approachable under ϵ.

c. For the game in which $u_i(a) = 0$ for $i = 1$, 2 and all $a \in A$, show that only the mixed strategy Nash equilibrium α in which $\alpha_i(a_i) = \alpha_i(b_i) = \frac{1}{2}$ for $i = 1$, 2 is approachable under ϵ. (Other equilibria are approachable under other perturbations.)

Thus Harsanyi's rationale for a mixed strategy equilibrium is that even if no player makes any effort to use his pure strategies with the required probabilities, the random variations in the payoff functions induce each player to choose his pure strategies with the right frequencies. The equilibrium behavior of the other players is such that a player who chooses the uniquely optimal pure strategy for each realization of his payoff function chooses his actions with the frequencies required by his equilibrium mixed strategy.

MJO Harsanyi's result is an elegant response to the claim that a player has no reason to choose her equilibrium mixed strategy since she is indifferent between all strategies with the same support. I argued above that for some games, including strictly competitive games, this criticism is muted, since there are other reasons for players to choose their equilibrium mixed strategies. Harsanyi's result shows that in almost *any* game the force of the criticism is limited, since almost any mixed strategy Nash equilibrium is close to a strict pure strategy equilibrium of *any* perturbation of the game in which the players' payoffs are subject to small random variations.

3.2.5 Mixed Strategies as Beliefs

Under another interpretation, upon which we elaborate in Section 5.4, a mixed strategy Nash equilibrium is a profile β of *beliefs*, in which β_i is the *common* belief of all the *other* players about player i's actions, with the property that for each player i each action in the support of β_i is optimal given β_{-i}. Under this interpretation each player chooses a single action rather than a mixed strategy. An equilibrium is a steady state of the players' beliefs, not their actions. These beliefs are required to satisfy two properties: they are common among all players and are consistent with the assumption that every player is an expected utility maximizer.

If we were to start from this idea, we would formulate the notion of equilibrium as follows.

▶ DEFINITION 44.1 A *mixed strategy Nash equilibrium* of a finite strategic game is a mixed strategy profile α^* with the property that for every player i every action in the support of α_i^* is a best response to α_{-i}^*.

Lemma 33.2 shows that this definition is equivalent to our previous definition (32.3) and thus guarantees that the idea is indeed an interpretation of mixed strategy equilibrium.

Note, however, that when we interpret mixed strategy equilibrium in this way the predictive content of an equilibrium is small: it predicts only that each player uses an action that is a best response to the equilibrium beliefs. The set of such best responses includes any action in the support of a player's equilibrium mixed strategy and may even include actions outside the support of this strategy.

3.3   Correlated Equilibrium

In Section 3.2.3 we discuss an interpretation of a mixed strategy Nash equilibrium as a steady state in which each player's action depends on a signal that he receives from "nature". In this interpretation the signals are private and independent.

What happens if the signals are not private and independent? Suppose, for example, that in BoS (see Figure 35.1) *both* players observe a random variable that takes each of the two values x and y with probability $\frac{1}{2}$. Then there is a new equilibrium, in which both players choose *Bach* if the realization is x and *Stravinsky* if the realization is y. Given each player's information, his action is optimal: if the realization is x then he knows that the other player chooses *Bach*, so that it is optimal for him to choose *Bach*, and symmetrically if the realization is y.

In this example the players observe the same random variable. More generally, their information may be less than perfectly correlated. Suppose, for example, that there is a random variable that takes the three values x, y, and z, and player 1 knows only that the realization is either x or that it is a member of $\{y, z\}$, while player 2 knows only that it is either a member of $\{x, y\}$ or that it is z. That is, player 1's *information partition* is $\{\{x\}, \{y, z\}\}$ and player 2's is $\{\{x, y\}, \{z\}\}$. Under these assumptions a strategy of player 1 consists of two actions: one that she uses when she knows that the realization is x and one that she uses when she knows that the realization is a member of $\{y, z\}$. Similarly, a strategy of player 2 consists of two actions, one for $\{x, y\}$ and one for z. A player's strategy is optimal if, given the strategy of the other player, for any realization of his information he can do no better by choosing an action different from that dictated by his strategy. To illustrate how a player uses his information in choosing an optimal action, suppose that the probabilities of y and z are η and ζ and player 2's strategy is to take the action a_2 if he knows that the realization is in $\{x, y\}$ and b_2 if he knows that the realization is z. Then if player 1 is informed that either y or z has occurred he chooses an action that is optimal given that player 2 chooses a_2 with probability $\eta/(\eta + \zeta)$ (the probability of y conditional on $\{y, z\}$) and b_2 with probability $\zeta/(\eta + \zeta)$.

These examples lead us to the following notion of equilibrium.

▶ DEFINITION 45.1 A **correlated equilibrium** of a strategic game $\langle N, (A_i), (u_i) \rangle$ consists of

- a finite probability space (Ω, π) (Ω is a set of **states** and π is a probability measure on Ω)

- for each player $i \in N$ a partition $\mathcal{P}_i$ of Ω (player i's **information partition**)

- for each player $i \in N$ a function $\sigma_i : \Omega \to A_i$ with $\sigma_i(\omega) = \sigma_i(\omega')$ whenever $\omega \in P_i$ and $\omega' \in P_i$ for some $P_i \in \mathcal{P}_i$ (σ_i is player i's **strategy**)

such that for every $i \in N$ and every function $\tau_i : \Omega \to A_i$ for which $\tau_i(\omega) = \tau_i(\omega')$ whenever $\omega \in P_i$ and $\omega' \in P_i$ for some $P_i \in \mathcal{P}_i$ (i.e. for every strategy of player i) we have

$$\sum_{\omega \in \Omega} \pi(\omega) u_i(\sigma_{-i}(\omega), \sigma_i(\omega)) \geq \sum_{\omega \in \Omega} \pi(\omega) u_i(\sigma_{-i}(\omega), \tau_i(\omega)). \qquad (45.2)$$

Note that the probability space and information partition are not exogenous but are part of the equilibrium. Note also that (45.2) is equivalent to the requirement that for every state ω that occurs with positive probability the action $\sigma_i(\omega)$ is optimal given the other players' strategies and player i's knowledge about ω. (This equivalence depends on the assumption that the players' preferences obey expected utility theory.)

We begin by showing that the set of correlated equilibria contains the set of mixed strategy Nash equilibria.

■ PROPOSITION 45.3 *For every mixed strategy Nash equilibrium α of a finite strategic game $\langle N, (A_i), (u_i) \rangle$ there is a correlated equilibrium $\langle (\Omega, \pi), (\mathcal{P}_i), (\sigma_i) \rangle$ in which for each player $i \in N$ the distribution on A_i induced by σ_i is α_i.*

Proof. Let $\Omega = A\ (= \times_{j \in N} A_j)$ and define π by $\pi(a) = \Pi_{j \in N} \alpha_j(a_j)$. For each $i \in N$ and $b_i \in A_i$ let $P_i(b_i) = \{a \in A : a_i = b_i\}$ and let $\mathcal{P}_i$ consist of the $|A_i|$ sets $P_i(b_i)$. Define σ_i by $\sigma_i(a) = a_i$ for each $a \in A$. Then $\langle(\Omega, \pi), (\mathcal{P}_i), (\sigma_i)\rangle$ is a correlated equilibrium since (45.2) is satisfied for every strategy τ_i: the left-hand side is player i's payoff in the mixed strategy Nash equilibrium α and the right-hand side is his payoff when he uses the mixed strategy in which he chooses the action $\tau_i(a)$ with probability $\alpha_i(a_i)$ and every other player j uses the mixed strategy α_j. Further, the distribution on A_i induced by σ_i is α_i. $\square$

The following example is a formal expression of the example with which we began this section.

◇ EXAMPLE 46.1 The three mixed strategy Nash equilibrium payoff profiles in BoS (see Example 34.1) are $(2, 1)$, $(1, 2)$, and $(\frac{2}{3}, \frac{2}{3})$. In addition one of the correlated equilibria yields the payoff profile $(\frac{3}{2}, \frac{3}{2})$: let $\Omega = \{x, y\}$, $\pi(x) = \pi(y) = \frac{1}{2}$, $\mathcal{P}_1 = \mathcal{P}_2 = \{\{x\}, \{y\}\}$, $\sigma_i(x) = Bach$, and $\sigma_i(y) = Stravinsky$ for $i = 1, 2$. One interpretation of this equilibrium is that the players observe the outcome of a public coin toss, which determines which of the two pure strategy Nash equilibria they play.

This example suggests the following result.

■ PROPOSITION 46.2 *Let $G = \langle N, (A_i), (u_i)\rangle$ be a strategic game. Any convex combination of correlated equilibrium payoff profiles of G is a correlated equilibrium payoff profile of G.*

Proof. Let $u^1, \ldots, u^K$ be correlated equilibrium payoff profiles and let $(\lambda^1, \ldots, \lambda^K) \in \mathbb{R}^K$ with $\lambda^k \geq 0$ for all k and $\sum_{k=1}^{K} \lambda^k = 1$. For each value of k let $\langle(\Omega^k, \pi^k), (\mathcal{P}_i^k), (\sigma_i^k)\rangle$ be a correlated equilibrium that generates the payoff profile u^k; without loss of generality assume that the sets Ω^k are disjoint. The following defines a correlated equilibrium for which the payoff profile is $\sum_{k=1}^{K} \lambda^k u^k$. Let $\Omega = \cup_k \Omega^k$, and for any $\omega \in \Omega$ define π by $\pi(\omega) = \lambda^k \pi^k(\omega)$ where k is such that $\omega \in \Omega^k$. For each $i \in N$ let $\mathcal{P}_i = \cup_k \mathcal{P}_i^k$ and define σ_i by $\sigma_i(\omega) = \sigma_i^k(\omega)$ where k is such that $\omega \in \Omega^k$. $\square$

We can interpret the correlated equilibrium constructed in this proof as follows: first a public random device determines which of the K correlated equilibria is to be played, and then the random variable corresponding to the kth correlated equilibrium is realized.

◇ EXAMPLE 46.3 Consider the game in the left-hand side of Figure 47.1. The Nash equilibrium payoff profiles are $(2, 7)$ and $(7, 2)$ (pure) and

	L	R
T	6,6	2,7
B	7,2	0,0

	L	R
T	y	z
B	x	–

Figure 47.1 An example of a correlated equilibrium. On the left is a strategic game. The table on the right gives the choices of the players as a function of the state in a correlated equilibrium of the game.

$(4\frac{2}{3}, 4\frac{2}{3})$ (mixed). The following correlated equilibrium yields a payoff profile that is outside the convex hull of these three profiles. Let $\Omega = \{x, y, z\}$ and $\pi(x) = \pi(y) = \pi(z) = \frac{1}{3}$; let player 1's partition be $\{\{x\}, \{y, z\}\}$ and player 2's be $\{\{x, y\}, \{z\}\}$. Define the strategies as follows: $\sigma_1(x) = B$ and $\sigma_1(y) = \sigma_1(z) = T$; $\sigma_2(x) = \sigma_2(y) = L$ and $\sigma_2(z) = R$. (The relation between the choices and the states is shown in the right-hand side of Figure 47.1.) Then player 1's behavior is optimal given player 2's: in state x, player 1 knows that player 2 plays L and thus it is optimal for her to play B; in states y and z she assigns equal probabilities to player 2 using L and R, so that it is optimal for her to play T. Symmetrically, player 2's behavior is optimal given player 1's, and hence we have a correlated equilibrium; the payoff profile is $(5, 5)$.

This example, in which we can identify the set of states with the set of outcomes, suggests the following result.

∎ PROPOSITION 47.1 *Let $G = \langle N, (A_i), (u_i) \rangle$ be a finite strategic game. Every probability distribution over outcomes that can be obtained in a correlated equilibrium of G can be obtained in a correlated equilibrium in which the set of states is A and for each $i \in N$ player i's information partition consists of all sets of the form $\{a \in A : a_i = b_i\}$ for some action $b_i \in A_i$.*

Proof. Let $\langle (\Omega, \pi), (\mathcal{P}_i), (\sigma_i) \rangle$ be a correlated equilibrium of G. Then $\langle (\Omega', \pi'), (\mathcal{P}'_i), (\sigma'_i) \rangle$ is also a correlated equilibrium, where $\Omega' = A$, $\pi'(a) = \pi(\{\omega \in \Omega : \sigma(\omega) = a\})$ for each $a \in A$, $\mathcal{P}'_i$ consists of sets of the type $\{a \in A : a_i = b_i\}$ for some $b_i \in A_i$, and σ'_i is defined by $\sigma'_i(a) = a_i$. □

This result allows us to confine attention, when calculating correlated equilibrium payoffs, to equilibria in which the set of states is the set of outcomes. Note however that such equilibria may have no natural interpretation.

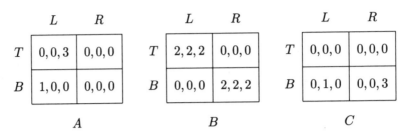

Figure 48.1 A three-player game. Player 1 chooses one of the two rows, player 2 chooses one of the two columns, and player 3 chooses one of the three tables.

In the definition of a correlated equilibrium we assume that the players share a common belief about the probabilities with which the states occur. If there is a random variable about which the players hold different beliefs then additional equilibrium payoff profiles are possible. Suppose, for example, that player 1 is sure that team T_1 will beat team T_2 in some contest, while player 2 is sure that team T_2 will win. Then there is an equilibrium of BoS (Example 34.1) in which the outcome is (*Bach, Bach*) if T_1 wins and (*Stravinsky, Stravinsky*) if team T_2 wins, which gives each player an expected payoff of 2! (In Section 5.3 we show that it cannot be common knowledge between two players that their beliefs differ in the way we have just assumed if they have the same priors.)

[?] EXERCISE 48.1   Consider the three-player game with the payoffs given in Figure 48.1. (Player 1 chooses one of the two rows, player 2 chooses one of the two columns, and player 3 chooses one of the three tables.)

  a. Show that the pure strategy equilibrium payoffs are $(1, 0, 0)$, $(0, 1, 0)$, and $(0, 0, 0)$.

  b. Show that there is a correlated equilibrium in which player 3 chooses B and players 1 and 2 play (T, L) and (B, R) with equal probabilities.

  c. Explain the sense in which player 3 prefers not to have the information that players 1 and 2 use to coordinate their actions.

3.4   Evolutionary Equilibrium

In this section we describe the basic idea behind a variant of the notion of Nash equilibrium called *evolutionary equilibrium*. This notion is designed to model situations in which the players' actions are determined by the forces of evolution. We confine the discussion to a simple case in which the members of a single population of organisms (animals, human

beings, plants, ...) interact with each other pairwise. In each match
each organism takes an action from a set B. The organisms do not
consciously choose actions; rather, they either inherit modes of behavior
from their forebears or are assigned them by mutation. We assume that
there is a function u that measures each organism's ability to survive: if
an organism takes the action a when it faces the distribution β of actions
in the population of its potential opponents, then its ability to survive
is measured by the expectation of $u(a, b)$ under β. This description corresponds to a two-player symmetric strategic game $\langle \{1, 2\}, (B, B), (u_i) \rangle$
where $u_1(a, b) = u(a, b)$ and $u_2(a, b) = u(b, a)$.

A candidate for an evolutionary equilibrium is an action in B. The
notion of equilibrium is designed to capture a steady state in which all
organisms take this action and no mutant can invade the population.
More precisely, the idea is that for every possible action $b \in B$ the evolutionary process occasionally transforms a small fraction of the population into mutants who follow b. In an equilibrium any such mutant must
obtain an expected payoff lower than that of the equilibrium action, so
that it dies out. Now, if the fraction $\epsilon > 0$ of the population consists of
mutants taking the action b while all other organisms take the action b^*,
then the average payoff of a mutant is $(1 - \epsilon)u(b, b^*) + \epsilon u(b, b)$ (since with
probability $1 - \epsilon$ it encounters a non-mutant and with probability ϵ it
encounters another mutant), while the average payoff of a non-mutant
is $(1 - \epsilon)u(b^*, b^*) + \epsilon u(b^*, b)$. Therefore for b^* to be an evolutionary
equilibrium we require

$$(1 - \epsilon)u(b, b^*) + \epsilon u(b, b) < (1 - \epsilon)u(b^*, b^*) + \epsilon u(b^*, b)$$

for all values of ϵ sufficiently small. This inequality is satisfied if and
only if for every $b \neq b^*$ either $u(b, b^*) < u(b^*, b^*)$, or $u(b, b^*) = u(b^*, b^*)$
and $u(b, b) < u(b^*, b)$, so that we can define an evolutionary equilibrium
as follows.

▷ DEFINITION 49.1 Let $G = \langle \{1, 2\}, (B, B), (u_i) \rangle$ be a symmetric strategic game, where $u_1(a, b) = u_2(b, a) = u(a, b)$ for some function u. An
evolutionarily stable strategy (ESS) of G is an action $b^* \in B$ for
which (b^*, b^*) is a Nash equilibrium of G and $u(b, b) < u(b^*, b)$ for every
best response $b \in B$ to b^* with $b \neq b^*$.

In the following example, as in much of the literature, the set B is
taken to be the set of mixed strategies over some finite set of actions.

◇ EXAMPLE 49.2 (*Hawk–Dove*) From time to time pairs of animals in
a population fight over a prey with value 1. Each animal can behave

	D	H
D	$\frac{1}{2},\frac{1}{2}$	$0,1$
H	$1,0$	$\frac{1}{2}(1-c),\frac{1}{2}(1-c)$

Figure 50.1 A Hawk–Dove game.

γ,γ	$1,-1$	$-1,1$
$-1,1$	γ,γ	$1,-1$
$1,-1$	$-1,1$	γ,γ

Figure 50.2 A game without an ESS. Each pure strategy yields a mutant a payoff higher than the unique symmetric equilibrium mixed strategy.

either like a dove (D) or like a hawk (H). If both animals in a match are dovish then they split the value of the prey; if they are both hawkish then the value of the prey is reduced by c and is split evenly; if one of them is hawkish and the other is dovish then the hawk gets 1 and the dove 0. The game is shown in Figure 50.1. (If $c > 1$, it has the same structure as that in Figure 17.2.) Let B be the set of all mixed strategies over $\{D, H\}$. If $c > 1$, the game has a unique symmetric mixed strategy Nash equilibrium, in which each player uses the strategy $(1 - 1/c, 1/c)$; this strategy is the only ESS. (In particular, in this case a population exclusively of hawks is not evolutionarily stable.) If $c < 1$, the game has a unique mixed strategy Nash equilibrium in which each player uses the pure strategy H; this strategy is the only ESS.

It is immediate from Definition 49.1 that if (b^*, b^*) is a symmetric Nash equilibrium and no strategy other than b^* is a best response to b^* (i.e. (b^*, b^*) is a *strict* equilibrium) then b^* is an ESS. A nonstrict equilibrium strategy may not be an ESS: consider the two-player symmetric game in which each player has two actions and $u(a, b) = 1$ for all $(a, b) \in B \times B$. For a more interesting example of a nonstrict equilibrium strategy that is not an ESS, consider the game in Figure 50.2 in which B consists of all mixed strategies over a set containing three members and $0 \leq \gamma \leq 1$. This game has a unique symmetric mixed strategy Nash equilibrium in which each player's mixed strategy is $(\frac{1}{3}, \frac{1}{3}, \frac{1}{3})$; in this equilibrium the expected payoff of each player is $\gamma/3$. A mutant who uses any of the three pure strategies obtains an expected payoff of $\gamma/3$ when it encounters a non-mutant, but the higher payoff γ when it encounters another mutant. Hence the equilibrium mixed strategy is not an ESS (from which it follows that not every game that has a Nash equilibrium has an ESS).

[?] EXERCISE 51.1 Show that in every two-player symmetric strategic game in which each player has two pure strategies and the payoffs to the four strategy profiles are different there is a mixed strategy that is an ESS.

Notes

The modern formulation of a mixed strategy is due to Borel (1921; 1924, pp. 204–221; 1927), although the idea dates back at least to the early eighteenth century (see Guilbaud (1961) and Kuhn (1968)). Borel establishes the existence of a mixed strategy Nash equilibrium for some special strictly competitive games; von Neumann (1928) proves the existence of an equilibrium for all strictly competitive games. The existence result (Proposition 33.1) that we prove (which covers all finite strategic games) is due to Nash (1950a, 1951). The notion of a correlated equilibrium is due to Aumann (1974), whose paper is also the basis for the other material in Section 3.3. The idea of an evolutionarily stable strategy is due to Maynard Smith and Price (see Maynard Smith (1972) and Maynard Smith and Price (1973); see also Maynard Smith (1974, 1982)).

The large population model mentioned in Section 3.2.2 is due to Rosenthal (1979). The idea of interpreting mixed strategies as pure strategies in an extended game discussed in Section 3.2.3 is due to Harsanyi (1973), as is the content of Section 3.2.4. The interpretation of a mixed strategy Nash equilibrium given in Section 3.2.5 is discussed in Aumann (1987a). Some of the criticism of mixed strategy Nash equilibrium given in Section 3.2 is taken from Rubinstein (1991). The examples in Section 3.3 are due to Aumann (1974).

Our proof of Proposition 33.1, due to Nash (1950a), appeals to Proposition 20.3, the proof of which uses Kakutani's fixed point theorem. Nash (1951) presents an alternative proof of Proposition 33.1 that uses the more basic fixed point theorem of Brouwer, which applies to point-valued functions.

The game in Exercise 35.1 is taken from Moulin (1986, p. 72). Exercise 36.3 is taken from Arrow, Barankin, and Blackwell (1953).

For a discussion of mixed strategy Nash equilibrium when the players' preferences do not satisfy the assumptions necessary to be represented by expected utility functions see Crawford (1990). The notion of ESS that we discuss in Section 3.4 has been extended in various directions; see van Damme (1991, Chapter 9).

We have not addressed the question of whether there is any dynamic adjustment process that leads to an equilibrium. One such process, called *fictitious play*, is suggested by Brown (1951), and has recently been reconsidered. In this process each player always chooses a best response to the statistical frequency of the other players' past actions. Robinson (1951) shows that the process converges to a mixed strategy Nash equilibrium in any strictly competitive game; Shapley (1964, Section 5) shows that this is not necessarily so in games that are not strictly competitive. Recent research focuses on models that explicitly capture the forces of evolution and learning; see Battigalli, Gilli, and Molinari (1992) for an introduction to this work.

4 Rationalizability and Iterated Elimination of Dominated Actions

In this chapter we examine the consequences of requiring that a player's choice be optimal given a belief consistent with the view that every player is rational, every player thinks that every player is rational, every player thinks that every player thinks that every player is rational, and so on.

4.1 Rationalizability

In Chapters 2 and 3 we discuss solution concepts for strategic games in which each player's choice is required to be optimal given his belief about the other players' behavior, a belief that is required to be correct. That is, we assume that each player knows the other players' equilibrium behavior. If the players participate repeatedly in the situation that the game models then they can obtain this knowledge from the steady state behavior that they observe. However, if the game is a one-shot event in which all players choose their actions simultaneously then it is not clear how each player can know the other players' equilibrium actions; for this reason game theorists have developed solution concepts that do not entail this assumption.

In this chapter we study some such solution concepts, in which the players' beliefs about each other's actions are not assumed to be correct, but are constrained by considerations of rationality: each player believes that the actions taken by every other player is a best response to some belief, and, further, each player assumes that every other player reasons in this way and hence thinks that every other player believes that every other player's action is a best response to some belief, and so on.

The solution concepts that we study are weaker than Nash equilibrium. In fact, in many games they do not exclude any action from being

used. Nevertheless we find the approach interesting in that it explores
the logical implications of assumptions about the players' knowledge
that are weaker than those in the previous chapters.

Fix a strategic game $\langle N, (A_i), (u_i) \rangle$ (in which the expectation of u_i
represents player i's preferences over lotteries on $A = \times_{j \in N} A_j$ for each
$i \in N$). In order to develop the ideas in this chapter it is not necessary
to assume that the action set A_i of each player is finite, though for
simplicity we adopt this assumption in some of the discussion. A *belief* of
player i (about the actions of the other players) is a probability measure
on A_{-i} $(= \times_{j \in N \setminus \{i\}} A_j)$. Note that this definition allows a player to
believe that the other players' actions are correlated: a belief is not
necessarily a product of independent probability measures on each of the
action sets A_j for $j \in N \setminus \{i\}$. As before, an action $a_i \in A_i$ of player i is
a *best response* to a belief if there is no other action that yields player i
a higher payoff given the belief. We frequently use the phrase "player i
thinks that some other player j is rational", which we take to mean that
player i thinks that whatever action player j chooses is a best response
to player j's belief about the actions of the players other than j.

If player i thinks that every other player j is rational then he must be
able to rationalize his belief μ_i about the other players' actions as follows:
every action of any other player j to which the belief μ_i assigns positive
probability must be a best response to a belief of player j. If player i
further thinks that every other player j thinks that every player $h \neq j$
(including player i) is rational then he, player i, must also have a view
about player j's view about player h's beliefs. If player i's reasoning has
unlimited depth, we are led to the following definition.

▶ DEFINITION 54.1 An action $a_i \in A_i$ is **rationalizable** in the strategic
game $\langle N, (A_i), (u_i) \rangle$ if there exists

- a collection $((X_j^t)_{j \in N})_{t=1}^{\infty}$ of sets with $X_j^t \subseteq A_j$ for all j and t,
- a belief μ_i^1 of player i whose support is a subset of X_{-i}^1, and
- for each $j \in N$, each $t \geq 1$, and each $a_j \in X_j^t$, a belief $\mu_j^{t+1}(a_j)$ of
  player j whose support is a subset of X_{-j}^{t+1}

such that

- a_i is a best response to the belief μ_i^1 of player i
- $X_i^1 = \varnothing$ and for each $j \in N \setminus \{i\}$ the set X_j^1 is the set of all $a_j' \in A_j$
  such that there is some a_{-i} in the support of μ_i^1 for which $a_j = a_j'$
- for every player $j \in N$ and every $t \geq 1$ every action $a_j \in X_j^t$ is a best
  response to the belief $\mu_j^{t+1}(a_j)$ of player j

- for each $t \geq 2$ and each $j \in N$ the set X_j^t is the set of all $a_j' \in A_j$ such that there is some player $k \in N \setminus \{j\}$, some action $a_k \in X_k^{t-1}$, and some a_{-k} in the support of $\mu_k^t(a_k)$ for which $a_j' = a_j$.

Note that formally the second and fourth conditions in the second part of this definition are superfluous; we include them so that the definition corresponds more closely to the motivation we gave. Note also that we include the set X_i^1 in the collection $((X_j^t)_{j \in N})_{t=1}^\infty$, even though it is required to be empty, merely to simplify the notation. If $|N| \geq 3$ then X_i^1 is the only such superfluous set, while if $|N| = 2$ there are many (X_i^t for any odd t and, for $j \neq i$, X_j^t for any even t).

The set X_j^1 for $j \in N \setminus \{i\}$ is interpreted to be the set of actions of player j that are assigned positive probability by the belief μ_i^1 of player i about the actions of the players other than i that justifies i choosing a_i. For any $j \in N$ the interpretation of X_j^2 is that it is the set of all actions a_j of player j such that there exists at least one action $a_k \in X_k^1$ of some player $k \neq j$ that is justified by the belief $\mu_k^2(a_k)$ that assigns positive probability to a_j.

To illustrate what the definition entails, suppose there are three players, each of whom has two possible actions, A and B. Assume that the action A of player 1 is rationalizable and that player 1's belief μ_1^1 used in the rationalization assigns positive probability to the choices of players 2 and 3 being either (A, A) or (B, B). Then $X_2^1 = X_3^1 = \{A, B\}$. The beliefs $\mu_2^2(A)$ and $\mu_2^2(B)$ of player 2 that justify his choices of A and B concern the actions of players 1 and 3; the beliefs $\mu_3^2(A)$ and $\mu_3^2(B)$ of player 3 concern players 1 and 2. These four beliefs do not have to induce the same beliefs about player 1 and do not have to assign positive probability to the action A. The set X_1^2 consists of all the actions of player 1 that are assigned positive probability by either $\mu_2^2(A)$, $\mu_3^2(A)$, $\mu_2^2(B)$, or $\mu_3^2(A)$.

This definition of rationalizability is equivalent to the following.

▶ DEFINITION 55.1 An action $a_i \in A_i$ is **rationalizable** in the strategic game $\langle N, (A_i), (u_i) \rangle$ if for each $j \in N$ there is a set $Z_j \subseteq A_j$ such that

- $a_i \in Z_i$

- every action $a_j \in Z_j$ is a best response to a belief $\mu_j(a_j)$ of player j whose support is a subset of Z_{-j}.

Note that if $(Z_j)_{j \in N}$ and $(Z_j')_{j \in N}$ satisfy this definition then so does $(Z_j \cup Z_j')_{j \in N}$, so that the set of profiles of rationalizable actions is the largest set $\times_{j \in N} Z_j$ for which $(Z_j)_{j \in N}$ satisfies the definition.

■ LEMMA 56.1 *Definitions 54.1 and 55.1 are equivalent.*

Proof. If $a_i \in A_i$ is rationalizable according to Definition 54.1 then define $Z_i = \{a_i\} \cup (\cup_{t=1}^{\infty} X_i^t)$ and $Z_j = (\cup_{t=1}^{\infty} X_j^t)$ for each $j \in N \setminus \{i\}$. If it is rationalizable according to Definition 55.1 then define $\mu_i^1 = \mu_i(a_i)$ and $\mu_j^t(a_j) = \mu_j(a_j)$ for each $j \in N$ and each integer $t \geq 2$. Then the sets X_j^t defined in the second and fourth parts of Definition 54.1 are subsets of Z_j and satisfy the conditions in the first and third parts.   □

It is clear from Definition 55.1 that in a finite game any action that a player uses with positive probability in some mixed strategy Nash equilibrium is rationalizable (take Z_j to be the support of player j's mixed strategy). The following result shows that the same is true for actions used with positive probability in some correlated equilibrium.

■ LEMMA 56.2 *Every action used with positive probability by some player in a correlated equilibrium of a finite strategic game is rationalizable.*

Proof. Denote the strategic game by $\langle N, (A_i), (u_i) \rangle$; choose a correlated equilibrium, and for each player $i \in N$ let Z_i be the set of actions that player i uses with positive probability in the equilibrium. Then any $a_i \in Z_i$ is a best response to the distribution over A_{-i} generated by the strategies of the players other than i, conditional on player i choosing a_i. The support of this distribution is a subset of Z_{-i} and hence by Definition 55.1 a_i is rationalizable.   □

In the Prisoner's Dilemma (Example 16.2) only the Nash equilibrium action *Confess* is rationalizable. In all the other games in Section 2.3 both actions of each player are rationalizable, since in each case both actions are used with positive probability in some mixed strategy Nash equilibrium. Thus rationalizability puts no restriction on the outcomes in these games. For many other games the restrictions that rationalizability imposes are weak. However, in some games rationalizability provides a sharp answer, as the following exercises demonstrate.

?⃞ EXERCISE 56.3 Find the set of rationalizable actions of each player in the two-player game in Figure 57.1.

?⃞ EXERCISE 56.4 (*Cournot duopoly*) Consider the strategic game $\langle \{1,2\}, (A_i), (u_i) \rangle$ in which $A_i = [0, 1]$ and $u_i(a_1, a_2) = a_i(1 - a_1 - a_2)$ for $i = 1$, 2. Show that each player's only rationalizable action is his unique Nash equilibrium action.

?⃞ EXERCISE 56.5 (*Guess the average*) In the game in Exercise 35.1 show that each player's equilibrium action is his unique rationalizable action.

	b_1	b_2	b_3	b_4
a_1	0, 7	2, 5	7, 0	0, 1
a_2	5, 2	3, 3	5, 2	0, 1
a_3	7, 0	2, 5	0, 7	0, 1
a_4	0, 0	0, -2	0, 0	10, -1

Figure 57.1 The two-player game in Exercise 56.3.

? EXERCISE 57.1 Suppose that two players choose locations a_1 and a_2 in the unit interval; each wishes to be as close as possible to the other, the payoff of each player being $-|a_1 - a_2|$. Show that every action of each player is rationalizable, while the set of Nash equilibria is $\{(a_1, a_2): a_1 = a_2\}$. Now assume that each player is informed of the distance to his opponent. Modify Definition 55.1 by adding the condition that the support of a belief that rationalizes a pair (a_i, d) consisting of an action a_i and a distance d be a subset of $\{a_j - d, a_j + d\}$. Show that for no $d > 0$ is there an action a_i for which (a_i, d) is rationalizable in this new sense, while $(a_i, 0)$ is rationalizable for every a_i.

Note that in Definitions 54.1 and 55.1 we take a belief of player i to be a probability distribution on A_{-i}, which allows each player to believe that his opponents' actions are correlated. In most of the literature, players are not allowed to entertain such beliefs: it is assumed that each player's belief is a product of independent probability distributions, one for each of the other players. (Such a restriction is obviously inconsequential in a two-player game.) This assumption is consistent with the motivation behind the notion of mixed strategy equilibrium. Our definition of rationalizability requires that at all levels of rationalization the players be rational; the alternative definition of rationalizability requires in addition that at all levels of rationalization the beliefs preserve the assumption of independence.

The two definitions have different implications, as the game in Figure 58.1 shows. In this game there are three players; player 1 chooses one of the two rows, player 2 chooses one of the two columns, and player 3 chooses one of the four tables. All three players obtain the same payoffs, given by the numbers in the boxes. We claim that the action M_2

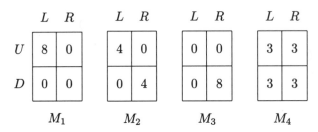

Figure 58.1 A three-player strategic game. Player 1 chooses one of the two rows, player 2 chooses one of the two columns, and player 3 chooses one of the four tables. All three players obtain the same payoffs, given by the numbers in the boxes.

of player 3 is rationalizable in the sense of Definitions 54.1 and 55.1, in which a player may believe that his opponent's actions are correlated, but is not rationalizable if players are restricted to beliefs that are products of independent probability distributions. To see this, note that the action U of player 1 is a best response to a belief that assigns probability one to (L, M_2) and the action D is a best response to the belief that assigns probability one to (R, M_2); similarly, both actions of player 2 are best responses to beliefs that assign positive probability only to U, D, and M_2. Further, the action M_2 of player 3 is a best response to the belief in which players 1 and 2 play (U, L) and (D, R) with equal probabilities. Thus M_2 is rationalizable in the sense that we have defined (take $Z_1 = \{U, D\}$, $Z_2 = \{L, R\}$, and $Z_3 = \{M_2\}$ in Definition 55.1). However, it is not a best response to any pair of (independent) mixed strategies and is thus not rationalizable under the modified definition in which each player's belief is restricted to be a product of independent beliefs. (In order for M_2 to be a best response we need $4pq + 4(1 - p)(1 - q) \geq \max\{8pq, 8(1 - p)(1 - q), 3\}$, where $(p, 1 - p)$ and $(q, 1 - q)$ are mixed strategies of players 1 and 2 respectively, an inequality that is not satisfied for any values of p and q.)

4.2   Iterated Elimination of Strictly Dominated Actions

Like the notion of rationalizability, the solution concept that we now study looks at a game from the point of view of a *single* player. Each player takes an action based on calculations that do not require knowledge of the actions taken by the other players. To define the solution we start by eliminating actions that a player should definitely *not* take. In a complicated game it is particularly attractive to assume that players, looking for ways to simplify the situation they confront, will adopt such a tack. We assume that players exclude from consideration actions that are not best responses *whatever* the other players do. A player who knows that the other players are rational can assume that they too will exclude such actions from consideration. Now consider the game G' obtained from the original game G by eliminating all such actions. Once again, a player who knows that the other players are rational should not choose an action that is not a best response whatever the other players do in G'. Further, a player who knows that the other players know that he is rational can argue that they too will not choose actions that are never best responses in G'. Continuing to argue in this way suggests that the outcome of G must survive an unlimited number of rounds of such elimination. We now formalize this idea and show that it is equivalent to the notion of rationalizability.

4.2.1 Never-Best Responses

▸ DEFINITION 59.1 An action of player i in a strategic game is a **never-best response** if it is not a best response to any belief of player i.

Clearly any action that is a never-best response is not rationalizable. If an action a_i of player i is a never-best response then for every belief of player i there is *some* action, which may depend on the belief, that is better for player i than a_i. We now show that if a_i is a never-best response in a finite game then there is a *mixed* strategy that, whatever belief player i holds, is better for player i than a_i. This alternative property is defined precisely as follows.

▸ DEFINITION 59.2 The action $a_i \in A_i$ of player i in the strategic game $\langle N, (A_i), (u_i) \rangle$ is **strictly dominated** if there is a mixed strategy α_i of player i such that $U_i(a_{-i}, \alpha_i) > u_i(a_{-i}, a_i)$ for all $a_{-i} \in A_{-i}$, where $U_i(a_{-i}, \alpha_i)$ is the payoff of player i if he uses the mixed strategy α_i and the other players' vector of actions is a_{-i}.

In fact, we show that in a game in which the set of actions of each player is finite an action is a never-best response *if and only if* it is strictly dominated. Thus in such games the notion of strict domination has a decision-theoretic basis that does not involve mixed strategies. It follows that even if one rejects the idea that mixed strategies can be objects of choice, one can still argue that a player will not use an action that is strictly dominated.

■ LEMMA 60.1 *An action of a player in a finite strategic game is a never-best response if and only if it is strictly dominated.*

Proof. Let the strategic game be $G = \langle N, (A_i), (u_i) \rangle$ and let $a_i^* \in A_i$. Consider the auxiliary strictly competitive game G' (see Definition 21.1) in which the set of actions of player 1 is $A_i \setminus \{a_i^*\}$, that of player 2 is A_{-i}, and the preferences of player 1 are represented by the payoff function v_1 given by $v_1(a_i, a_{-i}) = u_i(a_{-i}, a_i) - u_i(a_{-i}, a_i^*)$. (Note that the argument (a_i, a_{-i}) of v_1 is a pair of actions in G' while the arguments (a_{-i}, a_i) and (a_{-i}, a_i^*) are action profiles in G.) For any given mixed strategy profile (m_1, m_2) in G' we denote by $v_1(m_1, m_2)$ the expected payoff of player 1.

The action a_i^* is a never-best response in G if and only if for any mixed strategy of player 2 in G' there is an action of player 1 that yields a positive payoff; that is, if and only if $\min_{m_2} \max_{a_i} v_1(a_i, m_2) > 0$. This is so if and only if $\min_{m_2} \max_{m_1} v_1(m_1, m_2) > 0$ (by the linearity of v_1 in m_1).

Now, by Proposition 33.1 the game G' has a mixed strategy Nash equilibrium, so from part (b) of Proposition 22.2 applied to the mixed extension of G' we have $\min_{m_2} \max_{m_1} v_1(m_1, m_2) > 0$ if and only if $\max_{m_1} \min_{m_2} v_1(m_1, m_2) > 0$; that is, if and only if there exists a mixed strategy m_1^* of player i in G' for which $v_1(m_1^*, m_2) > 0$ for all m_2 (that is, for all beliefs on A_{-i}). Since m_1^* is a probability measure on $A_i \setminus \{a_i^*\}$ it is a mixed strategy of player 1 in G; the condition $v_1(m_1^*, m_2) > 0$ for all m_2 is equivalent to $U_i(a_{-i}, m_1^*) - U_i(a_{-i}, a_i^*) > 0$ for all $a_{-i} \in A_{-i}$, which is equivalent to a_i^* being strictly dominated. $\square$

Note that the argument in this proof depends upon our assumption that the players' preferences over lotteries satisfy the assumptions of von Neumann and Morgenstern; if the preferences do not satisfy these assumptions then the properties of being a never-best response and being strictly dominated are not equivalent in general.

4.2.2 Iterated Elimination of Strictly Dominated Actions

We now define formally the procedure that we described at the beginning of the section.

▶ DEFINITION 60.2 The set $X \subseteq A$ of outcomes of a finite strategic game $\langle N, (A_i), (u_i) \rangle$ **survives iterated elimination of strictly dominated actions** if $X = \times_{j \in N} X_j$ and there is a collection $((X_j^t)_{j \in N})_{t=0}^T$ of sets that satisfies the following conditions for each $j \in N$.

• $X_j^0 = A_j$ and $X_j^T = X_j$.

$$
\begin{array}{c|c|c|}
 & \multicolumn{1}{c}{L} & \multicolumn{1}{c}{R} \\
\cline{2-3}
T & 3,0 & 0,1 \\
\cline{2-3}
M & 0,0 & 3,1 \\
\cline{2-3}
B & 1,1 & 1,0 \\
\cline{2-3}
\end{array}
$$

Figure 61.1 A two-player strategic game. The only rationalizable action of player 1 is M and the only rationalizable action of player 2 is R.

- $X_j^{t+1} \subseteq X_j^t$ for each $t = 0, \ldots, T - 1$.
- For each $t = 0, \ldots, T - 1$ every action of player j in $X_j^t \setminus X_j^{t+1}$ is strictly dominated in the game $\langle N, (X_i^t), (u_i^t) \rangle$, where u_i^t for each $i \in N$ is the function u_i restricted to $\times_{j \in N} X_j^t$.
- No action in X_j^T is strictly dominated in the game $\langle N, (X_i^T), (u_i^T) \rangle$.

◇ EXAMPLE 61.1 In the game in Figure 61.1 the action B is dominated by the mixed strategy in which T and M are each used with probability $\frac{1}{2}$. After B is eliminated from the game, L is dominated by R; after L is eliminated T is dominated by M. Thus (M, R) is the only outcome that survives iterated elimination of strictly dominated actions.

We now show that in a finite game a set of outcomes that survives iterated elimination of dominated actions exists and is the set of profiles of rationalizable actions.

■ PROPOSITION 61.2 If $X = \times_{j \in N} X_j$ survives iterated elimination of strictly dominated actions in a finite strategic game $\langle N, (A_i), (u_i) \rangle$ then X_j is the set of player j's rationalizable actions for each $j \in N$.

Proof. Suppose that $a_i \in A_i$ is rationalizable and let $(Z_j)_{j \in N}$ be the profile of sets in Definition 55.1 that supports a_i. For any value of t we have $Z_j \subseteq X_j^t$ since each action in Z_j is a best response to some belief over Z_{-j} and hence is not strictly dominated in the game $\langle N, (X_i^t), (u_i^t) \rangle$ (by Lemma 60.1). Hence $a_i \in X_i$.

We now show that for every $j \in N$ every member of X_j is rationalizable. By definition, no action in X_j is strictly dominated in the game in which the set of actions of each player i is X_i, so by Lemma 60.1 every action in X_j is a best response among the members of X_j to some belief on X_{-j}. We need to show that every action in X_j is a best response among all the members of the set A_j to some belief on X_{-j}. If $a_j \in X_j$

is not a best response among all the members of A_j then there is a value
of t such that a_j is a best response among the members of X_j^t to a belief
μ_j on X_{-j}, but is not a best response among the members of X_j^{t-1}.
Then there is an action $b_j \in X_j^{t-1} \setminus X_j^t$ that is a best response among
the members of X_j^{t-1} to μ_j, contradicting the fact that b_j is eliminated
at the tth stage of the procedure. $\square$

Note that the procedure in Definition 60.2 does not require that *all*
strictly dominated strategies be eliminated at any stage. Thus the result
shows that the order and speed of elimination have no effect on the set
of outcomes that survive.

Lemma 60.1 and the equivalence of the notions of iterated elimination
of strictly dominated actions and rationalizability fail if we modify the
definition of rationalizability to require the players to believe that their
opponents' actions are independent. To see this, consider the game in
Figure 58.1. The action M_2 is a best response to the belief of player 3
in which players 1 and 2 play (U, L) and (D, R) with equal probabilities
and is thus not strictly dominated. However, as we saw before, it is not
a best response to any pair of (independent) mixed strategies and is thus
not rationalizable under the modified definition in which each player's
belief is restricted to be a product of independent beliefs.

4.3 Iterated Elimination of Weakly Dominated Actions

We say that a player's action is weakly dominated if the player has
another action at least as good no matter what the other players do and
better for at least some vector of actions of the other players.

▸ DEFINITION 62.1 The action $a_i \in A_i$ of player i in the strategic game
$\langle N, (A_i), (u_i) \rangle$ is **weakly dominated** if there is a mixed strategy α_i
of player i such that $U_i(a_{-i}, \alpha_i) \geq u_i(a_{-i}, a_i)$ for all $a_{-i} \in A_{-i}$ and
$U_i(a_{-i}, \alpha_i) > u_i(a_{-i}, a_i)$ for some $a_{-i} \in A_{-i}$, where $U_i(a_{-i}, \alpha_i)$ is the
payoff of player i if he uses the mixed strategy α_i and the other players'
vector of actions is a_{-i}.

By Lemma 60.1 an action that is weakly dominated but not strictly
dominated is a best response to some belief. This fact makes the argument against using a weakly dominated action weaker than that against
using a strictly dominated action. Yet since there is no advantage to using a weakly dominated action, it seems very natural to eliminate such
actions in the process of simplifying a complicated game.

$$\begin{array}{c|c|c|} & L & R \\\hline T & 1,1 & 0,0 \\\hline M & 1,1 & 2,1 \\\hline B & 0,0 & 2,1 \\\hline \end{array}$$

Figure 63.1 A two-player game in which the set of actions that survive iterated elimination of weakly dominated actions depends on the order in which actions are eliminated.

The notion of weak domination leads to a procedure analogous to iterated elimination of strictly dominated actions (Definition 60.2). However, this procedure is less compelling since the set of actions that survive iterated elimination of weakly dominated actions may depend on the order in which actions are eliminated, as the two-player game in Figure 63.1 shows. The sequence in which we first eliminate T (weakly dominated by M) and then L (weakly dominated by R) leads to an outcome in which player 2 chooses R and the payoff profile is $(2,1)$. On the other hand, the sequence in which we first eliminate B (weakly dominated by M) and then R (weakly dominated by L) leads to an outcome in which player 2 chooses L and the payoff profile is $(1,1)$. We discuss further the procedure of iterated elimination of weakly dominated actions in Section 6.6.

? EXERCISE 63.1 Consider a variant of the game in Example 18.6 in which there are two players, the distribution of the citizens' favorite positions is uniform, and each player is restricted to choose a position of the form ℓ/m for some $\ell \in \{0, \ldots, m\}$, where m is even. Show that the only outcome that survives iterated elimination of weakly dominated actions is that in which both players choose the position $\frac{1}{2}$.

? EXERCISE 63.2 (*Dominance solvability*) A strategic game is *dominance solvable* if all players are indifferent between all outcomes that survive the iterative procedure in which *all* the weakly dominated actions of each player are eliminated at each stage. Give an example of a strategic game that is dominance solvable but for which it is not the case that all players are indifferent between all outcomes that survive iterated elimination of weakly dominated actions (a procedure in which not all weakly dominated actions may be eliminated at each stage).

? EXERCISE 64.1 Each of two players announces a nonnegative integer
equal to at most 100. If $a_1 + a_2 \leq 100$, where a_i is the number announced
by player i, then each player i receives payoff of a_i. If $a_1 + a_2 > 100$
and $a_i < a_j$ then player i receives a_i and player j receives $100 - a_i$; if
$a_1 + a_2 > 100$ and $a_i = a_j$ then each player receives 50. Show that the
game is dominance solvable (see the previous exercise) and find the set
of surviving outcomes.

Lemma 60.1 shows that in a finite game an action that is not strictly
dominated is a best response to some belief. The following exercise
strengthens this conclusion for an action (or mixed strategy) that is not
weakly dominated.

? EXERCISE 64.2 Show that in a finite strategic game any mixed strategy
of a player that is not weakly dominated is a best response to a belief
that assigns positive probability to every vector of actions of the other
players. [Hint: Let $\langle N, (A_i), (u_i) \rangle$ be the game and let U be the set of
all vectors of the form $(u_1(a^1_{-1}, m_1), \ldots, u_1(a^k_{-1}, m_1))$, where m_1 ranges
over the mixed strategies of player 1 and $\{a^1_{-1}, \ldots, a^k_{-1}\}$ is the set of
all vectors of actions for the players other than player 1. Let $u^* \in U$
correspond to a mixed strategy of player 1 that is not weakly dominated.
You need to show that there exists a positive vector p^* with $p^* \cdot u^* \geq p^* \cdot u$
for all $u \in U$. To do so, let $u^* = 0$ without loss of generality, and for
any $\epsilon > 0$ let $P(\epsilon) = \{p \in \mathbb{R}^k : p_i \geq \epsilon \text{ for all } i \text{ and } \sum_{i=1}^k p_i = 1\}$. Use
the result of Exercise 36.3 for the sets $P(\epsilon)$ and U and let $\epsilon \to 0$; use
also the fact that U is the convex hull of a finite number of vectors.]

Notes

The notion of rationalizability originated with Bernheim (1984) and
Pearce (1984) (both of whom restrict players to believe that the actions
of their opponents are independent). (Spohn (1982) discusses the idea,
but does not formalize it.) Versions of the procedure of iterated elimination of dominated strategies were first studied in detail by Gale (1953)
and Luce and Raiffa (1957, pp. 108–109, 173); the formulation that we
give is due to Moulin (1979). Lemma 60.1 is due to Pearce (1984); it is
closely related to Lemma 3.2.1 of van Damme (1983). Proposition 61.2
is due to Pearce (1984, p. 1035).

The result in Exercise 56.4 is due to Gabay and Moulin (1980),
Bernheim (1984), and Moulin (1984). Exercise 56.5 is taken from
Moulin (1986, p. 72). Exercise 57.1 is taken from Rubinstein and Wolin-

sky (1994). The notion of dominance solvability in Exercise 63.2 is due to Moulin (1979); it is closely related to the notion of "solvability in the complete weak sense" of Luce and Raiffa (1957, p. 109). Exercise 64.1 is due to Brams and Taylor (1994) and Exercise 64.2 is due to Arrow, Barankin, and Blackwell (1953).

For a family of games in which rationalizability gives a sharp answer see Vives (1990) and Milgrom and Roberts (1990).

5 Knowledge and Equilibrium

In this chapter we describe a model of knowledge and use it to formalize the idea that an event is "common knowledge", to ask if it is possible for people to "agree to disagree", and to express formally the assumptions on players' knowledge that lie behind the concepts of Nash equilibrium and rationalizability.

5.1 A Model of Knowledge

A strategic game models interaction between players. Consequently we are interested not only in a player's knowledge of an exogenous parameter but also in his knowledge about the knowledge of another player. We begin by giving a brief introduction to a model of the knowledge of a single decision-maker.

At the basis of the model is a set Ω of *states*. The notion of a state is given two interpretations in the literature. At one extreme, a state is viewed as a description of the contingencies that the decision-maker perceives to be relevant in the context of a certain decision problem. This is the interpretation used in standard economic models of uncertainty. At the other extreme a state is viewed as a full description of the world, including not only the decision-maker's information and beliefs but also his actions.

5.1.1 The Information Function

One way to define the extent of a decision-maker's knowledge of the state is to specify an *information function* P that associates with every state $\omega \in \Omega$ a nonempty subset $P(\omega)$ of Ω. The interpretation is that when the state is ω the decision-maker knows only that the state is in the set $P(\omega)$. That is, he considers it possible that the true state could be any state in $P(\omega)$ but not any state outside $P(\omega)$.

▶ DEFINITION 68.1 An **information function** for the set Ω of states is a function P that associates with every state $\omega \in \Omega$ a nonempty subset $P(\omega)$ of Ω.

When we use an information function to model a decision-maker's knowledge we usually assume that the pair $\langle \Omega, P \rangle$ consisting of the set of states and the information function satisfies the following two conditions:

P1 $\omega \in P(\omega)$ for every $\omega \in \Omega$.

P2   If $\omega' \in P(\omega)$ then $P(\omega') = P(\omega)$.

P1 says that the decision-maker never excludes the true state from the set of states he regards as feasible: he is never certain that the state is different from the true state. P2 says that the decision-maker uses the consistency or inconsistency of states with his information to make inferences about the state. Suppose, contrary to P2, that $\omega' \in P(\omega)$ and there is a state $\omega'' \in P(\omega')$ with $\omega'' \notin P(\omega)$. Then if the state is ω the decision-maker can argue that since ω'' is inconsistent with his information the true state cannot be ω'. Similarly, if there is a state $\omega'' \in P(\omega)$ with $\omega'' \notin P(\omega')$ then when the state is ω the decision-maker can argue that since ω'' is consistent with his information the true state cannot be ω'.

The following condition is equivalent to P1 and P2.

▶ DEFINITION 68.2 An information function P for the set Ω of states is **partitional** if there is a partition of Ω such that for any $\omega \in \Omega$ the set $P(\omega)$ is the element of the partition that contains ω.

■ LEMMA 68.3 *An information function is partitional if and only if it satisfies P1 and P2.*

Proof. If P is partitional then it clearly satisfies P1 and P2. Now suppose that P satisfies P1 and P2. If $P(\omega)$ and $P(\omega')$ intersect and $\omega'' \in P(\omega) \cap P(\omega')$ then by P2 we have $P(\omega) = P(\omega') = P(\omega'')$; by P1 we have $\cup_{\omega \in \Omega} P(\omega) = \Omega$. Thus P is partitional.   □

Given this result, an information function that satisfies P1 and P2 may be specified by the *information partition* that it induces.

◇ EXAMPLE 68.4 Let $\Omega = [0, 1)$ and assume that the decision-maker observes only the first four digits of the decimal expansion of a number. Then for each $\omega \in \Omega$ the set $P(\omega)$ is the set of all states $\omega' \in \Omega$ such that the first four digits of ω' are the same as those of ω. This information function is partitional.

☐ EXERCISE 69.1 Let Q be a set of questions to which the answer is either "Yes" or "No". A state is a list of answers to all the questions in Q. Suppose that the information function P has the property that for some state ω_1 the set $P(\omega_1)$ consists of all states in which the answers to the first two questions are the same as in ω_1, while for some other state ω_2 the set $P(\omega_2)$ consists of all states in which the answers to the first three questions are the same as in ω_2. Is P necessarily partitional?

☐ EXERCISE 69.2 A decision-maker is told an integer n but remembers only that the number is either $n - 1$, n, or $n + 1$. Model the decision-maker's knowledge by an information function and determine if the function is partitional.

5.1.2 The Knowledge Function

We refer to a set of states (a subset of Ω) as an **event**. Given our interpretation of an information function, a decision-maker for whom $P(\omega) \subseteq E$ knows, in the state ω, that some state in the event E has occurred. In this case we say that in the state ω the decision-maker **knows** E. Given P we now define the decision maker's **knowledge function** K by

$$K(E) = \{\omega \in \Omega : P(\omega) \subseteq E\}. \tag{69.3}$$

For any event E the set $K(E)$ is the set of all states in which the decision-maker knows E. The knowledge function K that is derived from any information function satisfies the following three properties.

K1 $K(\Omega) = \Omega$.

This says that in all states the decision-maker knows that some state in Ω has occurred.

K2 If $E \subseteq F$ then $K(E) \subseteq K(F)$.

This says that if F occurs whenever E occurs and the decision-maker knows E then he knows F: if E implies F then knowledge of E implies knowledge of F.

K3 $K(E) \cap K(F) = K(E \cap F)$.

The interpretation of this property is that if the decision-maker knows both E and F then he knows $E \cap F$.

If P satisfies P1 then the associated knowledge function K satisfies the following additional property.

K4 (Axiom of Knowledge) $K(E) \subseteq E$.

This says that whenever the decision-maker knows E then indeed some member of E is the true state: the decision-maker does not know anything that is false. The axiom is derived from P1 as follows: if $\omega \in K(E)$ then $P(\omega) \subseteq E$, so that by P1 we have $\omega \in E$.

If P is partitional (i.e. satisfies both P1 and P2) then $K(E)$ is the union of all the members of the partition that are subsets of E. (If E does not contain any member of the partition then $K(E)$ is empty.) In this case the knowledge function K satisfies the following two additional properties.

K5 (Axiom of Transparency) $K(E) \subseteq K(K(E))$.

Given our interpretation of $K(E)$ as the event in which the decision-maker knows E, we interpret $K(K(E))$ to be the event in which the decision-maker knows that he knows E. Thus K5 says that if the decision-maker knows E then he knows that he knows E. As we remarked above, if P satisfies P1 and P2 then the set $K(E)$ is a union of members of the partition induced by P; K5 follows from the observation that if F is a union of members of the partition then $K(F) = F$.

K6 (Axiom of Wisdom) $\Omega \setminus K(E) \subseteq K(\Omega \setminus K(E))$.

The interpretation of this axiom is that the decision-maker is aware of what he does not know: if he does not know E then he knows that he does not know E. Since P is partitional, $K(E)$ is a union of members of the partition induced by P; thus $\Omega \setminus K(E)$ also is such a union, and K6 follows.

Note that given that K satisfies K4 the properties in K5 and K6 in fact hold with equality.

We have taken an information function as the primitive and derived from it a knowledge function. Alternatively we can start by defining a **knowledge function** for the set Ω to be a function K that associates a subset of Ω with each event $E \subseteq \Omega$. We can then derive from it an information function P as follows: for each state ω let

$$P(\omega) = \cap \{E \subseteq \Omega \colon K(E) \ni \omega\}. \qquad (70.1)$$

(If there is no event E for which $\omega \in K(E)$ then we take the intersection to be Ω.)

? EXERCISE 71.1

    a. Given an information function P, let K be the knowledge function defined by (69.3) and let P' be the information function derived from K in (70.1). Show that $P' = P$.

    b. Given a knowledge function K that satisfies K1, K2, and K3, let P be the information function defined by (70.1) and let K' be the knowledge function derived from P in (69.3). Show that $K' = K$.

? EXERCISE 71.2 Using the framework we have described, we can formulate an individual's decision problem as follows. Let A be a set of actions, Ω a set of states, P a partitional information function, π a probability measure on Ω, and $u: A \times \Omega \rightarrow \mathbb{R}$ a function whose expected value represents the individual's preferences over lotteries on A. The individual's problem is to choose a function $a: \Omega \rightarrow A$ (called an *act*) for which $a(\omega) = a(\omega')$ whenever $\omega \in P(\omega)$ and $\omega' \in P(\omega)$ to solve $\max_a \mathrm{E}_\pi u(a(\omega), \omega)$ (where E is the expectation operator). Define the partitional information function P' to be *coarser* than the information function P if $P(\omega) \subseteq P'(\omega)$ for all $\omega \in \Omega$ (i.e. if each member of the partition induced by P' is a union of members of the partition induced by P). Show that if P' is coarser than P then the best act under the information function P' is no better than the best act under the information function P. Contrast this result with that of Exercise 28.2.

5.1.3 *An Illustrative Example: The Puzzle of the Hats*

The following puzzle, which "swept Europe" some time in the first half of the twentieth century (Littlewood (1953, p. 3)), illustrates the concepts that we have defined. Each of n "perfectly rational" individuals, seated around a table, is wearing a hat that is either white or black. Each individual can see the hats of the other $n - 1$ individuals, but not his own. An observer announces: "Each of you is wearing a hat that is either white or black; at least one of the hats is white. I will start to count slowly. After each number you will have the opportunity to raise a hand. You may do so only when you know the color of your hat." When, for the first time, will any individual raise his hand?

We can answer this question by using the formal model we have introduced, as follows. Initially, after the observer's announcement, the set of states is the set of all configurations $c = (c_1, \ldots, c_n)$ of colors for the hats, where each c_i is either W or B and at least one c_i is W. These $2^n - 1$ states constitute the set

$$\Omega = \{c \in \{B, W\}^n : |\{i : c_i = W\}| \geq 1\}.$$

The initial information function P_i^1 of any individual i is given as follows: in any state c the set $P_i^1(c)$ consists of all the states that accord with i's observations, and thus contains at most two states, which differ only in the color of i's hat. Precisely, if c is a state in which a player different from i has a white hat then $P_i^1(c) = \{(c_{-i}, W), (c_{-i}, B)\}$, and if c is the state in which all the other hats are black then $P_i^1(c) = \{c\}$ (since not all the hats are black).

What does it mean for an individual i with information function P_i to "know the color c_i of his hat"? It means either he knows that the event $\{c : c_i = W\}$ occurs or he knows that the event $\{c : c_i = B\}$ occurs. Thus the event "i knows the color of his hat" is

$$E_i = \{c : P_i(c) \subseteq \{c : c_i = B\} \text{ or } P_i(c) \subseteq \{c : c_i = W\}\}.$$

It is only in a state c in which there is exactly one individual i for whom $c_i = W$ that $P_j^1(c) \subseteq E_j$ for some j, and in this case $P_i^1(c) \subseteq E_i$, so that i raises his hand.

Now let $F^1 = \{c : |\{i : c_i = W\}| = 1\}$, the set of states for which someone raises a hand at the first stage. If nobody raises a hand at the first stage then all individuals obtain the additional information that the state is not in F^1, and thus for all i and for all $c \notin F^1$ we have $P_i^2(c) = P_i^1(c) \backslash F^1$. That is, in any such state every individual concludes that at least two individuals have white hats. We have $P_i^2(c) = P_i^1(c) = \{(c_{-i}, W), (c_{-i}, B)\}$ unless $c_j = W$ for exactly one individual $j \neq i$, in which case $P_i^2(c_{-i}, W) = \{(c_{-i}, W)\}$ (and $P_j^2(c_{-j}, W) = \{(c_{-j}, W)\}$). In other words, in any state c for which $c_j = W$ and $c_h = W$ for precisely two individuals j and h we have $P_j^2(c) \subseteq E_j$ and $P_h^2(c) \subseteq E_h$, and hence j and h each raises a hand at the second stage. Now let $F^2 = \{c : |\{i : c_i = W\}| = 2\}$, the set of states in which the process ends at the second stage. In states for which no hand is raised after the observer counts 2 ($c \notin F^1 \cup F^2$) all individuals conclude that at least three hats are white and the process continues with $P_i^3(c) = P_i^2(c) \backslash F^2$. It is easy to see that if k hats are white then no one raises a hand until the observer counts k, at which point the k individuals with white hats do so.

5.2 Common Knowledge

We say that an event is "mutual knowledge" in some state if in that state each individual knows the event. We say that an event is "common knowledge" if not only is it mutual knowledge but also each individual knows that all other individuals know it, each individual knows that all other individuals know that all the individuals know it, and so on. Restricting for simplicity to the case of two individuals the notion of common knowledge is formalized in the following definition.

▶ DEFINITION 73.1 Let K_1 and K_2 be the knowledge functions of individuals 1 and 2 for the set Ω of states. An event $E \subseteq \Omega$ is **common knowledge between 1 and 2 in the state** $\omega \in \Omega$ if ω is a member of every set in the infinite sequence $K_1(E)$, $K_2(E)$, $K_1(K_2(E))$, $K_2(K_1(E))$, ...

Another definition of common knowledge (which we show in Proposition 74.2 is equivalent) is stated in terms of the individuals' information functions.

▶ DEFINITION 73.2 Let P_1 and P_2 be the information functions of individuals 1 and 2 for the set Ω of states. An event $F \subseteq \Omega$ is **self-evident between 1 and 2** if for all $\omega \in F$ we have $P_i(\omega) \subseteq F$ for $i = 1, 2$. An event $E \subseteq \Omega$ is **common knowledge between 1 and 2 in the state** $\omega \in \Omega$ if there is a self-evident event F for which $\omega \in F \subseteq E$.

In words, an event E is self-evident between two individuals if whenever it occurs both individuals know that it occurs (i.e. whenever it occurs it is mutual knowledge between the individuals), and is common knowledge in the state ω if there is a self-evident event containing ω whose occurrence implies E.

◇ EXAMPLE 73.3 Let $\Omega = \{\omega_1, \omega_2, \omega_3, \omega_4, \omega_5, \omega_6\}$, let P_1 and P_2 be the partitional information functions of individuals 1 and 2, and let K_1 and K_2 be the associated knowledge functions. Let the partitions induced by the information functions be

$$\mathcal{P}_1 = \{\{\omega_1, \omega_2\}, \{\omega_3, \omega_4, \omega_5\}, \{\omega_6\}\}$$
$$\mathcal{P}_2 = \{\{\omega_1\}, \{\omega_2, \omega_3, \omega_4\}, \{\omega_5\}, \{\omega_6\}\}.$$

The event $E = \{\omega_1, \omega_2, \omega_3, \omega_4\}$ does not contain any event that is self-evident between 1 and 2 and hence in no state is E common knowledge between 1 and 2 in the sense of the second definition (73.2). The event E is also not common knowledge in any state in the sense of the first definition (73.1) since $K_1(K_2(K_1(E))) = \varnothing$, as the following calculation demonstrates:

$$K_1(E) = \{\omega_1, \omega_2\}, \qquad\qquad K_2(E) = E,$$
$$K_2(K_1(E)) = \{\omega_1\}, \qquad\qquad K_1(K_2(E)) = \{\omega_1, \omega_2\},$$
$$K_1(K_2(K_1(E))) = \varnothing, \qquad\qquad K_2(K_1(K_2(E))) = \{\omega_1\}.$$

The event $F = \{\omega_1, \omega_2, \omega_3, \omega_4, \omega_5\}$ is self-evident between 1 and 2 and hence is common knowledge between 1 and 2 in any state in F in the sense of the second definition. Since $K_1(F) = K_2(F) = F$ the event F is also common knowledge between 1 and 2 at any state in F in the sense of the first definition.

Before showing that the two definitions of common knowledge are equivalent we establish the following.

■ LEMMA 74.1 *Let P_1 and P_2 be the partitional information functions of individuals 1 and 2 for the set Ω of states, let K_1 and K_2 be the associated knowledge functions, and let E be an event. Then the following three conditions are equivalent.*

a. $K_i(E) = E$ *for $i = 1, 2$.*

b. *E is self-evident between 1 and 2.*

c. *E is a union of members of the partition induced by P_i for $i = 1, 2$.*

Proof. Assume (a). Then for every $\omega \in E$ we have $P_i(\omega) \subseteq E$ for $i = 1$, 2, and hence (b) is satisfied. Assume (b). Then $E = \cup_{\omega \in E} P_i(\omega)$ for $i = 1$, 2, and thus E is a union of members of both partitions, so that (c) is satisfied. Finally (c) immediately implies (a).   □

We now show that Definitions 73.1 and 73.2 are equivalent.

■ PROPOSITION 74.2 *Let Ω be a finite set of states, let P_1 and P_2 be the partitional information functions of individuals 1 and 2, and let K_1 and K_2 be the associated knowledge functions. Then an event $E \subseteq \Omega$ is common knowledge between 1 and 2 in the state $\omega \in \Omega$ according to Definition 73.1 if and only if it is common knowledge between 1 and 2 in the state ω according to Definition 73.2.*

Proof. Assume that the event E is common knowledge between 1 and 2 in the state ω according to Definition 73.1. For each $i \in \{1, 2\}$ and $j \neq i$ we have $E \supseteq K_i(E) \supseteq K_j(K_i(E)) \supseteq K_i(K_j(K_i(E))) \supseteq \ldots$ and ω is a member of all these sets, which are hence nonempty. Thus since Ω is finite there is a set $F_i = K_i(K_j(K_i \cdots K_i(E) \cdots))$ for which $K_j(F_i) = F_i$; since P_i is partitional, K_i satisfies K4 and K5, so that we have also

$K_i(F_i) = F_i$. Thus by Lemma 74.1 the event F_i is self-evident between 1 and 2, so that E is common knowledge in ω according to Definition 73.2.

Now assume that $E \subseteq \Omega$ is common knowledge between 1 and 2 in the state ω according to Definition 73.2. Then there is a self-evident event F with $\omega \in F \subseteq E$. By Lemma 74.1 every set of the type $K_i(K_j(K_i \cdots K_i(F) \cdots))$ coincides with F. It follows from K2 that ω is a member of every set of the type $K_i(K_j(K_i \cdots K_i(E) \cdots))$ and thus E is common knowledge in ω according to Definition 73.1. $\qquad \square$

5.3 Can People Agree to Disagree?

An interesting question that can be addressed within the framework we have described is the following. Can it be common knowledge between two individuals with the same prior belief that individual 1 assigns probability η_1 to some event and individual 2 assigns probability $\eta_2 \neq \eta_1$ to the same event? It seems that the answer could be positive: the individuals might "agree to disagree" in this way when they possess different information. However, we now show that if the individuals' information functions are partitional then the answer is negative.

One of the contexts in which this result is of interest is that of a Bayesian game (Section 2.6). An assumption often made in the literature is that the players in such a game have identical prior beliefs. The result implies that under this assumption it cannot be common knowledge between the players that they assign different posterior probabilities to the same event. Thus if we want to model a situation in which such differences in beliefs are common knowledge, we must assume that the players' prior beliefs are different.

Let ρ be a probability measure on the set Ω of states, interpreted as the individuals' common prior belief, and let P_1 and P_2 be the individuals' information functions. If E is an event and $\rho(E|P_i(\omega)) = \eta_i$ (where $\rho(E|P_i(\omega))$ is the probability of E conditional on $P_i(\omega)$) then, given his information in the state ω, individual i assigns the probability η_i to the event E. Thus the event "individual i assigns the probability η_i to E" is $\{\omega \in \Omega : \rho(E|P_i(\omega)) = \eta_i\}$.

● PROPOSITION 75.1 *Suppose that the set Ω of states is finite and individuals 1 and 2 have the same prior belief. If each individual's information function is partitional and it is common knowledge between 1 and 2 in some state $\omega^* \in \Omega$ that individual 1 assigns probability η_1 to some event E and individual 2 assigns probability η_2 to E then $\eta_1 = \eta_2$.*

Proof. If the assumptions are satisfied then there is a self-evident event $F \ni \omega^*$ that is a subset of the intersection of $\{\omega \in \Omega: \rho(E|P_1(\omega)) = \eta_1\}$ and $\{\omega \in \Omega: \rho(E|P_2(\omega)) = \eta_2\}$, and hence a subset of each of these sets, where ρ is the common prior belief. By Lemma 74.1, for each individual i the event F is a union of members of i's information partition. Since Ω is finite, so is the number of sets in each union; let $F = \cup_k A_k = \cup_k B_k$. Now, for any nonempty disjoint sets C and D with $\rho(E|C) = \eta_i$ and $\rho(E|D) = \eta_i$ we have $\rho(E|C \cup D) = \eta_i$. Thus, since for each k we have $\rho(E|A_k) = \eta_1$, it follows that $\rho(E|F) = \eta_1$; similarly $\rho(E|F) = \eta_2$. Hence $\eta_1 = \eta_2$.   □

? EXERCISE 76.1 Show that if two individuals with partitional information functions have the same prior then it can be common knowledge between them that they assign different probabilities to some event. Show, however, that it can*not* be common knowledge that the probability assigned by individual 1 exceeds that assigned by individual 2.

? EXERCISE 76.2 Show that if two individuals with partitional information functions have the same prior then it *cannot* be common knowledge between them that individual 1 believes the expectation of some lottery to exceed some number η while individual 2 believes this expectation to be less than η. Show by an example that this result depends on the assumption that the individuals' information functions are partitional.

5.4   Knowledge and Solution Concepts

In the previous chapters we discussed the concepts of Nash equilibrium and rationalizability. When motivating these concepts we appealed informally to assumptions about what the players know. In this section we use the model described above to examine formally assumptions about the players' knowledge that lie behind the solution concepts.

Throughout we fix attention on a given strategic game $G = \langle N, (A_i), (\succsim_i) \rangle$ (see Definition 11.1).

Let Ω be a set of states, each of which is a description of the environment relevant to the game: that is, a description of each player's knowledge, action, and belief. Formally, each state $\omega \in \Omega$ consists of a specification for each player i of

- $P_i(\omega) \subseteq \Omega$, which describes player i's knowledge in state ω (where P_i is a partitional information function)

- $a_i(\omega) \in A_i$, the action chosen by player i in state ω

- $\mu_i(\omega)$, a probability measure on $A_{-i} = \times_{j \in N \setminus \{i\}} A_j$, the belief of player i in state ω about the actions of the other players. (Note that this allows a player to believe that the other players' actions are correlated.)

Note that the notion of a state, since it consists of a specification of the knowledge, action, and belief of each player, may be self-referential: if in state ω_1 some player does not know whether the state is ω_1 or ω_2 then the description of ω_1 refers to itself.

Our implicit assumption in this definition of the set of states is that it is common knowledge among the players that the game is G. Thus we ignore the possibility, for example, that some player does not know his own action set or the action set of another player, or some player i does not know whether player j knows player i's preferences. This assumption is stronger than we need for some of the results. To formalize weaker assumptions about the players' knowledge of the game we would need to extend the definition of the set of states, requiring that each state include a specification of the game that is played.

We now isolate properties of a state that imply that the actions in that state are consistent with various solution concepts. Our first result is that if in some state each player is rational, knows the other players' actions, and has a belief consistent with his knowledge, then the profile of actions chosen in that state is a Nash equilibrium of the game.

■ PROPOSITION 77.1 *Suppose that in the state* $\omega \in \Omega$ *each player* $i \in N$

  a. *knows the other players' actions:* $P_i(\omega) \subseteq \{\omega' \in \Omega : a_{-i}(\omega') = a_{-i}(\omega)\}$;

  b. *has a belief that is consistent with his knowledge: the support of* $\mu_i(\omega)$ *is a subset of* $\{a_{-i}(\omega') \in A_{-i} : \omega' \in P_i(\omega)\}$;

  c. *is rational:* $a_i(\omega)$ *is a best response of player* i *to* $\mu_i(\omega)$.

Then $(a_i(\omega))_{i \in N}$ *is a Nash equilibrium of* G.

Proof. By (c) the action $a_i(\omega)$ is a best response of player i to his belief, which by (b) assigns probability one to the set $\{a_{-i}(\omega') \in A_{-i} : \omega' \in P_i(\omega)\}$; by (a) this set is $\{a_{-i}(\omega)\}$. □

The assumption that each player knows the actions of all the other players is very strong. We now show that in a two-player game we can replace it with the assumption that each player knows the belief of the other player if we strengthen (c) to require not only that each player be rational but also that each player know that the other player is rational. Since the result involves mixed strategies we now let the strategic game under consideration be $G = \langle N, (A_i), (u_i) \rangle$, where for each $i \in N$ the expected value of the function u_i represents player i's preferences over lotteries on A.

∎ PROPOSITION 78.1 *Suppose that $|N| = 2$ and that in the state $\omega \in \Omega$ each player $i \in N$*

a. *knows the other player's belief: $P_i(\omega) \subseteq \{\omega' \in \Omega : \mu_j(\omega') = \mu_j(\omega)\}$ for $j \neq i$;*

b. *has a belief that is consistent with his knowledge: the support of $\mu_i(\omega)$ is a subset of $\{a_j(\omega') \in A_j : \omega' \in P_i(\omega)\}$ for $j \neq i$;*

c. *knows that the other is rational: for any $\omega' \in P_i(\omega)$ the action $a_j(\omega')$ is a best response of player j to $\mu_j(\omega')$ for $j \neq i$.*

Then the mixed strategy profile $(\alpha_1, \alpha_2) = (\mu_2(\omega), \mu_1(\omega))$ is a mixed strategy Nash equilibrium of G.

Proof. Let a_i^* be an action of player i that is in the support of $\alpha_i = \mu_j(\omega)$. By (b) there is a state $\omega' \in P_j(\omega)$ such that $a_i(\omega') = a_i^*$. It follows from (c) that the action a_i^* is a best response of player i to $\mu_i(\omega')$, which by (a) is equal to $\mu_i(\omega)$.   □

Note that neither proposition requires the players to derive their beliefs from some common prior on Ω. In particular, note that in (b) we require only that each player's belief be consistent with his knowledge. Note also that the assumption that the game is common knowledge can be weakened in both results: in Proposition 77.1 it is sufficient to assume each player knows his own action set and preferences, and in Proposition 78.1 it is sufficient to assume the game is mutual knowledge.

The following example demonstrates that Proposition 78.1 does not have an analog when there are more than two players. Consider the game at the top of Figure 79.1. (Note that player 3's payoff is always 0.) Let the set of states be $\Omega = \{\alpha, \beta, \gamma, \delta, \epsilon, \xi\}$ and let the players' action functions and information functions be those given in the table at the bottom of the figure; assume that the players' beliefs are derived from the same prior, which is given in the first row of the table.

Consider the state δ. We claim that the three conditions of the proposition are satisfied. Condition (b) is satisfied since each player's belief at δ is defined from the common prior. It remains to verify that in this state each player knows the beliefs of the other players and knows that the other players are rational. Consider player 1. She knows that the state is either δ or ϵ, so that she knows that player 2's information is either $\{\gamma, \delta\}$ or $\{\epsilon, \xi\}$. In both cases player 2 believes that with prob-

	L	R
U	2,3,0	2,0,0
D	0,3,0	0,0,0

	L	R
U	0,0,0	0,2,0
D	3,0,0	3,2,0

A B

State	α	β	γ	δ	ϵ	ξ
Probability $\times 63$	32	16	8	4	2	1
1's action	U	D	D	D	D	D
2's action	L	L	L	L	L	L
3's action	A	B	A	B	A	B
1's partition	$\{\alpha\}$	$\{\beta$	$\gamma\}$	$\{\delta$	$\epsilon\}$	$\{\xi\}$
2's partition	$\{\alpha$	$\beta\}$	$\{\gamma$	$\delta\}$	$\{\epsilon$	$\xi\}$
3's partition	$\{\alpha\}$	$\{\beta\}$	$\{\gamma\}$	$\{\delta\}$	$\{\epsilon\}$	$\{\xi\}$

Figure 79.1 At the top is a three-player game in which player 1 chooses one of the two rows, player 2 chooses one of the two columns, and player 3 chooses one of the two tables. At the bottom are action functions, information functions, and beliefs for the players in the game.

ability $\frac{2}{3}$ the pair of actions chosen by players 1 and 3 is (D, A) and that with probability $\frac{1}{3}$ it is (D, B). Given this belief, the action L that player 2 takes is optimal. Thus player 1 knows that player 2 is rational.

Similarly player 2 knows that player 1's information is either $\{\beta, \gamma\}$ or $\{\delta, \epsilon\}$. In both cases player 1 believes that with probability $\frac{2}{3}$ players 2 and 3 will choose (L, B) and that with probability $\frac{1}{3}$ they will choose (L, A). Given this belief, the action D that she takes is optimal, so that player 2 knows that player 1 is rational.

Player 3 knows that player 1's information is $\{\delta, \epsilon\}$ and that player 2's information is $\{\gamma, \delta\}$. Thus, as argued above, player 3 knows that players 1 and 2 are rational.

In the three states γ, δ, and ϵ, player 3's belief is that the pair of actions of players 1 and 2 is (D, L), and thus in the state δ players 1 and 2 know player 3's belief. They also know she is rational since her payoffs are always zero.

However in the state δ the beliefs do not define a Nash equilibrium. In fact, the players' beliefs about each other's behavior do not even coincide: Player 1 believes that player 3 chooses A with probability $\frac{1}{3}$

while player 2 believes that she does so with probability $\frac{2}{3}$. Neither of these beliefs together with the actions D and L forms a mixed strategy Nash equilibrium of the game.

What makes this example work is that in state δ player 1 does not know that player 2 knows her belief: player 1 thinks that the state might be ϵ, in which player 2 does not know whether player 1 believes that player 3 plays B or that player 3 plays B with probability $\frac{2}{3}$ and A with probability $\frac{1}{3}$.

Aumann and Brandenburger (1995) show that if all players share a common prior and in some state rationality is mutual knowledge and the players' beliefs are *common knowledge* then the beliefs at that state form a mixed strategy Nash equilibrium even if there are more than two players. The key point is that if the beliefs of players 1 and 2 about player 3's action are common knowledge and if all the players share the same prior, then the beliefs must be the same (by an argument like that in the proof of Proposition 75.1).

The following result formalizes the arguments in Chapter 4 to the effect that the notion of rationalizability rests on weaker assumptions about the players' knowledge than that of Nash equilibrium, requiring only that it be common knowledge among the players that all players are rational. (The result does not depend essentially on the assumption that there are two players, though the statement is simpler in this case.)

■ PROPOSITION 80.1 *Suppose that $|N| = 2$ and that in the state $\omega \in \Omega$ it is common knowledge between the players that each player's belief is consistent with his knowledge and that each player is rational. That is, suppose that there is a self-evident event $F \ni \omega$ such that for every $\omega' \in F$ and each $i \in N$*

    a. *the support of $\mu_i(\omega')$ is a subset of $\{a_j(\omega'') \in A_j : \omega'' \in P_i(\omega')\}$ for $j \neq i$;*

    b. *the action $a_i(\omega')$ is a best response of player i to $\mu_i(\omega')$.*

Then for each $i \in N$ the action $a_i(\omega)$ is rationalizable in G.

Proof. For each $i \in N$ let $Z_i = \{a_i(\omega') \in A_i : \omega' \in F\}$. By (b) we know that for any $\omega' \in F$ the action $a_i(\omega')$ is a best response to $\mu_i(\omega')$, whose support, by (a), is a subset of $\{a_j(\omega'') \in A_j : \omega'' \in P_i(\omega')\}$. Since F is self-evident we have $P_i(\omega') \subseteq F$, and thus $\{a_j(\omega'') \in A_j : \omega'' \in P_i(\omega')\} \subseteq Z_j$. Hence (using Definition 55.1) $a_i(\omega)$ is rationalizable.   □

The three results in this section derive implications for the players' actions or beliefs in a *particular* state from assumptions about their

	A	B
A	M, M	$1, -L$
B	$-L, 1$	$0, 0$

G_a (probability $1 - p$)

	A	B
A	$0, 0$	$1, -L$
B	$-L, 1$	M, M

G_b (probability p)

Figure 81.1 The component games of the Electronic Mail game. The parameters satisfy $L > M > 1$ and $p < \frac{1}{2}$.

knowledge in that state. The result in the following exercise is based on an assumption of a different type—that in *every* state the players' rationality is common knowledge. If this assumption is satisfied and the players' beliefs are derived from a common prior then the distribution of the players' actions over Ω is a correlated equilibrium.

[?] EXERCISE 81.1 Suppose that for all $\omega \in \Omega$ all players are rational (and hence their rationality is common knowledge in every state, since any fact that is true in all states is common knowledge in every state). Show that if each player's belief in every state is derived from a common prior ρ on Ω for which $\rho(P_i(\omega)) > 0$ for all $i \in N$ and all $\omega \in \Omega$, and $a_i(\omega') = a_i(\omega)$ for each $i \in N$ and each $\omega' \in P_i(\omega)$, then $\langle (\Omega, \rho), (P_i), (a_i) \rangle$, where $\mathcal{P}_i$ is the partition induced by P_i, is a correlated equilibrium of G. (The proof is very simple; the main task is to understand the content of the result.)

5.5 The Electronic Mail Game

In this section we study a game that illustrates the concepts introduced in this chapter. Each of two players has to choose one of the actions A or B. With probability $p < \frac{1}{2}$ the game in which the players are involved is G_b; with probability $1 - p$ it is G_a. In both G_a and G_b it is mutually beneficial for the players to choose the same action, but the action that is best depends on the game: in G_a the outcome (A, A) is best, while in game G_b the outcome (B, B) is best. The payoffs are shown in Figure 81.1, where $L > M > 1$. Note that even if a player is sure that the game is G_b, it is risky for him to choose B unless he is sufficiently confident that his partner is going to choose B as well.

Which is the true game is known initially only to player 1. Assume first that player 2 cannot obtain this information. Then we can model the situation as a Bayesian game (Definition 25.1) in which there are two states a and b and the information structures induced by the signal functions are $\{\{a\}, \{b\}\}$ for player 1 and $\{\{a, b\}\}$ for player 2. This game has a unique Nash equilibrium, in which both players always choose A; the expected payoff of each player is $(1 - p)M$.

Now assume that player 1 can communicate with player 2 in such a way that the game becomes common knowledge between the two players. In this case each player's information structure is $\{\{a\}, \{b\}\}$ and the (degenerate) Bayesian game has a Nash equilibrium in which each player chooses A in state a and B in state b; the payoff of each player is M.

In the situation we study in this section, the players can communicate, but the means that is open to them does not allow the game to become common knowledge. Specifically, the players are restricted to communicate via computers under the following protocol. If the game is G_b then player 1's computer *automatically* sends a message to player 2's computer; if the game is G_a then no message is sent. If a computer receives a message then it *automatically* sends a confirmation; this is so not only for the original message but also for the confirmation, the confirmation of the confirmation, and so on. The protocol is designed to send confirmations because the technology has the property that there is a small probability $\epsilon > 0$ that any given message does not arrive at its intended destination. If a message does not arrive then the communication stops. At the end of the communication phase each player's screen displays the number of messages that his machine has sent.

To discuss the players' knowledge in this situation we need to specify a set of states and the players' information functions. Define the set of states to be $\Omega = \{(Q_1, Q_2): Q_1 = Q_2 \text{ or } Q_1 = Q_2 + 1\}$. In the state (q, q) player 1's computer sends q messages, all of which arrive at player 2's computer, and the qth message sent by player 2's computer goes astray. In the state $(q+1, q)$ player 1's computer sends $q+1$ messages, and all but the last arrive at player 2's computer. Player 1's information function is defined by $P_1(q, q) = \{(q, q), (q, q - 1)\}$ if $q \geq 1$ and $P_1(0, 0) = \{(0, 0)\}$; player 2's information function is defined by $P_2(q, q) = \{(q, q), (q + 1, q)\}$ for all q. Denote by $G(Q_1, Q_2)$ the game that is played in the state (Q_1, Q_2); that is, $G(0, 0) = G_a$ and $G(Q_1, Q_2) = G_b$ otherwise. Player 1 knows the game in all states. Player 2 knows the game in all states except $(0, 0)$ and $(1, 0)$. In each of the states $(1, 0)$ and $(1, 1)$ player 1 knows that the game is G_b but does not know that player 2 knows it. Similarly in each of the states $(1, 1)$ and $(2, 1)$ player 2 knows that the game is G_b but does not know whether player 1 knows that player 2 knows that the game is G_b. And so on. In any state (q, q) or $(q + 1, q)$ the larger the value of q the more statements of the type "player i knows that player j knows that player i knows that ... the game is G_b" are correct, but in no state is it common knowledge that the game is G_b.

If ϵ is small then with high probability each player sees a very high number on his screen. When player 1 sees "1" on her screen, she is not sure whether player 2 knows that the game is G_b, and consequently may hesitate to play B. But if the number on her screen is, for example, 17 then it seems to be "almost" common knowledge that the game is G_b, and thus it may seem that she will adhere to the more desirable equilibrium (B, B) of the game G_b. Her decision will depend on her belief about what player 2 will do if the number on his screen is 16 or 17. In turn, player 2's decision depends on his belief about what player 1 will do if the number on her screen is 16. And so on. To study these considerations we now define the following Bayesian game, referred to as the *electronic mail game*.

- The set of states is $\Omega = \{(Q_1, Q_2): Q_1 = Q_2 \text{ or } Q_1 = Q_2 + 1\}$.
- The signal function τ_i of each player i is defined by $\tau_i(Q_1, Q_2) = Q_i$.
- Each player's belief on Ω is the same, derived from the technology (characterized by ϵ) and the assumption that the game is G_a with probability $1 - p$: $p_i(0, 0) = 1 - p$, $p_i(q + 1, q) = p\epsilon(1 - \epsilon)^{2q}$, and $p_i(q + 1, q + 1) = p\epsilon(1 - \epsilon)^{2q+1}$ for any nonnegative integer q.
- In each state (Q_1, Q_2) the payoffs are determined by the game $G(Q_1, Q_2)$.

■ PROPOSITION 83.1 *The electronic mail game has a unique Nash equilibrium, in which both players always choose A.*

Proof. In the state $(0, 0)$ the action A is strictly dominant for player 1, so that in any Nash equilibrium player 1 chooses A when receiving the signal 0. If player 2 gets no message (i.e. his signal is 0) then he knows that either player 1 did not send a message (an event with probability $1 - p$) or the message that player 1 sent did not arrive (an event with probability $p\epsilon$). If player 2 chooses A then, since player 1 chooses A in the state $(0, 0)$, player 2's expected payoff is at least $(1 - p)M/[(1 - p) + p\epsilon]$ whatever player 1 chooses in the state $(1, 0)$; if player 2 chooses B then his payoff is at most $[-L(1 - p) + p\epsilon M]/[(1 - p) + p\epsilon]$. Therefore it is strictly optimal for player 2 to choose A when his signal is 0.

Assume now that we have shown that for all (Q_1, Q_2) with $Q_1 + Q_2 < 2q$ players 1 and 2 both choose A in any equilibrium. Consider player 1's decision when she sends q messages. In this case player 1 is uncertain whether $Q_2 = q$ or $Q_2 = q - 1$. Given that she did not receive a confirmation of her qth message, the probability that she assigns to $Q_2 = q - 1$ is $z = \epsilon/[\epsilon + (1 - \epsilon)\epsilon] > \frac{1}{2}$. Thus she believes that it is more likely that her last message did not arrive than that player 2 got the message. (This is the key point in the argument.) If she chooses B then her expected payoff is at most $z(-L) + (1 - z)M$ (since under the induction assumption, she knows that if $Q_2 = q - 1$ then player 2 chooses A). If she chooses A then her payoff is at least 0. Given that $L > M$ and $z > \frac{1}{2}$, her best action is thus A. By a similar argument, if players 1 and 2 both choose A in any equilibrium for all (Q_1, Q_2) with $Q_1 + Q_2 < 2q + 1$ then player 2 chooses A when his signal is q. Hence each player chooses A in response to every possible signal.   □

Thus even if both players know that the game is G_b and even if the noise in the network (the probability ϵ) is arbitrarily small, the players act as if they had no information and play A, as they do in the absence of an electronic mail system!

What would *you* do if the number on your screen were 17? It is hard to imagine that when L slightly exceeds M and ϵ is small a player who sees the number 17 on his screen will not choose B. The contrast between our intuition and the game theoretic analysis makes the equilibrium paradoxical. In this respect the example joins a long list of games (like the finitely repeated Prisoner's Dilemma (see Proposition 155.1), the chain-store game (see Section 6.5.1), and the centipede game (see Section 6.5.2)) in which it seems that the source of the discrepancy between our intuition and the analysis lies in the fact that mathematical induction is not part of the reasoning process of human beings.

Notes

The basic model of knowledge described in Section 5.1 was formulated in the 1950s and 1960s; Hintikka (1962) is seminal. The concept of common knowledge is due to Lewis (1969) and Aumann (1976). Lewis gives an informal definition (and discusses the philosophical background for Sections 5.1 and 5.2); Aumann gives a formal definition and proves Proposition 75.1. Section 5.4 is based on Brandenburger (1992) and Aumann and Brandenburger (1995). (Spohn (1982) contains a result that is a precursor to Proposition 78.1.) The electronic mail game of Section 5.5 is studied by Rubinstein (1989); it is close in spirit to the "coordinated attack problem" studied by computer scientists (see, for example, Halpern (1986)).

The origin of the puzzle of the hats in Section 5.1.3 is unclear; see Littlewood (1953, p. 3). Exercise 76.2 is based on Milgrom and Stokey (1982) and Exercise 81.1 is based on Aumann (1987a).

For discussions of models of interactive knowledge in which the players' information functions are not partitional see Bacharach (1985) and Samet (1990). For surveys of the literature see Binmore and Brandenburger (1990) and Geanakoplos (1992, 1994).

II Extensive Games with Perfect Information

An *extensive game* is an explicit description of the sequential structure of the decision problems encountered by the players in a strategic situation. The model allows us to study solutions in which each player can consider his plan of action not only at the beginning of the game but also at any point of time at which he has to make a decision. By contrast, the model of a strategic game restricts us to solutions in which each player chooses his plan of action once and for all; this plan can cover unlimited contingencies, but the model of a strategic game does not allow a player to reconsider his plan of action after some events in the game have unfolded.

A general model of an extensive game allows each player, when making his choices, to be imperfectly informed about what has happened in the past. We study such a model in Part III. In this part we investigate a simpler model in which each player is perfectly informed about the players' previous actions at each point in the game. In Chapter 6 we describe the basic model. In the next three chapters we study two interesting classes of extensive games with perfect information: bargaining games of alternating offers (Chapter 7) and repeated games (Chapters 8 and 9). In Chapter 10 we present some of the main results of implementation theory (using the models of both strategic and extensive games).

6 Extensive Games with Perfect Information

In this chapter we study the model of an extensive game with perfect information. We argue that the solution concept of Nash equilibrium is unsatisfactory in this model since it ignores the sequential structure of the decision problems. We define the alternative notion of subgame perfect equilibrium, in which a player is required to reassess his plans as play proceeds. At the end of the chapter we compare this solution concept with that of iterated elimination of weakly dominated actions.

6.1 Extensive Games with Perfect Information

6.1.1 Definition

An extensive game is a detailed description of the sequential structure of the decision problems encountered by the players in a strategic situation. There is perfect information in such a game if each player, when making any decision, is perfectly informed of all the events that have previously occurred. For simplicity we initially restrict attention to games in which no two players make decisions at the same time and all relevant moves are made by the players (no randomness ever intervenes). (We remove these two restrictions in Section 6.3.)

▶ DEFINITION 89.1 An **extensive game with perfect information** has the following components.

- A set N (the set of **players**).

- A set H of sequences (finite or infinite) that satisfies the following three properties.

  ∘ The empty sequence $\varnothing$ is a member of H.

○ If $(a^k)_{k=1,\ldots,K} \in H$ (where K may be infinite) and $L < K$ then $(a^k)_{k=1,\ldots,L} \in H$.

○ If an infinite sequence $(a^k)_{k=1}^{\infty}$ satisfies $(a^k)_{k=1,\ldots,L} \in H$ for every positive integer L then $(a^k)_{k=1}^{\infty} \in H$.

(Each member of H is a **history**; each component of a history is an **action** taken by a player.) A history $(a^k)_{k=1,\ldots,K} \in H$ is **terminal** if it is infinite or if there is no a^{K+1} such that $(a^k)_{k=1,\ldots,K+1} \in H$. The set of terminal histories is denoted Z.

- A function P that assigns to each nonterminal history (each member of $H \setminus Z$) a member of N. (P is the **player function**, $P(h)$ being the player who takes an action after the history h.)

- For each player $i \in N$ a preference relation $\succsim_i$ on Z (the **preference relation** of player i).

Sometimes it is convenient to specify the structure of an extensive game without specifying the players' preferences. We refer to a triple $\langle N, H, P \rangle$ whose components satisfy the first three conditions in the definition as an **extensive game form with perfect information**.

If the set H of possible histories is finite then the game is *finite*. If the longest history is finite then the game has a *finite horizon*. Let h be a history of length k; we denote by (h, a) the history of length $k + 1$ consisting of h followed by a.

Throughout this chapter we refer to an extensive game with perfect information simply as an "extensive game". We interpret such a game as follows. After any nonterminal history h player $P(h)$ chooses an action from the set

$$A(h) = \{a : (h, a) \in H\}.$$

The empty history is the starting point of the game; we sometimes refer to it as the *initial history*. At this point player $P(\varnothing)$ chooses a member of $A(\varnothing)$. For each possible choice a^0 from this set player $P(a^0)$ subsequently chooses a member of the set $A(a^0)$; this choice determines the next player to move, and so on. A history after which no more choices have to be made is terminal. Note that a history may be an infinite sequence of actions. Implicit in the definition of a history as a sequence (rather than as a more complex mathematical object, like a string of sequences) is the assumption that no action may be taken after any infinite history, so that each such history is terminal. As in the case of a strategic game we often specify the players' preferences over terminal histories by giving payoff functions that represent the preferences.

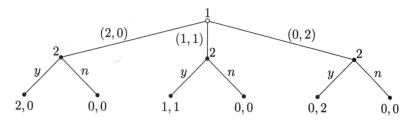

Figure 91.1 An extensive game that models the procedure for allocating two identical indivisible objects between two people described in Example 91.1.

◇ EXAMPLE 91.1 Two people use the following procedure to share two desirable identical indivisible objects. One of them proposes an allocation, which the other then either accepts or rejects. In the event of rejection, neither person receives either of the objects. Each person cares only about the number of objects he obtains.

An extensive game that models the individuals' predicament is $\langle N, H, P, (\succsim_i) \rangle$ where

- $N = \{1, 2\}$;
- H consists of the ten histories $\varnothing$, $(2,0)$, $(1,1)$, $(0,2)$, $((2,0), y)$, $((2,0), n)$, $((1,1), y)$, $((1,1), n)$, $((0,2), y)$, $((0,2), n)$;
- $P(\varnothing) = 1$ and $P(h) = 2$ for every nonterminal history $h \neq \varnothing$.
- $((2,0), y) \succ_1 ((1,1), y) \succ_1 ((0,2), y) \sim_1 ((2,0), n) \sim_1 ((1,1), n) \sim_1 ((0,2), n)$ and $((0,2), y) \succ_2 ((1,1), y) \succ_2 ((2,0), y) \sim_2 ((0,2), n) \sim_2 ((1,1), n) \sim_2 ((2,0), n)$.

A convenient representation of this game is shown in Figure 91.1. The small circle at the top of the diagram represents the initial history $\varnothing$ (the starting point of the game). The 1 above this circle indicates that $P(\varnothing) = 1$ (player 1 makes the first move). The three line segments that emanate from the circle correspond to the three members of $A(\varnothing)$ (the possible actions of player 1 at the initial history); the labels beside these line segments are the names of the actions, $(k, 2 - k)$ being the proposal to give k of the objects to player 1 and the remaining $2 - k$ to player 2. Each line segment leads to a small disk beside which is the label 2, indicating that player 2 takes an action after any history of length one. The labels beside the line segments that emanate from these disks are the names of player 2's actions, y meaning "accept" and n meaning "reject". The numbers below the terminal histories are payoffs that represent the players' preferences. (The first number in each pair is the payoff of player 1 and the second is the payoff of player 2.)

Figure 91.1 suggests an alternative definition of an extensive game in which the basic component is a tree (a connected graph with no cycles). In this formulation each node corresponds to a history and any pair of nodes that are connected corresponds to an action; the names of the actions are not part of the definition. This definition is more conventional, but we find Definition 89.1, which takes the players' actions as primitives, to be more natural.

6.1.2 Strategies

A strategy of a player in an extensive game is a plan that specifies the action chosen by the player for *every* history after which it is his turn to move.

▶ DEFINITION 92.1 A **strategy of player** $i \in N$ in an extensive game with perfect information $\langle N, H, P, (\succsim_i) \rangle$ is a function that assigns an action in $A(h)$ to each nonterminal history $h \in H \setminus Z$ for which $P(h) = i$.

Note that the notion of a strategy of a player in a game $\langle N, H, P, (\succsim_i) \rangle$ depends only on the game form $\langle N, H, P \rangle$.

To illustrate the notion of a strategy consider the game in Figure 91.1. Player 1 takes an action only after the initial history $\varnothing$, so that we can identify each of her strategies with one of the three possible actions that she can take after this history: $(2,0)$, $(1,1)$, and $(0,2)$. Player 2 takes an action after each of the three histories $(2,0)$, $(1,1)$, and $(0,2)$, and in each case he has two possible actions. Thus we can identify each of his strategies with a triple $a_2 b_2 c_2$ where a_2, b_2, and c_2 are the actions that he chooses after the histories $(2,0)$, $(1,1)$, and $(0,2)$. The interpretation of player 2's strategy $a_2 b_2 c_2$ is that it is a contingency plan: *if* player 1 chooses $(2,0)$ *then* player 2 will choose a_2; *if* player 1 chooses $(1,1)$ *then* player 2 will choose b_2; and *if* player 1 chooses $(0,2)$ *then* player 2 will choose c_2.

The game in Figure 93.1 illustrates an important point: a strategy specifies the action chosen by a player for *every* history after which it is his turn to move, *even for histories that, if the strategy is followed, are never reached.* In this game player 1 has *four* strategies AE, AF, BE, and BF. That is, her strategy specifies an action after the history (A, C) even if it specifies that she chooses B at the beginning of the game. In this sense a strategy differs from what we would naturally consider to be a plan of action; we return to this point in Section 6.4. As we shall see in a moment, for some purposes we can regard BE and

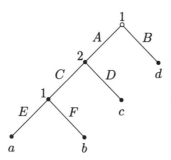

Figure 93.1 An extensive game in which player 1 moves both before and after player 2.

BF as the same strategy; however, in other cases it is important to keep them distinct.

For each strategy profile $s = (s_i)_{i \in N}$ in the extensive game $\langle N, H, P, (\succsim_i) \rangle$ we define the **outcome** $O(s)$ of s to be the terminal history that results when each player $i \in N$ follows the precepts of s_i. That is, $O(s)$ is the (possibly infinite) history $(a^1, \ldots, a^K) \in Z$ such that for $0 \leq k < K$ we have $s_{P(a^1, \ldots, a^k)}(a^1, \ldots, a^k) = a^{k+1}$.

As in a strategic game we can define a *mixed strategy* to be a probability distribution over the set of (pure) strategies. In extensive games with perfect information little is added by considering such strategies. Thus we postpone discussing them until Chapter 11, where we study extensive games in which the players are not perfectly informed when taking actions; in such games the notion of a mixed strategy has more significance.

6.1.3 Nash Equilibrium

The first solution concept we define for an extensive game ignores the sequential structure of the game; it treats the strategies as choices that are made once and for all before play begins.

▸ DEFINITION 93.1 A **Nash equilibrium of an extensive game with perfect information** $\langle N, H, P, (\succsim_i) \rangle$ is a strategy profile s^* such that for every player $i \in N$ we have

$$O(s^*_{-i}, s^*_i) \succsim_i O(s^*_{-i}, s_i) \text{ for every strategy } s_i \text{ of player } i.$$

Alternatively, we can define a Nash equilibrium of an extensive game Γ as a Nash equilibrium of the strategic game derived from Γ defined as follows.

▶ DEFINITION 94.1 The **strategic form of the extensive game with perfect information** $\Gamma = \langle N, H, P, (\succsim_i) \rangle$ is the strategic game $\langle N, (S_i), (\succsim_i') \rangle$ in which for each player $i \in N$

- S_i is the set of strategies of player i in Γ
- $\succsim_i'$ is defined by $s \succsim_i' s'$ if and only if $O(s) \succsim_i O(s')$ for every $s \in \times_{i \in N} S_i$ and $s' \in \times_{i \in N} S_i$.

? EXERCISE 94.2 Let G be a two-player strategic game $\langle \{1,2\}, (A_i), (\succsim_i) \rangle$ in which each player has two actions: $A_i = \{a_i', a_i''\}$ for $i = 1, 2$. Show that G is the strategic form of an extensive game with perfect information if and only if either for some $a_1 \in A_1$ we have $(a_1, a_2') \sim_i (a_1, a_2'')$ for $i = 1, 2$ or for some $a_2 \in A_2$ we have $(a_1', a_2) \sim_i (a_1'', a_2)$ for $i = 1, 2$.

If Nash equilibrium were the only solution we defined for extensive games, we could define a strategy more restrictively than we have done so: we could require that a strategy specify a player's action only after histories that are not inconsistent with the actions that it specifies at earlier points in the game. This is so because the outcome $O(s)$ of the strategy profile s is not affected by the actions that the strategy s_i of any player i specifies after contingencies that are inconsistent with s_i. Precisely, we can define a *reduced strategy* of player i to be a function f_i whose domain is a subset of $\{h \in H : P(h) = i\}$ and has the following properties: (i) it associates with every history h in the domain of f_i an action in $A(h)$ and (ii) a history h with $P(h) = i$ is in the domain of f_i if and only if all the actions of player i in h are those dictated by f_i (that is, if $h = (a^k)$ and $h' = (a^k)_{k=1,\ldots,L}$ is a subsequence of h with $P(h') = i$ then $f_i(h') = a^{L+1}$). Each reduced strategy of player i corresponds to a set of strategies of player i; for each vector of strategies of the other players each strategy in this set yields the same outcome (that is, the strategies in the set are *outcome-equivalent*). The set of Nash equilibria of an extensive game corresponds to the Nash equilibria of the strategic game in which the set of actions of each player is the set of his *reduced* strategies. (The full definition of a strategy is needed for the concept of subgame perfect equilibrium, which we define in the next section.)

As an example of the set of reduced strategies of a player in an extensive game, consider the game in Figure 93.1. Player 1 has three reduced strategies: one defined by $f_i(\varnothing) = B$ (with domain $\{\varnothing\}$), one defined by $f_i(\varnothing) = A$ and $f_i(A, C) = E$ (with domain $\{\varnothing, (A, C)\}$), and one defined by $f_i(\varnothing) = A$ and $f_i(A, C) = F$ (with domain $\{\varnothing, (A, C)\}$).

For some games some of a player's *reduced* strategies are equivalent in the sense that, regardless of the strategies of the other players, they generate the same *payoffs* for all players (though not the same outcome). That is, for some games there is a redundancy in the definition of a strategy, from the point of view of the players' payoffs, beyond that captured by the notion of a reduced strategy. For example, if $a = b$ in the game in Figure 93.1 then player 1's two reduced strategies in which she chooses A at the start of the game are equivalent from the point of view of payoffs. To capture this further redundancy, together with the redundancy captured by the notion of a reduced strategy, we can define the following variant of the strategic form.

▶ DEFINITION 95.1 Let $\Gamma = \langle N, H, P, (\succsim_i) \rangle$ be an extensive game with perfect information and let $\langle N, (S_i), (\succsim_i') \rangle$ be its strategic form. For any $i \in N$ define the strategies $s_i \in S_i$ and $s_i' \in S_i$ of player i to be *equivalent* if for each $s_{-i} \in S_{-i}$ we have $(s_{-i}, s_i) \sim_j' (s_{-i}, s_i')$ for all $j \in N$. The **reduced strategic form of** Γ is the strategic game $\langle N, (S_i'), (\succsim_i'') \rangle$ in which for each $i \in N$ each set S_i' contains one member of each set of equivalent strategies in S_i and $\succsim_i''$ is the preference ordering over $\times_{j \in N} S_j'$ induced by $\succsim_i'$.

(Note that this definition specifies the names of the actions in the reduced strategic form; every choice of such actions defines a different reduced strategic form. However, the names of the actions do not matter in any conventional game theoretic analysis, so that we refer to *the* reduced strategic form of a game.)

The strategic and reduced strategic forms of the game in Figure 93.1 are shown in Figure 96.1. If $a = b$ then the strategies AE and AF of player 1 are equivalent, so that player 1 has only two actions in the reduced strategic form of the game.

The next example illustrates the notion of Nash equilibrium and points to an undesirable feature that equilibria may possess.

◇ EXAMPLE 95.2 The game in Figure 96.2 has two Nash equilibria: (A, R) and (B, L), with payoff profiles $(2, 1)$ and $(1, 2)$. The strategy profile (B, L) is a Nash equilibrium because *given* that player 2 chooses L after the history A, it is optimal for player 1 to choose B at the start of the game (if she chooses A instead, then given player 2's choice she obtains 0 rather than 1), and *given* player 1's choice of B it is optimal for player 2 to choose L (since his choice makes no difference to the outcome).

Our interpretation of a nonterminal history as a point at which a player may reassess his plan of action leads to an argument that the Nash equilibrium (B, L) in this game lacks plausibility. *If the history A*

$$
\begin{array}{c|c|c}
 & C & D \\
\hline
AE & a & c \\
\hline
AF & b & c \\
\hline
BE & d & d \\
\hline
BF & d & d \\
\end{array}
\qquad
\begin{array}{c|c|c}
 & C & D \\
\hline
AE & a & c \\
\hline
AF & b & c \\
\hline
B & d & d \\
\end{array}
$$

Figure 96.1 The strategic form (left) and reduced strategic form (right) of the extensive game in Figure 93.1.

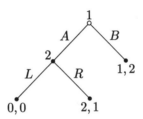

Figure 96.2 An example of a two-player extensive game.

were to occur then player 2 would, it seems, choose R over L, since he obtains a higher payoff by doing so. The equilibrium (B, L) is sustained by the "threat" of player 2 to choose L if player 1 chooses A. This threat is not credible since player 2 has no way of committing himself to this choice. Thus player 1 can be confident that if she chooses A then player 2 will choose R; since she prefers the outcome (A, R) to the Nash equilibrium outcome (B, L), she thus has an incentive to deviate from the equilibrium and choose A. In the next section we define a notion of equilibrium that captures these considerations.

◇ EXAMPLE 96.1 The Nash equilibria of the game in Figure 91.1 are $((2, 0), yyy)$, $((2, 0), yyn)$, $((2, 0), yny)$, $((2, 0), ynn)$, $((1, 1), nyy)$, $((1, 1), nyn)$, $((0, 2), nny)$, $((2, 0), nny)$, and $((2, 0), nnn)$. The first four result in the division $(2, 0)$, the next two result in the division $(1, 1)$, the next one results in the division $(0, 2)$, and the last two result in the division $(0, 0)$. All of these equilibria except $((2, 0), yyy)$ and $((1, 1), nyy)$ involve an action of player 2 that is implausible after some history (since he rejects a proposal that gives him at least one of the objects); like the equilibrium (B, L) in Example 95.2, they are ruled out by the notion of equilibrium we now define.

6.2 Subgame Perfect Equilibrium

Motivated by the discussion at the end of the previous section we now define the notion of subgame perfect equilibrium. We begin by defining the notion of a subgame.

▶ DEFINITION 97.1 The **subgame of the extensive game with perfect information** $\Gamma = \langle N, H, P, (\succsim_i) \rangle$ **that follows the history** h is the extensive game $\Gamma(h) = \langle N, H|_h, P|_h, (\succsim_i|_h) \rangle$, where $H|_h$ is the set of sequences h' of actions for which $(h, h') \in H$, $P|_h$ is defined by $P|_h(h') = P(h, h')$ for each $h' \in H|_h$, and $\succsim_i|_h$ is defined by $h' \succsim_i|_h h''$ if and only if $(h, h') \succsim_i (h, h'')$.

The notion of equilibrium we now define requires that the action prescribed by each player's strategy be optimal, given the other players' strategies, after *every* history. Given a strategy s_i of player i and a history h in the extensive game Γ, denote by $s_i|_h$ the strategy that s_i induces in the subgame $\Gamma(h)$ (i.e. $s_i|_h(h') = s_i(h, h')$ for each $h' \in H|_h$); denote by O_h the outcome function of $\Gamma(h)$.

▶ DEFINITION 97.2 A **subgame perfect equilibrium of an extensive game with perfect information** $\Gamma = \langle N, H, P, (\succsim_i) \rangle$ is a strategy profile s^* such that for every player $i \in N$ and every nonterminal history $h \in H \setminus Z$ for which $P(h) = i$ we have

$$O_h(s^*_{-i}|_h, s^*_i|_h) \succsim_i|_h O_h(s^*_{-i}|_h, s_i)$$

for every strategy s_i of player i in the subgame $\Gamma(h)$.

Equivalently, we can define a subgame perfect equilibrium to be a strategy profile s^* in Γ for which for any history h the strategy profile $s^*|_h$ is a Nash equilibrium of the subgame $\Gamma(h)$.

The notion of subgame perfect equilibrium eliminates Nash equilibria in which the players' threats are not credible. For example, in the game in Figure 96.2 the only subgame perfect equilibrium is (A, R) and in the game in Figure 91.1 the only subgame perfect equilibria are $((2, 0), yyy)$ and $((1, 1), nyy)$.

◇ EXAMPLE 97.3 (*Stackelberg games*) A Stackelberg game is a two-player extensive game with perfect information in which a "leader" chooses an action from a set A_1 and a "follower", informed of the leader's choice, chooses an action from a set A_2. The solution usually applied to such games in economics is that of subgame perfect equilibrium (though this terminology is not always used). Some (but not all) subgame perfect equilibria of a Stackelberg game correspond to solutions of the maximization problem

$$\max_{(a_1,a_2)\in A_1 \times A_2} u_1(a_1, a_2) \quad \text{subject to } a_2 \in \arg\max_{a_2' \in A_2} u_2(a_1, a_2'),$$

where u_i is a payoff function that represents player i's preferences. If the set A_i of actions of each player i is compact and the payoff functions u_i are continuous then this maximization problem has a solution.

? EXERCISE 98.1 Give an example of a subgame perfect equilibrium of a Stackelberg game that does not correspond to a solution of the maximization problem above.

To verify that a strategy profile s^* is a subgame perfect equilibrium, Definition 97.2 requires us to check, for every player i and every subgame, that there is no strategy that leads to an outcome that player i prefers. The following result shows that in a game with a finite horizon we can restrict attention, for each player i and each subgame, to alternative strategies that differ from s_i^* in the actions they prescribe after just *one* history. Specifically, a strategy profile is a subgame perfect equilibrium if and only if for each subgame the player who makes the first move cannot obtain a better outcome by changing only his initial action. For an extensive game Γ denote by $\ell(\Gamma)$ the length of the longest history in Γ; we refer to $\ell(\Gamma)$ as the *length of* Γ.

■ LEMMA 98.2 (The one deviation property) *Let* $\Gamma = \langle N, H, P, (\succsim_i) \rangle$ *be a finite horizon extensive game with perfect information. The strategy profile* s^* *is a subgame perfect equilibrium of* Γ *if and only if for every player* $i \in N$ *and every history* $h \in H$ *for which* $P(h) = i$ *we have*

$$O_h(s_{-i}^*|_h, s_i^*|_h) \succsim_i|_h O_h(s_{-i}^*|_h, s_i)$$

for every strategy s_i *of player* i *in the subgame* $\Gamma(h)$ *that differs from* $s_i^*|_h$ *only in the action it prescribes after the initial history of* $\Gamma(h)$.

Proof. If s^* is a subgame perfect equilibrium of Γ then it satisfies the condition. Now suppose that s^* is not a subgame perfect equilibrium; suppose that player i can deviate profitably in the subgame $\Gamma(h')$. Then there exists a profitable deviant strategy s_i of player i in $\Gamma(h')$ for which $s_i(h) \neq (s_i^*|_{h'})(h)$ for a number of histories h not larger than the length of $\Gamma(h')$; since Γ has a finite horizon this number is finite. From among all the profitable deviations of player i in $\Gamma(h')$ choose a strategy s_i for which the number of histories h such that $s_i(h) \neq (s_i^*|_{h'})(h)$ is minimal. Let h^* be the longest history h of $\Gamma(h')$ for which $s_i(h) \neq (s_i^*|_{h'})(h)$. Then the initial history of $\Gamma(h^*)$ is the only history in $\Gamma(h^*)$ at which the action prescribed by s_i differs from that prescribed by $s_i^*|_{h'}$. Further, $s_i|_{h^*}$ is a profitable deviation in $\Gamma(h^*)$, since otherwise there would be a profitable deviation in $\Gamma(h')$ that differs from $s_i^*|_{h'}$ after fewer histories than does s_i. Thus $s_i|_{h^*}$ is a profitable deviation in $\Gamma(h^*)$ that differs from $s_i^*|_{h^*}$ only in the action that it prescribes after the initial history of $\Gamma(h^*)$. $\qquad\square$

? EXERCISE 99.1 Give an example of an infinite horizon game for which the one deviation property does not hold.

We now prove that every finite extensive game with perfect information has a subgame perfect equilibrium. Our proof is constructive: for each of the longest nonterminal histories in the game we choose an optimal action for the player whose turn it is to move and replace each of these histories with a terminal history in which the payoff profile is that which results when the optimal action is chosen; then we repeat the procedure, working our way back to the start of the game. (The following result is known as Kuhn's theorem.)

■ PROPOSITION 99.2 *Every finite extensive game with perfect information has a subgame perfect equilibrium.*

Proof. Let $\Gamma = \langle N, H, P, (\succsim_i) \rangle$ be a finite extensive game with perfect information. We construct a subgame perfect equilibrium of Γ by induction on $\ell(\Gamma(h))$; at the same time we define a function R that associates a terminal history with every history $h \in H$ and show that this history is a subgame perfect equilibrium outcome of the subgame $\Gamma(h)$.

If $\ell(\Gamma(h)) = 0$ (i.e. h is a terminal history of Γ) define $R(h) = h$. Now suppose that $R(h)$ is defined for all $h \in H$ with $\ell(\Gamma(h)) \leq k$ for some $k \geq 0$. Let h^* be a history for which $\ell(\Gamma(h^*)) = k+1$ and let $P(h^*) = i$. Since $\ell(\Gamma(h^*)) = k+1$ we have $\ell(\Gamma(h^*, a)) \leq k$ for all $a \in A(h^*)$. Define $s_i(h^*)$ to be a $\succsim_i$-maximizer of $R(h^*, a)$ over $a \in A(h^*)$, and define $R(h^*) = R(h^*, s_i(h^*))$. By induction we have now defined a strategy profile s in Γ; by Lemma 98.2 this strategy profile is a subgame perfect equilibrium of Γ. $\qquad\square$

The procedure used in this proof is often referred to as *backwards induction*. In addition to being a means by which to prove the proposition, this procedure is an algorithm for calculating the set of subgame perfect equilibria of a finite game. Part of the appeal of the notion of subgame perfect equilibrium derives from the fact that the algorithm describes what appears to be a natural way for players to analyze such a game so long as the horizon is relatively short.

One conclusion we may draw from the result relates to chess. Under the rule that a game is a draw once a position is repeated three times, chess is finite, so that Proposition 99.2 implies that it has a subgame perfect equilibrium and hence also a Nash equilibrium. (Under the (official) rule that a player has the *option* of declaring a draw under these circumstances, chess is not finite.) Because chess is strictly competitive, Proposition 22.2 implies that the equilibrium payoff is unique and that any Nash equilibrium strategy of a player guarantees the player his equilibrium payoff. Thus either White has a strategy that guarantees that it wins, or Black has a strategy that guarantees that it wins, or each player has a strategy that guarantees that the outcome of the game is either a win for it or a draw.

[?] EXERCISE 100.1 Show that the requirement in Kuhn's theorem (Proposition 99.2) that the game be finite cannot be replaced by the requirement that it have a finite horizon, nor by the requirement that after any history each player have finitely many possible actions.

Note that Kuhn's theorem makes no claim of uniqueness. Indeed, the game in Figure 91.1 has two subgame perfect equilibria $((2,0), yyy)$ and $((1,1), nyy))$ that are not equivalent in terms of either player's preferences. However, it is clear that a finite game in which no player is indifferent between any two outcomes has a unique subgame perfect equilibrium. Further, if all players are indifferent between any two outcomes whenever any one player is indifferent, then even though there may be more than one subgame perfect equilibrium, all players are indifferent between all subgame perfect equilibria. This result is demonstrated in the following exercise.

[?] EXERCISE 100.2 Say that a finite extensive game with perfect information satisfies the *no indifference condition* if

$$z \sim_j z' \text{ for all } j \in N \text{ whenever } z \sim_i z' \text{ for some } i \in N,$$

where z and z' are terminal histories. Show, using induction on the length of subgames, that every player is indifferent among all subgame perfect equilibrium outcomes of such a game. Show also that if s and s' are subgame perfect equilibria then so is s'', where for each player i the strategy s_i'' is equal to either s_i or s_i' (i.e. the equilibria of the game are *interchangeable*).

? EXERCISE 101.1 Show that a subgame perfect equilibrium of an extensive game Γ is also a subgame perfect equilibrium of the game obtained from Γ by deleting a subgame not reached in the equilibrium and assigning to the terminal history thus created the outcome of the equilibrium in the deleted subgame.

? EXERCISE 101.2 Let s be a strategy profile in an extensive game with perfect information Γ; suppose that $P(h) = i$, $s_i(h) = a$, and $a' \in A(h)$ with $a' \neq a$. Consider the game Γ' obtained from Γ by deleting all histories of the form (h, a', h') for some sequence of actions h' and let s' be the strategy profile in Γ' that is induced by s. Show that if s is a subgame perfect equilibrium of Γ then s' is a subgame perfect equilibrium of Γ'.

? EXERCISE 101.3 Armies 1 and 2 are fighting over an island initially held by a battalion of army 2. Army 1 has K battalions and army 2 has L. Whenever the island is occupied by one army the opposing army can launch an attack. The outcome of the attack is that the occupying battalion and one of the attacking battalions are destroyed; the attacking army wins and, so long as it has battalions left, occupies the island with one battalion. The commander of each army is interested in maximizing the number of surviving battalions but also regards the occupation of the island as worth more than one battalion but less than two. (If, after an attack, neither army has any battalions left, then the payoff of each commander is 0.) Analyze this situation as an extensive game and, using the notion of subgame perfect equilibrium, predict the winner as a function of K and L.

6.3 Two Extensions of the Definition of a Game

The model of an extensive game with perfect information, as given in Definition 89.1, can easily be extended in two directions.

6.3.1 Exogenous Uncertainty

First we extend the model to cover situations in which there is some exogenous uncertainty. An *extensive game with perfect information and chance moves* is a tuple $\langle N, H, P, f_c, (\succsim_i) \rangle$ where, as before, N is a finite set of players and H is a set of histories, and

- P is a function from the nonterminal histories in H to $N \cup \{c\}$.
  (If $P(h) = c$ then *chance* determines the action taken after the
  history h.)

- For each $h \in H$ with $P(h) = c$, $f_c(\cdot|h)$ is a probability measure on
  $A(h)$; each such probability measure is assumed to be independent
  of every other such measure. ($f_c(a|h)$ is the probability that a
  occurs after the history h.)

- For each player $i \in N$, $\succsim_i$ is a preference relation on lotteries over
  the set of terminal histories.

A strategy for each player $i \in N$ is defined as before. The outcome of a
strategy profile is a probability distribution over terminal histories. The
definition of a subgame perfect equilibrium is the same as before (see
Definition 97.2).

☐ EXERCISE 102.1 Show that both the one deviation property (Lemma
98.2) and Kuhn's theorem (Proposition 99.2) hold for an extensive game
with perfect information and chance moves.

6.3.2 Simultaneous Moves

To model situations in which players move simultaneously after certain
histories, each of them being fully informed of all past events when
making his choice, we can modify the definition of an extensive game
with perfect information (Definition 89.1) as follows. An *extensive game
with perfect information and simultaneous moves* is a tuple $\langle N, H, P,$
$(\succsim_i) \rangle$ where N, H, and $\succsim_i$ for each $i \in N$ are the same as in Definition 89.1, P is a function that assigns to each nonterminal history a *set*
of players, and H and P jointly satisfy the condition that for every nonterminal history h there is a collection $\{A_i(h)\}_{i \in P(h)}$ of sets for which
$A(h) = \{a : (h, a) \in H\} = \times_{i \in P(h)} A_i(h)$.

A history in such a game is a sequence of vectors; the components of
each vector a^k are the actions taken by the players whose turn it is to
move after the history $(a^\ell)_{\ell=1}^{k-1}$. The set of actions among which each
player $i \in P(h)$ can choose after the history h is $A_i(h)$; the interpretation
is that the choices of the players in $P(h)$ are made simultaneously.

A *strategy* of player $i \in N$ in such a game is a function that assigns
an action in $A_i(h)$ to every nonterminal history h for which $i \in P(h)$.
The definition of a subgame perfect equilibrium is the same as that
in Definition 97.2 with the exception that "$P(h) = i$" is replaced by
"$i \in P(h)$".

[?] EXERCISE 103.1 Suppose that three players share a pie by using the following procedure. First player 1 proposes a division, then players 2 and 3 simultaneously respond either "yes" or "no". If players 2 and 3 both say "yes" then the division is implemented; otherwise no player receives anything. Each player prefers more of the pie to less. Formulate this situation as an extensive game with simultaneous moves and find its subgame perfect equilibria.

[?] EXERCISE 103.2 Consider the following two-player game. First player 1 can choose either *Stop* or *Continue*. If she chooses *Stop* then the game ends with the pair of payoffs $(1, 1)$. If she chooses *Continue* then the players simultaneously announce nonnegative integers and each player's payoff is the product of the numbers. Formulate this situation as an extensive game with simultaneous moves and find its subgame perfect equilibria.

[?] EXERCISE 103.3 Show that the one deviation property (Lemma 98.2) holds for an extensive game with simultaneous moves but that Kuhn's theorem (Proposition 99.2) does not.

6.4 The Interpretation of a Strategy

As we have noted, the definition of a strategy (92.1) does not correspond to a plan of action since it requires a player to specify his actions after histories that are impossible if he carries out his plan. For example, as we saw before, a strategy of player 1 in the game in Figure 104.1 specifies both the action she takes at the beginning of the game and the action she takes after the history (A, C), *even if* the action she takes at the beginning of the game is B.

One interpretation for the components of a player's strategy corresponding to histories that are not possible if the strategy is followed is that they are the *beliefs* of the other players about what the player will do in the event he does not follow his plan. For example, in the game in Figure 104.1, player 1's action after the history (A, C) can be thought of as player 2's belief about the choice that player 1 will make after this history, a belief that player 2 needs to hold in order to rationally choose an action. If player 1 plans to choose A then player 2's belief coincides with player 1's planned action after the history (A, C). However, if player 1 plans to choose B then such a belief cannot be derived from player 1's plan of action. In this case player 1's strategy nevertheless supplies such a belief. Note that the belief of player 2 about player 1 is

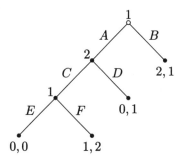

Figure 104.1 An extensive game in which player 1 moves both before and after player 2.

relevant to the analysis of the game even if player 1 plans to choose B, since to rationalize the choice of B player 1 needs to form a belief about player 2's plan after the history A.

This interpretation has a number of implications. First, it becomes problematic to speak of the "choice of a strategy", since a player does not choose the other players' beliefs. Second, in any equilibrium of a game with more than two players there is an implicit assumption that all the players other than any given player i hold the *same* beliefs about player i's behavior, not only if he follows his plan of action but also if he deviates from this plan. Third, one has to be careful if one imposes constraints on the strategies since one is then making assumptions not only about the players' plans of action, but also about their beliefs regarding each others' intentions when these plans of action are violated.

This interpretation of a strategy also diminishes the attraction of the notion of subgame perfect equilibrium. Consider again the game in Figure 104.1. There is no way, within the structure of the game, for player 2 to rationalize a choice of A by player 1 (since player 1 prefers the history B to *every* history that can result when she chooses A). Thus if she observes that player 1 chooses A, player 2 must give up a basic assumption about the game: she must believe either that player 1 is not rational, that player 1 perceives the game to differ from that in Figure 104.1, or that player 1 chose A by "mistake" (although such mistakes are not envisaged in the specification of the game). Yet the notion of subgame perfect equilibrium requires that, whatever history he observes, player 2 continue to maintain his original assumptions that player 1 is rational, knows the game, and does not make mistakes.

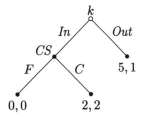

Figure 105.1 The structure of the players' choices in city k in the chain-store game. The first number in each pair is the chain-store's payoff and the second number is player k's payoff.

6.5 Two Notable Finite Horizon Games

In this section we demonstrate some of the strengths and weaknesses of the concept of subgame perfect equilibrium by examining two well-known games. It is convenient to describe each of these games by introducing a variable *time* that is discrete and starts at period 1. This variable is not an addition to the formal model of an extensive game; it is merely a device to simplify the description of the games and highlight their structures.

6.5.1 The Chain-Store Game

A chain-store (player CS) has branches in K cities, numbered $1, \ldots, K$. In each city k there is a single potential competitor, player k. In each period one of the potential competitors decides whether or not to compete with player CS; in period k it is player k's turn to do so. If player k decides to compete then the chain-store can either fight (F) or cooperate (C). The chain-store responds to player k's decision before player $k + 1$ makes its decision. Thus in period k the set of possible outcomes is $Q = \{Out, (In, C), (In, F)\}$. If challenged in any given city the chain-store prefers to cooperate rather than fight, but obtains the highest payoff if there is no entry. Each potential competitor is better off staying out than entering and being fought, but obtains the highest payoff when it enters and the chain-store is cooperative. The structure of the players' choices and their considerations in a single period are summarized in Figure 105.1.

Two assumptions complete the description of the game. First, at every point in the game all players know all the actions previously chosen. This allows us to model the situation as an extensive game with perfect information, in which the set of histories is $(\cup_{k=0}^{K} Q^k) \cup (\cup_{k=0}^{K-1} (Q^k \times \{In\}))$, where Q^k is the set of all sequences of k members of Q, and the player function is given by $P(h) = k+1$ if $h \in Q^k$ and $P(h) = CS$ if $h \in Q^k \times \{In\}$, for $k = 0, \ldots, K-1$. Second, the payoff of the chain-store in the game is the sum of its payoffs in the K cities.

The game has a multitude of Nash equilibria: every terminal history in which the outcome in any period is either Out or (In, C) is the outcome of a Nash equilibrium. (In any equilibrium in which player k chooses Out the chain-store's strategy specifies that it will fight if player k enters.)

In contrast, the game has a unique subgame perfect equilibrium; in this equilibrium every challenger chooses In and the chain-store always chooses C. (In city K the chain-store must choose C, regardless of the history, so that in city $K-1$ it must do the same; continuing the argument one sees that the chain-store must always choose C.)

For small values of K the Nash equilibria that are not subgame perfect are intuitively unappealing while the subgame perfect equilibrium is appealing. However, when K is large the subgame perfect equilibrium loses some of its appeal. The strategy of the chain-store in this equilibrium dictates that it cooperate with every entrant, regardless of its past behavior. Given our interpretation of a strategy (see the previous section), this means that even a challenger who has observed the chain-store fight with many entrants still believes that the chain-store will cooperate with it. Although the chain-store's unique subgame perfect equilibrium strategy does indeed specify that it cooperate with every entrant, it seems more reasonable for a competitor who has observed the chain-store fight repeatedly to believe that its entry will be met with an aggressive response, especially if there are many cities still to be contested. If a challenger enters then it is in the myopic interest of the chain-store to be cooperative, but intuition suggests that it may be in its long-term interest to build a reputation for aggressive behavior, in order to deter future entry. In Section 12.3.2 we study a perturbation of the chain-store game, in which the challengers are imperfectly informed about the motives of the chain-store, that attempts to capture this idea.

6.5.2   The Centipede Game

Two players are involved in a process that they alternately have the opportunity to stop. Each prefers the outcome when he stops the process in any period t to that in which the other player does so in period $t+1$. However, better still is any outcome that can result if the process is not

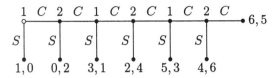

Figure 107.1 A six-period version of the centipede game.

stopped in either of these periods. After T periods, where T is even, the process ends. For $T = 6$ the game is shown in Figure 107.1. (The name "centipede" comes from the shape of the diagram.)

Formally, the set of histories in the game consists of all sequences $C(t) = (C, \ldots, C)$ of length t, for $0 \leq t \leq T$, and all sequences $S(t) = (C, \ldots, C, S)$ consisting of $t - 1$ repetitions of C followed by a single S, for $1 \leq t \leq T$. The player function is defined by $P(C(t)) = 1$ if t is even and $t \leq T - 2$ and $P(C(t)) = 2$ if t is odd. Player $P(C(t))$ prefers $S(t + 3)$ to $S(t + 1)$ to $S(t + 2)$ for $t \leq T - 3$, player 1 prefers $C(T)$ to $S(T - 1)$ to $S(T)$, and player 2 prefers $S(T)$ to $C(T)$.

The game has a unique subgame perfect equilibrium; in this equilibrium each player chooses S in every period. The outcome of this equilibrium is the same as the outcome of every Nash equilibrium. To see this, first note that there is no equilibrium in which the outcome is $C(T)$. Now assume that there is a Nash equilibrium that ends with player i choosing S in period t (i.e. after the history $C(t - 1)$). If $t \geq 2$ then player j can increase his payoff by choosing S in period $t - 1$. Hence in any equilibrium player 1 chooses S in the first period. In order for this to be optimal for player 1, player 2 must choose S in period 2. The notion of Nash equilibrium imposes no restriction on the players' choices in later periods: any pair of strategies in which player 1 chooses S in period 1 and player 2 chooses S in period 2 is a Nash equilibrium. (Note however that the *reduced* strategic form of the game has a unique Nash equilibrium.)

In the unique subgame perfect equilibrium of this game each player believes that the other player will stop the game at the next opportunity, even after a history in which that player has chosen to continue many times in the past. As in the subgame perfect equilibrium of the chain-store game such a belief is not intuitively appealing; unless T is very small it seems unlikely that player 1 would immediately choose S at the start of the game. The intuition in the centipede game is slightly different from that in the chain-store game in that after any long history *both* players have repeatedly violated the precepts of rationality enshrined in the notion of subgame perfect equilibrium. After a history in which both a player and his opponent have chosen to continue many times in the past, the basis on which the player should form a belief about his opponent's action in the next period is far from clear.

☐ EXERCISE 108.1 For any $\epsilon > 0$ define an *ϵ-equilibrium* of a strategic game to be a profile of actions with the property that no player has an alternative action that increases his payoff by more than ϵ. Show that for any positive integer k and any $\epsilon > 0$ there is a horizon T long enough that the strategic form of the modification of the centipede game in which all payoffs are divided by T has an ϵ-equilibrium in which the first player to stop the game does so in period k.

6.6 Iterated Elimination of Weakly Dominated Strategies

6.6.1 *Relation with Subgame Perfect Equilibrium*

In Section 4.3 we define the procedure of iterated elimination of weakly dominated actions for a strategic game and argue that though it is less appealing than the procedure of iterated elimination of strictly dominated actions (since a weakly dominated action is a best response to *some* belief), it is a natural method for a player to use to simplify a game. In the proof of Kuhn's theorem (Proposition 99.2) we define the procedure of backwards induction for finite extensive games with perfect information and show that it yields the set of subgame perfect equilibria of the game.

The two procedures are related. Let Γ be a finite extensive game with perfect information in which no player is indifferent between any two terminal histories. Then Γ has a unique subgame perfect equilibrium. We now define a sequence for eliminating weakly dominated actions in the strategic form G of Γ (weakly dominated strategies in Γ) with the property that all the action profiles of G that remain at the end of the procedure generate the unique subgame perfect equilibrium outcome of Γ.

Let h be a history of Γ with $P(h) = i$ and $\ell(\Gamma(h)) = 1$ and let $a_i^* \in A(h)$ be the unique action selected by the procedure of backwards induction for the history h. Backwards induction eliminates every strategy of player i that chooses an action different from a_i^* after the history h. Among these strategies, those consistent with h (i.e. that choose the component of h that follows h' whenever h' is a subhistory of

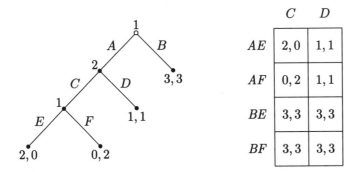

Figure 109.1 An extensive game (left) and its strategic form (right). There is an order of elimination of weakly dominated actions in the strategic form that eliminates the unique subgame perfect equilibrium of the extensive game.

h with $P(h') = i$) are weakly dominated actions in G. In the sequence of eliminations that we define, all of these weakly dominated actions are eliminated from G at this stage. Having performed this elimination for each history h with $\ell(\Gamma(h)) = 1$, we turn to histories h with $\ell(\Gamma(h)) = 2$ and perform an analogous elimination; we continue back to the beginning of the game in this way. Every strategy of player i that remains at the end of this procedure chooses the action selected by backwards induction after any history that is consistent with player i's subgame perfect equilibrium strategy. Thus in particular the subgame perfect equilibrium remains and every strategy profile that remains generates the unique subgame perfect equilibrium outcome.

Note, however, that other orders of elimination may remove all subgame perfect equilibria. In the game in Figure 109.1, for example, the unique subgame perfect equilibrium is (BE, D), while if in the strategic form the weakly dominated action AE is eliminated then D is weakly dominated in the remaining game; if AF is eliminated after D then neither of the two remaining action profiles $((BE, C)$ and $(BF, C))$ are subgame perfect equilibria of the extensive game.

Note also that if some player is indifferent between two terminal histories then there may be (i) an order of elimination that eliminates a subgame perfect equilibrium outcome and (ii) no order of elimination for which all surviving strategy profiles generate subgame perfect equilibrium outcomes. The game in Figure 110.1 demonstrates (i): the strategies AC, AD, and BD of player 1 are all weakly dominated by BC; after they are eliminated no remaining pair of actions yields the subgame perfect equilibrium outcome (A, R). If the payoff $(1, 2)$ is replaced by $(2, 0)$

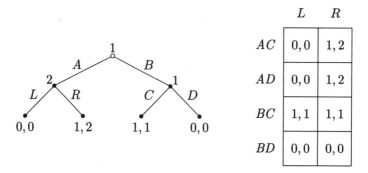

Figure 110.1 An extensive game (left) and its strategic form (right). There is an order of elimination of weakly dominated actions in the strategic form that eliminates a subgame perfect equilibrium outcome of the extensive game.

then the modified game demonstrates (ii): the outcome (A, L), which is not even a Nash equilibrium outcome, survives any order of elimination.

6.6.2 Forward Induction

We now present two examples that show that the iterated elimination of weakly dominated strategies captures some interesting features of players' reasoning in extensive games.

◇ EXAMPLE 110.1 (*BoS with an outside option*) Consider the extensive game with perfect information and simultaneous moves shown in Figure 111.1. In this game player 1 first decides whether to stay at home and read a book or to go to a concert. If she decides to read a book then the game ends; if she decides to go to a concert then she is engaged in the game BoS (Example 15.3) with player 2. (After the history *Concert* the players choose actions simultaneously.) Each player prefers to hear the music of his favorite composer in the company of the other player rather than either go to a concert alone or stay at home, but prefers to stay at home rather than either go out alone or hear the music of his less-preferred composer.

In the reduced strategic form of this game S is strictly dominated for player 1 by *Book*. If it is eliminated then S is weakly dominated for player 2 by B. Finally, *Book* is strictly dominated by B for player 1. The outcome that remains is (B, B).

This sequence of eliminations corresponds to the following argument for the extensive game. If player 2 has to make a decision he knows that player 1 has *not* chosen *Book*. Such a choice makes sense for player 1

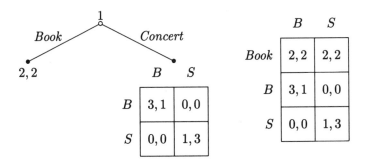

Figure 111.1 BoS with an outside option (left; an extensive game with perfect information and simultaneous moves) and its reduced strategic form (right).

only if she plans to choose B. Thus player 2 should choose B also. The logic of such an argument is referred to in the literature as "forward induction".

In the following example the iterated elimination of weakly dominated strategies leads to a conclusion that is more striking.

◇ EXAMPLE 111.1 (*Burning money*) Consider the game at the top of Figure 112.1. Two individuals are going to play BoS with monetary payoffs as in the left-hand table in the figure. Before doing so player 1 can discard a dollar (take the action D) or refrain from doing so (take the action 0); her move is observed by player 2. Both players are risk-neutral. (Note that the two subgames that follow player 1's initial move are strategically identical.)

The reduced strategic form of the game is shown in the bottom of Figure 112.1. Weakly dominated actions can be eliminated iteratively as follows.

1. DS is weakly dominated for player 1 by $0B$

2. SS is weakly dominated for player 2 by SB

3. BS is weakly dominated for player 2 by BB

4. $0S$ is strictly dominated for player 1 by DB

5. SB is weakly dominated for player 2 by BB

6. DB is strictly dominated for player 1 by $0B$

The single strategy pair that remains is $(0B, BB)$: the fact that player 1 can throw away a dollar implies, under iterated elimination of weakly dominated actions, that the outcome is player 1's favorite.

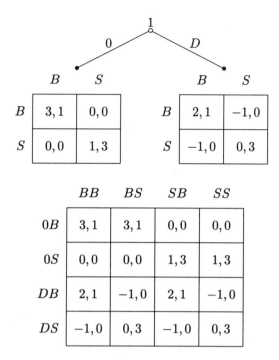

Figure 112.1 An extensive game with perfect information and simultaneous moves in which player 1 can choose to destroy a dollar before playing the game BoS. The extensive form is given at the top and the reduced strategic form at the bottom.

An intuitive argument that corresponds to this sequence of eliminations is the following. Player 1 must anticipate that if she chooses 0 then she will obtain an expected payoff of at least $\frac{3}{4}$, since for every belief about the behavior of player 2 she has an action that yields her at least this expected payoff. Thus if player 2 observes that player 1 chooses D then he must expect that player 1 will subsequently choose B (since the choice of S cannot possibly yield player 1 a payoff in excess of $\frac{3}{4}$). Given this, player 2 should choose B if player 1 chooses D; player 1 knows this, so that she can expect to obtain a payoff of 2 if she chooses D. But now player 2 can rationalize the choice 0 by player 1 only by believing that player 1 will choose B (since S can yield player 1 no more than 1), so that the best action of player 2 after observing 0 is B. This makes 0 the best action for player 1.

We now discuss these two examples in light of the distinction we made in Section 1.5 between the "steady state" and "deductive" approaches to game theory. From the point of view of the steady state interpretation the two examples share the same argument: the beliefs of player 2 in the equilibria in which the outcome is *Book* in the first example or $(0, (S, S))$ in the second example are both unreasonable in the sense that if player 1 deviates (to *Concert* or D) then the only sensible conclusion for player 2 to reach is that player 1 intends to play B subsequently, which means that player 2 should play B, making the deviation profitable for player 1. From the point of view of the deductive interpretation the two games differ, at least to the extent that the argument in the second example is more complex. In the first example player 1 has to reason about how player 2 will interpret an action (*Concert*) that she takes. In the second example player 1's reasoning about player 2's interpretation of her intended action 0 involves her belief about how player 2 would rationalize an action (D) that she does *not* take.

The second example raises a question about how to specify a game that captures a given situation. The arguments we have made are obviously based on the supposition that the game in Figure 112.1 reflects the situation as perceived by the players. In particular, they presume that the players perceive the possibility of disposing of a dollar to be relevant to the play of BoS. We believe this to be an implausible presumption: no reasonable person would consider the possibility of disposing a dollar to be relevant to the choice of which concert to attend. Thus we argue that a game that models the situation should simply exclude the possibility of disposal. (AR argues this to be so even if the game, including the move in which player 1 can burn money, is presented explicitly to the players by a referee, since before a player analyzes a situation strategically he "edits" the description of the situation, eliminating "irrelevant" factors.) On what principles do we base the claim that the possibility of disposing of a dollar is irrelevant? The answer is far from clear; some ideas follow. (a) The disposal does not affect the players' payoffs in BoS. (b) If the disposal is informative about the rationality of player 1, a sensible conclusion might be that a player who destroys a dollar is simply irrational. (In contrast, spending money on advertising, for example, may signal useful information.) (c) The dissimilarity between the two parts of the game makes it unlikely that player 2 will try to deduce from player 1's behavior in the first stage how she will behave in the second stage.

One interpretation of the arguments in this section is that each player accompanies his actions by messages explaining his future intentions. Thus to investigate the arguments further it may seem natural to augment the games by adding moves that have such explicit meaning. However, if we do so then we face difficulties, as the following example shows.

$$
\begin{array}{cc}
 & A \quad\ B \\
\begin{array}{c} A \\ \\ B \end{array} &
\begin{array}{|c|c|}
\hline
2,2 & 0,0 \\
\hline
0,0 & 1,1 \\
\hline
\end{array}
\end{array}
$$

Figure 114.1 The game relevant to Exercise 114.2.

Suppose that BoS is to be played and that player 1 is able, before BoS starts, to send a message (any string of symbols) to player 2. Assume further that each player cares only about the outcome of BoS, not about the content of any message that is sent or about the relationship between the action that he takes and the message. This extensive game has subgame perfect equilibrium outcomes in which both (B, B) and (S, S) are played in BoS; in particular, there is an equilibrium in which player 2 completely ignores player 1's message. This is so because player 2 is not forced to interpret the message sent by player 1 as meaningful, even if the message is "I am about to play B". The fact that a message can be sent does not affect the outcome because the names of the actions do not play any role in the concept of Nash equilibrium. A reasonable conclusion appears to be that a modification of the model of an extensive game is required if we wish to model communication between players.

[?] EXERCISE 114.1 Examine the variant of the game at the top of Figure 112.1 in which player 1 first has the option of burning a dollar, then player 2, having observed player 1's action, is also allowed to burn a dollar, and finally players 1 and 2 engage in BoS. Find the set of outcomes that survive iterated elimination of weakly dominated actions and compare it with the outcome that does so in the game in Figure 112.1.

[?] EXERCISE 114.2 Consider the game that differs from that at the top of Figure 112.1 only in that the game in which the players engage after player 1 has the option to burn a dollar is that shown in Figure 114.1. Find the set of outcomes that survives iterated elimination of weakly dominated actions.

Notes

The notion of an extensive game is due to von Neumann and Morgenstern (1944); Kuhn (1950, 1953) suggested the formulation we describe. The notion of subgame perfect equilibrium is due to Selten (1965).

The one deviation property (Lemma 98.2) is closely related to a principle of dynamic programming. Proposition 99.2 is due to Kuhn (1950, 1953). The idea of regarding games with simultaneous moves as games with perfect information is due to Dubey and Kaneko (1984). Some of our discussion of the interpretation of a strategy in Section 6.4 is based on Rubinstein (1991). The chain-store game studied in Section 6.5.1 is due to Selten (1978) and the centipede game studied in Section 6.5.2 is due to Rosenthal (1981). Some of the issues that these games raise are studied by Reny (1993). (See Section 12.3.2 for a variant of the chain-store game due to Kreps and Wilson (1982a) and Milgrom and Roberts (1982).) Moulin (1986) gives results that relate the procedure of iterated elimination of weakly dominated actions and the solution of subgame perfect equilibrium. The game in Figure 109.1 is taken (with modification) from Reny (1992). The idea of forward induction (together with the game in Example 110.1) is due to Kohlberg; it is discussed in Kohlberg and Mertens (1986). The game in Example 111.1 is due to van Damme (1989); see also Ben-Porath and Dekel (1992) and Osborne (1990). (For more discussion of the issues that arise in this game see Rubinstein (1991).)

Exercise 103.2 is based on an idea of Kreps; Exercise 108.1 is due to Radner (1980) (see also Radner (1986)).

7 Bargaining Games

Groups of people often have to choose collectively an outcome in a situation in which unanimity about the best outcome is lacking. Here we study a model, based on an extensive game with perfect information, that captures some of the features of such a situation.

7.1 Bargaining and Game Theory

Game theory deals with situations in which people's interests conflict. The people involved may try to resolve the conflict by *committing* themselves *voluntarily* to a course of action that is beneficial to all of them. If there is more than one course of action more desirable than disagreement for all individuals and there is conflict over which course of action to pursue then some form of negotiation over how to resolve the conflict is necessary. The negotiation process may be modeled using the tools of game theory; the model in this chapter is an example of such an analysis.

Since the presence of a conflict of interest is central to game theoretic situations, the theory of bargaining is more than just an application of game theory; models of bargaining lie at the heart of the subject and have attracted a great deal of attention since its inception. Most of the early work uses the axiomatic approach initiated by John Nash, whose work we discuss in Chapter 15. In this chapter we use the model of an extensive game with perfect information to study some features of bargaining, in particular the influence of the participants' impatience and risk aversion on the outcome.

7.2   A Bargaining Game of Alternating Offers

Consider a situation in which two bargainers have the opportunity to reach agreement on an outcome in some set X and perceive that if they fail to do so then the outcome will be some fixed event D. The set X may, for example, be the set of feasible divisions of a desirable pie and D may be the event in which neither party receives any of the pie. To model such a situation as an extensive game we have to specify the procedure that the parties follow when negotiating.

The procedure we study is one in which the players alternate offers. It can be described conveniently by introducing the variable "time", the values of which are the nonnegative integers. The first move of the game occurs in period 0, when player 1 makes a proposal (a member of X), which player 2 then either accepts or rejects. Acceptance ends the game while rejection leads to period 1, in which player 2 makes a proposal, which player 1 has to accept or reject. Again, acceptance ends the game; rejection leads to period 2, in which it is once again player 1's turn to make a proposal. The game continues in this fashion: so long as no offer has been accepted, in every even period player 1 makes a proposal that player 2 must either accept or reject, and in every odd period player 2 makes a proposal that player 1 must either accept or reject. There is no bound on the number of rounds of negotiation: the game has an infinite horizon. (See Section 8.2 for a discussion of the choice between a finite and infinite horizon when modeling a situation as a game.) The fact that some offer is rejected places no restrictions on the offers that may subsequently be made. In particular, a player who rejects a proposal x may subsequently make a proposal that is worse for him than x. If no offer is ever accepted then the outcome is the disagreement event D.

We now give a formal description of the situation as an extensive game with perfect information (see Definition 89.1). The set of players is $N = \{1,2\}$. Let X, the set of possible *agreements*, be a compact connected subset of a Euclidian space, and let T be the set of nonnegative integers. The set of histories H is the set of all sequences of one of the following types, where $t \in T$, $x^s \in X$ for all s, A means "accept", and R means "reject".

I. $\varnothing$ (the initial history), or $(x^0, R, x^1, R, \ldots, x^t, R)$

II. $(x^0, R, x^1, R, \ldots, x^t)$

III. $(x^0, R, x^1, R, \ldots, x^t, A)$

IV. $(x^0, R, x^1, R, \ldots)$

It follows from this description of the histories that the player whose turn it is to move chooses a member of X after a history of type I and a member of $\{A, R\}$ after a history of type II. Histories of type III and IV are terminal; those of type III are finite, while those of type IV are infinite. The player function is defined as follows: $P(h) = 1$ if h is of type I or type II and t is odd or if h is empty; $P(h) = 2$ if h is of type I or type II and t is even.

To complete the description of the game we need to specify the players' preferences over terminal histories. We assume that each player cares only about whether agreement is reached and the time and content of the agreement, not about the path of proposals that preceded the agreement. Precisely, the set of terminal histories is partitioned as follows: for each $x \in X$ and $t \in T$ the set of all histories of type III for which $x^t = x$ is a member of the partition, denoted by (x, t), and the set of all histories of type IV is a member of the partition, denoted by D. The preference relation of each player i over histories is induced from a preference relation $\succsim_i$ over the set $(X \times T) \cup \{D\}$ of members of this partition (that is, each player is indifferent between any two histories that lie in the same member of the partition). We assume that each player i's preference relation $\succsim_i$ satisfies the following conditions.

- No agreement is worse than disagreement: $(x, t) \succsim_i D$ for all $(x, t) \in X \times T$.

- Time is valuable: $(x, t) \succsim_i (x, t + 1)$ for every period $t \in T$ and every agreement $x \in X$, with strict preference if $(x, 0) \succ_i D$.

- Preferences are stationary: $(x, t) \succsim_i (y, t+1)$ if and only if $(x, 0) \succsim_i (y, 1)$, and $(x, t) \succsim_i (y, t)$ if and only if $(x, 0) \succsim_i (y, 0)$.

- Preferences are continuous: if $x_n \in X$ and $y_n \in X$ for all n, $\{x_n\}$ converges to $x \in X$, $\{y_n\}$ converges to $y \in X$, and $(x_n, t) \succsim_i (y_n, s)$ for all n, then $(x, t) \succsim_i (y, s)$.

These assumptions imply that for *any* $\delta \in (0, 1)$ there is a continuous function $u_i \colon X \to \mathbb{R}$ such that the preference relation $\succsim_i$ is represented on $X \times T$ by the function $\delta^t u_i(x)$ in the sense that $(x, t) \succsim_i (y, s)$ if and only if $\delta^t u_i(x) \geq \delta^s u_i(y)$. (This follows from Fishburn and Rubinstein (1982, Theorems 1 and 2).) Note that if $\delta^t u_i(x)$ represents $\succsim_i$ then for *any* $\epsilon \in (0, 1)$ the function $\epsilon^t v_i(x)$ where v_i is defined by $v_i(x) = (u_i(x))^{(\ln \epsilon)/(\ln \delta)}$ also represents $\succsim_i$. Thus if $\delta^t u_i(x)$ and $\epsilon^t v_i(x)$ are representations of two preference relations and $\delta > \epsilon$ then we *cannot* conclude that the first preference relation is more "patient" than the second unless $v_i = u_i$.

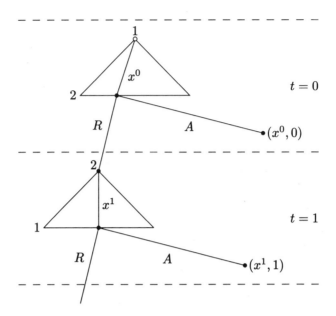

Figure 120.1 A representation of the first two periods of a bargaining game of alternating offers.

We refer to the extensive game with perfect information $\langle N, H, P, (\succsim_i)\rangle$ thus defined as the **bargaining game of alternating offers** $\langle X, (\succsim_i)\rangle$.

The first two periods of such a game are illustrated in Figure 120.1. (Note that x^0 is only one of the proposals available to player 1 at the start of the game, and x^1 is only one of the proposals available to player 2 after he rejects x^0.)

An important example of a bargaining game of alternating offers is the following.

◇ EXAMPLE 120.1 (*Split-the-pie*) The set of possible agreements X is the set of all divisions of a desirable pie:

$$X = \{(x_1, x_2): x_i \geq 0 \text{ for } i = 1, 2 \text{ and } x_1 + x_2 = 1\}.$$

The preference relation $\succsim_i$ of each player i over $(X \times T) \cup \{D\}$ has the properties that $(x, t) \succsim_i (y, t)$ if and only if $x_i \geq y_i$ (pie is desirable) and $D \sim_1 ((0, 1), 0)$ and $D \sim_2 ((1, 0), 0)$ (in the event of disagreement both players receive nothing). Thus $\succsim_i$ can be represented on $X \times T$ by a function of the form $\delta_i^t w_i(x_i)$ in which $0 < \delta_i < 1$ and w_i is increasing and continuous, with $w_i(0) = 0$.

The set of Nash equilibria of a bargaining game of alternating offers is very large. In particular, for any $x^* \in X$ there is a Nash equilibrium in which the players immediately agree on x^* (i.e. player 1's equilibrium strategy assigns x^* to the initial history and player 2's strategy assigns A to the history x^*). One such equilibrium is that in which both players *always* propose x^* and *always* accept a proposal x if and only if $x = x^*$. (Alternatively, each player i could accept a proposal x in period t if and only if $(x, t) \succsim_i (x^*, t)$.) In addition, for many specifications of the players' preferences there are Nash equilibria in which an agreement is not reached immediately. For example, for any agreement x and period t in a split-the-pie game there is a Nash equilibrium for which the outcome is the acceptance of x in period t. One such equilibrium is that in which through period $t - 1$ each player demands the whole pie and rejects all proposals, and from period t on proposes x and accepts only x.

These Nash equilibria illustrate the point we made at the end of Section 6.1.3: the notion of Nash equilibrium does not exclude the use of "incredible threats". Consider the Nash equilibrium of a split-the-pie game in which both players always propose x^* and player i accepts a proposal x in period t if and only if $(x, t) \succsim_i (x^*, t)$. If $(x^*, 0) \succ_2 (x^*, 1)$ then by the continuity of the players' preferences there is an agreement x in which x_2 is slightly less than x_2^* for which $(x, 0) \succ_1 (x^*, 0)$ and $(x, 0) \succ_2 (x^*, 1)$. In the equilibrium player 2's strategy dictates that in any period he reject such a proposal x; this "threat" induces player 1 to propose x^*. Player 2's threat is incredible, given player 1's strategy: the best outcome that can occur if player 2 carries out his threat to reject x is that there is agreement on x^* in the next period, an outcome that player 2 likes less than agreement on x in period 0, which he can achieve by accepting x. As we explained in the previous chapter, the notion of subgame perfect equilibrium is designed to isolate equilibria in which no player's strategy has this unattractive property.

7.3 Subgame Perfect Equilibrium

7.3.1 Characterization

We now show that under some additional assumptions a bargaining game of alternating offers has an essentially unique subgame perfect equilibrium, which we characterize. The first assumption is designed to avoid redundancies.

A1 For no two agreements x and y is it the case that $(x, 0) \sim_i (y, 0)$ for both $i = 1$ and $i = 2$.

The next assumption simplifies the analysis.

A2 $(b^i, 1) \sim_j (b^i, 0) \sim_j D$ for $i = 1$, 2, $j \neq i$ where b^i is the best agreement for player i.

To state the next two assumptions we define the *Pareto frontier* of the set X of agreements to be the set of agreements x for which there is no agreement y with $(y, 0) \succ_i (x, 0)$ for $i = 1$, 2. We refer to a member of the Pareto frontier as an *efficient* agreement.

A3 The Pareto frontier of X is strictly monotone: if an agreement x is efficient then there is no other agreement y such that $(y, 0) \succsim_i (x, 0)$ for both players i.

A4 There is a unique pair (x^*, y^*) of agreements for which $(x^*, 1) \sim_1 (y^*, 0)$, $(y^*, 1) \sim_2 (x^*, 0)$, and both x^* and y^* are efficient.

The most important of these assumptions is A4. In a split-the-pie game a sufficient condition for it to be satisfied is that each player's preference relation exhibit "increasing loss to delay": $x_i - f_i(x)$ is an increasing function of x_i for each player i, where $f(x)$ is the agreement for which $(f(x), 0) \sim_i (x, 1)$. Another case in which assumption A4 is satisfied is that in which the Pareto frontier of the set X of agreements is the set $\{x \in \mathbb{R}^2 : x_2 = g(x_1)\}$ for some decreasing concave function g and the preference relation of each player i is represented by the function $\delta_i^t x_i$ for some $0 < \delta_i < 1$.

■ PROPOSITION 122.1 *A bargaining game of alternating offers* $\langle X, (\succsim_i) \rangle$ *that satisfies A1 through A4 has a subgame perfect equilibrium. Let* (x^*, y^*) *be the unique pair of efficient agreements for which*

$$(x^*, 1) \sim_1 (y^*, 0) \qquad and \qquad (y^*, 1) \sim_2 (x^*, 0). \qquad (122.2)$$

In every subgame perfect equilibrium player 1 always proposes x^*, *accepts* y^* *and any proposal* x *for which* $(x, 0) \succ_1 (y^*, 0)$, *and rejects any proposal* x *for which* $(x, 0) \prec_1 (y^*, 0)$; *player 2 always proposes* y^*, *accepts* x^* *and any proposal* x *for which* $(x, 0) \succ_2 (x^*, 0)$, *and rejects any proposal* x *for which* $(x, 0) \prec_2 (x^*, 0)$.

Proof. First we verify that the pair of strategies defined in the proposition is a subgame perfect equilibrium. To do so we use the fact that the game has the "one deviation property": a pair of strategies is a subgame perfect equilibrium if and only if for every history h the player whose turn it is to move cannot deviate profitably by changing his action after h alone. This property holds for every game with a finite horizon, as Lemma 98.2 shows. The proof that it holds also for a bargaining game of alternating offers is left as an exercise.

? EXERCISE 123.1 Show that every bargaining game of alternating offers satisfies the one deviation property.

We now need to check the optimality only of the action of each player after any possible nonterminal history. The most interesting case is a history of type II. Suppose that it is player 2's turn to respond to a proposal x^t in period t. If he accepts this proposal then the outcome is (x^t, t) while if he rejects it the outcome is $(y^*, t+1)$. Since his preferences are stationary it follows from (122.2) that $(x^t, t) \succsim_2 (y^*, t+1)$ if and only if $(x^t, 0) \succsim_2 (x^*, 0)$ and thus his acceptance rule is optimal.

We now turn to the more difficult part of the proof: the argument that the subgame perfect equilibrium is essentially unique. (The only indeterminacy is in each player's response to a proposal that he regards as indifferent to the equilibrium proposal of the other player and the other player regards as worse; note that no such proposal is efficient.)

Given the stationarity of the players' preferences, for $i = 1$, 2 all subgames that begin with a proposal of player i are identical. Let G_i be such a subgame (G_1 is the game itself). Choose $\delta \in (0,1)$ and for $i = 1, 2$, let $u_i \colon X \to \mathbb{R}$ be such that $\delta^t u_i(x)$ represents $\succsim_i$ on $X \times T$. Let $M_i(G_i)$ be the supremum of the set of subgame perfect equilibrium (SPE) payoffs of player i in G_i:

$$M_i(G_i) = \sup\{\delta^t u_i(x) \colon \text{ there is a SPE of } G_i \text{ with outcome } (x, t)\}.$$

Let $m_i(G_i)$ be the corresponding infimum.

Step 1. $M_1(G_1) = m_1(G_1) = u_1(x^*)$ and $M_2(G_2) = m_2(G_2) = u_2(y^*)$. (That is, in every SPE of G_1 the payoff of player 1 is $u_1(x^*)$ and in every SPE of G_2 the payoff of player 2 is $u_2(y^*)$.)

Proof. Describe the pairs of payoffs on the Pareto frontier of X by the function ϕ: if x is efficient then $u_2(x) = \phi(u_1(x))$. By the connectedness of X and the continuity of the preference relations the domain of ϕ is an interval and ϕ is continuous; by A3 it is one-to-one and decreasing.

We first show that

$$m_2(G_2) \geq \phi(\delta M_1(G_1)). \tag{123.2}$$

If player 1 rejects a proposal of player 2 in the first period of G_2 then her payoff is not more than $\delta M_1(G_1)$. Hence in any SPE of G_2 she must accept any proposal that gives her more than $\delta M_1(G_1)$. Thus player 2's payoff is not less than $\phi(\delta M_1(G_1))$ in any SPE of G_2.

We now show that

$$M_1(G_1) \leq \phi^{-1}(\delta m_2(G_2)). \tag{124.1}$$

In any SPE of G_1 player 2's payoff is not less than $\delta m_2(G_2)$ since player 2 can always reject the opening proposal of player 1. Thus the payoff that player 1 can obtain in any SPE of G_1 does not exceed $\phi^{-1}(\delta m_2(G_2))$.

Finally, we argue that $M_1(G_1) = u_1(x^*)$. Since there is an SPE of G_1 in which immediate agreement is reached on x^* we have $M_1(G_1) \geq u_1(x^*)$. We now show that our assumptions about the set of agreements and the uniqueness of the solution of (122.2) imply that $M_1(G_1) \leq u_1(x^*)$.

By A2 we have $\delta u_2(b^1) = u_2(b^1)$, so that $u_2(b^1) = 0$; by A3 and the definition of ϕ we have $\delta\phi(\delta u_1(b^1)) > 0 = u_2(b^1) = \phi(u_1(b^1))$. Since ϕ is decreasing we conclude that $u_1(b^1) > \phi^{-1}(\delta\phi(\delta u_1(b^1)))$. Now, by (123.2) and (124.1) we have $M_1(G_1) \leq \phi^{-1}(\delta\phi(\delta M_1(G_1)))$. Thus by the continuity of ϕ there exists $U_1 \in [M_1(G_1), u_1(b^1))$ satisfying $U_1 = \phi^{-1}(\delta\phi(\delta U_1))$. If $M_1(G_1) > u_1(x^*)$ then $U_1 \neq u_1(x^*)$, so that taking a^* and b^* to be efficient agreements for which $u_1(a^*) = U_1$ and $u_1(b^*) = \delta u_1(a^*)$ it follows that (a^*, b^*) is a solution to (122.2) that differs from (x^*, y^*), contradicting A4. Thus $M_1(G_1) \leq u_1(x^*)$.

Similarly we can show that $m_1(G_1) = u_1(x^*)$, $M_2(G_2) = u_2(y^*)$, and $m_2(G_2) = u_2(y^*)$, completing the proof of this step.

Step 2. In every SPE of G_1 player 1's initial proposal is x^*, which player 2 immediately accepts.

Proof. In every SPE of G_1 player 1's payoff is $u_1(x^*)$ (by Step 1) and player 2's payoff is at least $\delta u_2(y^*) = u_2(x^*)$, since the rejection of player 1's proposal leads to the subgame G_2, in which player 2's SPE payoff is $u_2(y^*)$. Thus by A1 and the fact that x^* is efficient, player 1's opening proposal is x^*, which is accepted by player 2.

Step 3. In every SPE of G_1 player 2's strategy accepts any proposal x for which $(x, 0) \succ_2 (x^*, 0)$ and rejects any proposal x for which $(x, 0) \prec_2 (x^*, 0)$.

Proof. A rejection by player 2 leads to G_2, in which player 2's payoff is $u_2(y^*)$ (by Step 1). Since $u_2(x^*) = \delta u_2(y^*)$, player 2 must accept any proposal x for which $(x, 0) \succ_2 (x^*, 0)$ and reject any x for which $(x, 0) \prec_2 (x^*, 0)$. (There is no restriction on player 2's response to a proposal $x \neq x^*$ for which $(x, 0) \sim_2 (x^*, 0)$.)

Finally, arguments analogous to those in Steps 3 and 4 apply to player 2's proposals and player 1's acceptance rule in every SPE of G_2. □

Note that if all the members of the set X of agreements are efficient (as, for example, in a split-the-pie game) then a bargaining game of alternating offers has a unique (not just essentially unique) subgame perfect equilibrium. The following example is frequently used in applications.

◇ EXAMPLE 125.1 Consider a split-the-pie game in which each player i's preferences are represented by the function $\delta_i^t x_i$ for some $\delta_i \in (0, 1)$. Then we have $x^* = (a^*, 1 - a^*)$ and $y^* = (1 - b^*, b^*)$, where a^* and b^* solve the pair of equations $1 - b^* = \delta_1 a^*$ and $1 - a^* = \delta_2 b^*$, so that $a^* = (1 - \delta_2)/(1 - \delta_1 \delta_2)$ and $b^* = (1 - \delta_1)/(1 - \delta_1 \delta_2)$.

An interesting case that is not covered by Proposition 122.1 is a variant of a split-the-pie game in which each player i incurs the cost $c_i > 0$ for every period in which agreement is not reached (and there is no upper bound on the total of these costs that a player can incur). That is, player i's payoff if the agreement x is concluded in period t is $x_i - c_i t$. This case violates A2, since $(x, 0) \succ_i (x, 1)$ for *every* agreement x. It also violates A4: if $c_1 \neq c_2$ then there is no pair of agreements satisfying the two conditions while if $c_1 = c_2$ then there are many such pairs of agreements.

? EXERCISE 125.2

    a. Show that if $c_1 < c_2$ then the game described in the previous paragraph has a unique subgame perfect equilibrium, and that this equilibrium has the same structure as that in Proposition 122.1 with $x^* = (1, 0)$ and $y^* = (1 - c_1, c_1)$.

    b. Show that if $c_1 = c_2 = c < 1$ then the game has many subgame perfect equilibrium outcomes including, if $c < \frac{1}{3}$, equilibria in which agreement is delayed.

7.3.2 Properties of Equilibrium

Efficiency The structure of a bargaining game of alternating offers allows bargaining to continue for ever, but, under assumptions A1 through A4, in all subgame perfect equilibria agreement is reached immediately on an agreement on the Pareto frontier of X (so that the outcome of the game is efficient).

Stationarity of Strategies   The subgame perfect equilibrium strategies are stationary: for any history after which it is player i's turn to propose an agreement he proposes the same agreement, and for any history after which it is his turn to respond to a proposal he uses the same criterion to choose his response. We have not *restricted* players to use stationary strategies; rather, such strategies emerge as a conclusion.

First Mover Advantage   Consider again Example 125.1. If $\delta_1 = \delta_2 = \delta$ then the amount of the pie that player 1 gets is $a^* = 1/(1+\delta) > \frac{1}{2}$. The only asymmetry in the game is that player 1 moves first; the fact that she obtains more than half of the pie indicates that there is an advantage to being the first to make a proposal. This first-mover advantage holds more generally: using A3 and the fact that x^* and y^* are efficient, we have $(x^*, 0) \succ_1 (y^*, 0)$ in any bargaining game of alternating offers that satisfies A1 through A4.

Comparative Statics of Impatience   The key feature of the players' preferences is that they exhibit impatience. It seems reasonable to expect that the more impatient a player the worse off he is in equilibrium. This is indeed so in the game in Example 125.1, since the values of a^* and b^* are increasing in δ_1 and δ_2 respectively. We now generalize this result to any bargaining game of alternating offers that satisfies A1 through A4.

Define $\succsim_i'$ to be *at least as impatient as* $\succsim_i$ if both induce the same ordering on $X \times \{0\}$ and $(x, 1) \precsim_i' (y, 0)$ whenever $(x, 1) \sim_i (y, 0)$.

■ PROPOSITION 126.1   Let $\langle X, (\succsim_i) \rangle$ and $\langle X, (\succsim_i') \rangle$ be bargaining games of alternating offers that satisfy A1 through A4 and suppose that $\succsim_1'$ is at least as impatient as $\succsim_1$ and $\succsim_2' = \succsim_2$. Let x^* be the agreement reached in every subgame perfect equilibrium of $\langle X, (\succsim_i) \rangle$ and let x' be the agreement reached in every subgame perfect equilibrium of $\langle X, (\succsim_i') \rangle$. Then $(x^*, 0) \succsim_1 (x', 0)$.

Proof. Assume not (so that, in particular, $x^* \neq x'$). Consider the subset S of $X \times X$ consisting of all pairs (x, y) such that x and y are efficient and $(y, 1) \sim_2 (x, 0)$. Let y' be the agreement for which $(x', y') \in S$. Since $(x', 1) \sim_1' (y', 0)$ (by (122.2)) it follows that $(x', 1) \succsim_1 (y', 0)$ and hence $(x', 1) \succ_1 (y', 0)$ by A4. By A2 we have $(b^1, b^1) \in S$, and by the assumption that time is valuable and A4 we have $(b^1, 1) \prec_1 (b^1, 0)$. Since X is compact and connected, the Pareto frontier of X is compact and connected, so that there is an agreement $\hat{x}$ on the path on the Pareto frontier that connects x' and b^1 such that $(\hat{x}, \hat{y}) \in S$ and $(\hat{x}, 1) \sim_1$ $(\hat{y}, 0)$. Since $(x', 0) \succ_1 (x^*, 0)$ and $(b^1, 0) \succsim_1 (x', 0)$ we have $\hat{x} \neq x^*$, contradicting A4.   □

7.4 Variations and Extensions

7.4.1 The Importance of the Procedure

To model bargaining as an extensive game we need to give an explicit description of the sequential structure of the decision problems encountered by the players: we need to specify a bargaining procedure. The variant of the model of a bargaining game of alternating offers considered in the next exercise demonstrates that the structure of the bargaining procedure plays an important role in determining the outcome.

☐ EXERCISE 127.1 Assume that player 1 makes all the offers (rather than the players alternating offers). Show that under A1 through A3 the resulting game has an essentially unique subgame perfect equilibrium, in which regardless of the players' preferences the agreement reached is b^1, the best possible agreement for player 1.

In a bargaining game of alternating offers the procedure treats the players almost symmetrically. The fact that the player to make the first offer is better off than his opponent in such a game is a vestige of the extreme advantage that a player enjoys if he is the only one to make offers.

7.4.2 Variants that Eliminate a Key Feature of the Model

A key feature of the model of a bargaining game of alternating offers is the ability of one player to force the other to choose between an agreement now and a more desirable agreement later. To illustrate this point, consider first the game in which the players make proposals simultaneously in each period, agreement being reached only if the proposals in any given period are compatible. In this case neither player can force the other to choose between an agreement now and a better agreement later; *every* efficient agreement is a subgame perfect equilibrium outcome.

To illustrate the point further, consider the case in which the set X of agreements contains finitely many elements, so that a player's ability to offer an agreement today that is slightly better than the agreement that the responder expects tomorrow is limited. In this case the range of subgame perfect equilibrium payoffs depends on the richness of the set X. The following exercise demonstrates this point in a specific case.

☐ EXERCISE 128.1 Consider a variant of a split-the-pie game in which the pie can be divided only into integral multiples of a basic indivisible unit $\epsilon > 0$ and the preferences of each player i are represented by the function $\delta^t x_i$. Denote this game by $\Gamma(\epsilon)$ and the game in which the pie is perfectly divisible by $\Gamma(0)$.

    a. Show that if δ is close enough to 1 then for every agreement $x \in X$ there is a subgame perfect equilibrium of $\Gamma(\epsilon)$ for which the outcome is $(x, 0)$.

    b. Show that if δ is close enough to 1 then for every outcome $z \in (X \times T) \cup \{D\}$ there is a subgame perfect equilibrium of $\Gamma(\epsilon)$ in which the outcome is z (use the equilibrium strategies in part a for $x = (1, 0)$ and $x = (0, 1)$ to deter deviations).

    c. Show conversely that for every $\delta \in (0, 1)$ and every $\eta > 0$ there exists $\bar{\epsilon} > 0$ such that if $\epsilon < \bar{\epsilon}$ then for $i = 1, 2$ the difference between player i's payoff in every subgame perfect equilibrium of $\Gamma(\epsilon)$ and his payoff in the unique subgame perfect equilibrium of $\Gamma(0)$ is less than η and agreement is reached immediately.

7.4.3 Opting Out

An interesting class of extensions of the model of a bargaining game of alternating offers is obtained by allowing one or both players, at various points in the game, to "opt out" (without requiring the approval of the other player) rather than continue bargaining. A simple case is that in which only one of the players, say player 2, can opt out, and can do so only when responding to an offer. Denote the outcome in which he does so in period t by (Out, t), and assume that $(Out, t) \sim_1 D$ for all $t \in T$.

Suppose first that $(Out, 0) \prec_2 (y^*, 1)$, where y^* is the unique subgame perfect equilibrium proposal of player 2 in the standard bargaining game of alternating offers. Then the ability of player 2 to opt out has no effect on the outcome: although player 2 has an additional "threat", it is worthless since he prefers to continue bargaining and obtain the outcome y^* with one period of delay.

Now suppose that $(Out, 0) \succ_2 (y^*, 1)$. Then player 2's threat is not worthless. In this case (under A1 through A4), in any subgame perfect equilibrium player 1 always proposes the efficient agreement $\hat{x}$ for which $(\hat{x}, 0) \sim_2 (Out, 0)$, which player 2 accepts, and player 2 always proposes the efficient agreement $\hat{y}$ for which $(\hat{y}, 0) \sim_1 (\hat{x}, 1)$, which player 1 accepts. Thus in this case the ability of player 2 to exercise an outside option causes the outcome of bargaining to be equivalent for him to the outcome that results if he opts out.

[?] EXERCISE 129.1 Prove the result that we have just described for a split-the-pie game in which each player i's preference relation over $(X \times T) \cup \{D\} \cup (\{Out\} \times T)$ is represented by u_i where $u_i(x, t) = \delta^t x_i$, $u_i(D) = 0$, $u_1(Out, t) = 0$ for all t, and $u_2(Out, t) = \delta^t b$ for some $b < 1$ and some $\delta \in (0, 1)$.

This result is sometimes called the "outside option principle". It is not robust to the assumptions about the points at which the players can exercise their outside options. For example, if one of the players can opt out at the end of *any* period, not just after he rejects an offer, then the game has a great multiplicity of subgame perfect equilibria (see Shaked (1994) and Osborne and Rubinstein (1990, Section 3.12)).

7.4.4 A Model in Which There Is a Risk of Breakdown

Finally, consider a modification of the model of a bargaining game of alternating offers in which at the end of each period a chance move ends the game with probability $\alpha \in (0, 1)$. (We consider this case again in Section 15.4.) Assume that the players do not care about the time at which agreement is reached; the pressure on each player to reach agreement is not the player's impatience but the risk that the negotiations will break down. In the extensive game (with perfect information and chance moves) that models this situation there are six types of history. Four of these types are analogs of types I through IV (see Section 7.2) in which each occurrence of R is replaced by (R, C), where C is the action of chance in which bargaining continues (rather than breaks down). A history of type V takes the form $(x^0, R, C, x^1, R, C, \ldots, x^t, R)$, after which it is the turn of chance to move, and a history of type VI is terminal and takes the form $(x^0, R, C, x^1, R, C, \ldots, x^t, R, B)$, where B stands for breakdown. We assume that the players are indifferent among all terminal histories in which no agreement is reached (i.e. among all histories of types IV and VI). Given the presence of chance moves, we need to specify the players' preferences over the set of lotteries over terminal histories. As before, we assume that these preferences depend only on the agreement finally reached (not on the path of rejected agreements). Further, we assume that the preference relation of each player i is represented by a von Neumann–Morgenstern utility function $u_i\colon X \cup \{B\} \to \mathbb{R}$. (Since histories of type IV do not occur with positive probability, whatever strategies the players employ, the players' preference relations do not have to rank D.) Finally, we assume that $u_i(B) = 0$, $u_i(x) \geq 0$ for all $x \in X$, and there is a unique pair (x^*, y^*) of efficient agreements satisfying

$$u_1(y^*) = (1 - \alpha)u_1(x^*) \quad \text{and} \quad u_2(x^*) = (1 - \alpha)u_2(y^*). \qquad (130.1)$$

(This is so, for example, if the players are splitting a pie, $u_i(x_1, x_2) = w_i(x_i)$ for some increasing concave function w_i, and $w_i(0) = 0$.)

▱ EXERCISE 130.2 Prove the analog of Proposition 122.1 for the variant of a bargaining game of alternating offers described in the previous paragraph.

7.4.5 More Than Two Players

Proposition 122.1 does not extend to the case in which there are more than two players, as the following three-player variant of a split-the-pie game (Example 120.1) demonstrates. The set of possible agreements is

$$X = \{(x_1, x_2, x_3)\colon x_i \geq 0 \text{ for } i = 1, 2, 3 \text{ and } x_1 + x_2 + x_3 = 1\}$$

and each player i's preferences are represented by $u_i(x, t) = \delta^t x_i$ for some $0 < \delta < 1$. The bargaining procedure is the following. Player 1 initially makes a proposal. A proposal x made by player j in period t is first considered by player $j + 1 \pmod 3$, who may accept or reject it. If he accepts it, then player $j + 2 \pmod 3$ may accept or reject it. If both accept it, then the game ends and x is implemented. Otherwise player $j + 1 \pmod 3$ makes the next proposal, in period $t + 1$.

Let $\frac{1}{2} \leq \delta < 1$. We claim that for every agreement x there is a subgame perfect equilibrium in which x is accepted immediately. Such an equilibrium may be described in terms of four commonly-held "states" x, e^1, e^2, and e^3, where e^i is the ith unit vector. In state y each player i makes the proposal y and accepts the proposal z if and only if $z_i \geq \delta y_i$. The initial state is x. Transitions between states occur only after a proposal has been made, before the response. If, in any state y, player i proposes z with $z_i > y_i$ then the state becomes e^j, where $j \neq i$ is the player with the lowest index for whom $z_j < \frac{1}{2}$. Such a player j exists, and the requirement that $\delta \geq \frac{1}{2}$ guarantees that it is optimal for him to reject player i's proposal. The main force holding this equilibrium together is that a player is rewarded for rejecting a deviant offer: after his rejection, he obtains all of the pie.

? EXERCISE 131.1 Show that if each player is restricted to use a stationary strategy (in which he makes the same proposal whenever he is the proposer, uses the same rule to accept proposals whenever he is the first responder, and uses the same rule to accept proposals whenever he is the second responder) then the unique subgame perfect equilibrium of the game described above assigns the fraction $\delta^{k-1}/(1 + \delta + \delta^2)$ of the pie to player k for $k = 1, 2, 3$.

Notes

The model in this chapter is due to Rubinstein (1982), as is Proposition 122.1. For an exposition and analysis of the model and its applications see Osborne and Rubinstein (1990).

Two precursors of the model that effectively restrict attention to finite-horizon games are found in Ståhl (1972) and Krelle (1976, pp. 607–632). For a discussion of time preferences see Fishburn and Rubinstein (1982). The proof of Proposition 122.1 is a modification of the original proof of Rubinstein (1982), following the ideas of Shaked and Sutton (1984a). The material in Section 7.4.2 is discussed in Muthoo (1991) and van Damme, Selten, and Winter (1990). The model in Section 7.4.3, in which a player can opt out, was suggested by Binmore, Shaked, and Sutton; see for example Shaked and Sutton (1984b) and Binmore (1985). The model in Section 7.4.4 is discussed in Binmore, Rubinstein, and Wolinsky (1986). The example discussed in Section 7.4.5 is due to Shaked; see Osborne and Rubinstein (1990, Section 3.13) for more detail. For another interpretation of the model of a bargaining game of alternating offers see Rubinstein (1995).

8 Repeated Games

The model of a repeated game is designed to examine the logic of long-term interaction. It captures the idea that a player will take into account the effect of his current behavior on the other players' future behavior, and aims to explain phenomena like cooperation, revenge, and threats.

8.1 The Basic Idea

The basic idea behind the theory is illustrated by the case in which two individuals repeatedly play the *Prisoner's Dilemma* (reproduced in Figure 134.1). Recall that this game has a unique Nash equilibrium, in which each player chooses D; further, for each player the action D strictly dominates the action C, so that the rationale behind the outcome (D, D) is very strong. Despite this, both players are better off if they "cooperate" and choose C. The main idea behind the theory of repeated games is that if the game is played repeatedly then the mutually desirable outcome in which (C, C) occurs in every period is stable if each player believes that a defection will terminate the cooperation, resulting in a subsequent loss for him that outweighs the short-term gain.

The primary achievement of the theory is to isolate types of strategies that support mutually desirable outcomes in any game. The theory gives us insights into the structure of behavior when individuals interact repeatedly, structure that may be interpreted in terms of a "social norm". The results that we describe show that the social norm needed to sustain mutually desirable outcomes involves each player's "punishing" any player whose behavior is undesirable. When we impose the requirement embedded in the notion of subgame perfect equilibrium that threats of punishment be credible, the social norm must also ensure that the punishers have an incentive to carry out the threats in circumstances

$$
\begin{array}{c c}
 & C \quad\ D \\
\begin{array}{c} C \\[1.5em] D \end{array} &
\begin{array}{|c|c|}
\hline
3,3 & 0,4 \\
\hline
4,0 & 1,1 \\
\hline
\end{array}
\end{array}
$$

Figure 134.1 The *Prisoner's Dilemma.*

in which the social norm requires them to do so. In this case the precise nature of the punishment depends on how the players value future outcomes. Sometimes it is sufficient that a punishment phase last for a limited amount of time, after which the players return to pursue the mutually desirable outcome; sometimes the social norm must entail future rewards for players who carry out costly punishments.

Although we regard these results about the *structure* of the equilibrium strategies to be the main achievement of the theory, most of the results in the literature focus instead on the set of *payoffs* that can be sustained by equilibria, giving conditions under which this set consists of nearly all reasonable payoff profiles. These "folk theorems" have two sides. On the one hand they demonstrate that socially desirable outcomes that cannot be sustained if players are short-sighted can be sustained if the players have long-term objectives. On the other hand they show that the set of equilibrium outcomes of a repeated game is huge, so that the notion of equilibrium lacks predictive power. "Folk theorems" are the focus of much of the formal development in this chapter. Nevertheless, we stress that in our opinion the main contribution of the theory is the discovery of interesting stable social norms (strategies) that support mutually desirable payoff profiles, and not simply the demonstration that equilibria *exist* that generate such profiles.

8.2 Infinitely Repeated Games vs. Finitely Repeated Games

The model of a repeated game has two versions: the horizon may be finite or infinite. As we shall see, the results in the two cases are different. An extreme (and far from general) case of the difference is that in which the constituent game is the *Prisoner's Dilemma*. We shall see below that in any *finite* repetition of this game the only Nash equilibrium outcome is that in which the players choose (D, D) in every period; on the other hand, in the *infinitely* repeated game the set of subgame perfect equilibrium payoff profiles is huge. Thus in applying the model of a repeated game in specific situations we may need to determine whether a finite or infinite horizon is appropriate.

In our view a model should attempt to capture the features of reality that the players *perceive*; it should not necessarily aim to describe the reality that an outside observer perceives, though obviously there are links between the two perceptions. Thus the fact that a situation has a horizon that is in some physical sense finite (or infinite) does not *necessarily* imply that the best model of the situation has a finite (or infinite) horizon. A model with an infinite horizon is appropriate if after each period the players believe that the game will continue for an additional period, while a model with a finite horizon is appropriate if the players clearly perceive a well-defined final period. The fact that players have finite lives, for example, does not imply that one should always model their strategic interaction as a finitely repeated game. If they play a game so frequently that the horizon approaches only very slowly then they may ignore the existence of the horizon entirely until its arrival is imminent, and until this point their strategic thinking may be better captured by a game with an infinite horizon.

AR In a situation that is objectively finite, a key criterion that determines whether we should use a model with a finite or an infinite horizon is whether the last period enters explicitly into the players' strategic considerations. For this reason, even some situations that involve a small number of repetitions are better analyzed as infinitely repeated games. For example, when laboratory subjects are instructed to play the *Prisoner's Dilemma* twenty times with payoffs as in Figure 134.1 (interpreted as dollars), I believe that their lines of reasoning are better modeled by an infinitely repeated game than by a 20-period repeated game, since except very close to the end of the game they are likely to ignore the existence of the final period.

MJO The behavior of experimental subjects who play the *Prisoner's Dilemma* repeatedly a finite number of times is inconsistent with the unique subgame perfect equilibrium of the finitely repeated game. The fact that it may be consistent with some subgame perfect equilibrium of the infinitely repeated game is uninteresting since the range of outcomes that are so-consistent is vast. Certainly the subgame perfect equilibria of the infinitely repeated game give no insights about the dependence of the subjects' behavior on the magnitude of the payoffs and the length of the game. (For a summary of the evidence see Rapoport (1987).) The experimental results definitely indicate that the notion of subgame per-

fect equilibrium in the finitely repeated *Prisoner's Dilemma* does not
capture human behavior. However, this deficiency appears to have more
to do with the backwards induction inherent in the notion of subgame
perfect equilibrium than with the finiteness of the horizon *per se*. A
model that will give us an understanding of the facts is likely to be a
variant of the finitely repeated game; some characteristics of the equilibria of the infinitely repeated game may be suggestive, but this model
itself appears unpromising as an explanatory tool. Moreover, in contexts in which the constituent game has multiple Nash equilibria, the
equilibria of finitely repeated games correspond well with the casual observation that people act cooperatively when the horizon is distant and
opportunistically when it is near; the equilibria of infinitely repeated
games can give us no insight into such behavior. Finally, in situations
in which people's discount factors decline to zero over time, even if they
never become zero (i.e. no fixed finite horizon is perceived), the equilibrium outcomes have more in common with those of finitely repeated
games than with those of infinitely repeated games.

AR   In much of the existing literature the fact that the set of equilibria in a long finitely repeated game may be very different from the
set of equilibria of an infinite repetition of the same constituent game
is regarded as "disturbing". In contrast, I find it attractive: the two
models capture a very realistic feature of life, namely the fact that the
existence of a prespecified finite period may crucially affect people's behavior (consider the last few months of a presidency or the fact that
religions attempt to persuade their believers that there is "life after
death").

MJO   First, for a large set of constituent games there is *no* discontinuity between the outcomes of the associated finitely and infinitely
repeated games (see Section 8.10). Second, in some cases in which the
discontinuity does exist it is indeed unappealing. If people who are faced
with a known fixed distant horizon behave as if the horizon is infinite
then this should be the prediction of a model with a fixed finite horizon;
if it is not then doubts are raised about the plausibility of the notion of
subgame perfect equilibrium in other contexts.

8.3   Infinitely Repeated Games: Definitions

The model of an infinitely repeated game captures a situation in which
players repeatedly engage in a strategic game G, which we refer to as the
constituent game. Throughout we restrict attention to games in which

*the action set of each player is compact and the preference relation of
each player is continuous.* There is no limit on the number of times that
G is played; on each occasion the players choose their actions simultaneously. When taking an action, a player knows the actions previously
chosen by all players. We model this situation as an extensive game with
perfect information (and simultaneous moves) as follows.

▶ DEFINITION 137.1 Let $G = \langle N, (A_i), (\succsim_i) \rangle$ be a strategic game; let
$A = \times_{i \in N} A_i$. An **infinitely repeated game** of G is an extensive
game with perfect information and simultaneous moves $\langle N, H, P, (\succsim_i^*) \rangle$
in which

- $H = \{\varnothing\} \cup (\cup_{t=1}^{\infty} A^t) \cup A^{\infty}$ (where $\varnothing$ is the initial history and A^{∞} is
  the set of infinite sequences $(a^t)_{t=1}^{\infty}$ of action profiles in G)

- $P(h) = N$ for each nonterminal history $h \in H$

- $\succsim_i^*$ is a preference relation on A^{∞} that extends the preference relation $\succsim_i$ in the sense that it satisfies the following condition of *weak
  separability*: if $(a^t) \in A^{\infty}$, $a \in A$, $a' \in A$, and $a \succsim_i a'$ then

$$(a^1, \ldots, a^{t-1}, a, a^{t+1}, \ldots) \succsim_i^* (a^1, \ldots, a^{t-1}, a', a^{t+1}, \ldots)$$

for all values of t.

A history is terminal if and only if it is infinite. After any nonterminal
history every player $i \in N$ chooses an action in A_i. Thus a strategy of
player i is a function that assigns an action in A_i to every finite sequence
of outcomes in G.

We now impose restrictions on the players' preference relations in addition to weak separability. We assume throughout that player i's preference relation $\succsim_i^*$ in the repeated game is based upon a payoff function u_i
that represents his preference relation $\succsim_i$ in G: we assume that whether
$(a^t) \succsim_i^* (b^t)$ depends only on the relation between the corresponding
sequences $(u_i(a^t))$ and $(u_i(b^t))$ of payoffs in G.

We consider three forms of the preference relations, the first of which
is defined as follows.

- **Discounting**: There is some number $\delta \in (0, 1)$ (the *discount factor*) such that the sequence (v_i^t) of real numbers is at least as good
  as the sequence (w_i^t) if and only if $\sum_{t=1}^{\infty} \delta^{t-1}(v_i^t - w_i^t) \geq 0$.

According to this criterion a player evaluates a sequence (v_i^t) of payoffs
by $\sum_{t=1}^{\infty} \delta^{t-1} v_i^t$ for some discount factor $\delta \in (0, 1)$. (Since we have
assumed that the values of the player's payoffs lie in a bounded set, this
sum is well-defined.) When the players' preferences take this form we

refer to the profile $((1-\delta)\sum_{t=1}^{T}\delta^{t-1}v_i^t)_{i\in N}$ as the **payoff profile** in the repeated game associated with the sequence $(v^t)_{t=1}^{\infty}$ of payoff profiles in the constituent game.

Preferences with discounting treat the periods differently: the value of a given gain diminishes with time. We now specify two alternative criteria that treat all periods symmetrically. In one criterion a player evaluates a sequence (v_i^t) of payoffs essentially by its limiting average $\lim_{T\to\infty}\sum_{t=1}^{T}v_i^t/T$. However, this limit does not always exist (the average payoff over t periods may continually oscillate as t increases); the criterion that we discuss is defined as follows. (It is convenient to define this criterion in terms of the strict preference relation.)

- **Limit of means**: The sequence (v_i^t) of real numbers is preferred to the sequence (w_i^t) if and only if $\liminf\sum_{t=1}^{T}(v_i^t-w_i^t)/T > 0$ (i.e. if and only if there exists $\epsilon > 0$ such that $\sum_{t=1}^{T}(v_i^t-w_i^t)/T > \epsilon$ for all but a finite number of periods T).

When the players' preferences take this form we refer to the profile $\lim_{T\to\infty}(\sum_{t=1}^{T}v_i^t/T)_{i\in N}$, if it exists, as the **payoff profile** in the repeated game associated with the sequence $(v^t)_{t=1}^{\infty}$ of payoff profiles in the constituent game.

Note that if the sequence (v_i^t) is preferred to the sequence (w_i^t) according to the limit of means then there is a discount factor δ close enough to 1 such that (v_i^t) is preferred to (w_i^t) by the discounting criterion.

Under the discounting criterion a change in the payoff in a single period can matter, whereas under the limit of means criterion payoff differences in any finite number of periods do not matter. A player whose preferences satisfy the limit of means is ready to sacrifice any loss in the first finite number of periods in order to increase the stream of payoffs he eventually obtains. For example, the stream $(0,\ldots,0,2,2,\ldots)$ of payoffs is preferred by the limit of means criterion to the constant stream $(1,1,\ldots)$ independent of the index of the period in which the player first gets 2 in the first stream. At first sight this property may seem strange. However, it is not difficult to think of situations in which decision makers put overwhelming emphasis on the long run at the expense of the short run (think of nationalist struggles).

We now introduce a criterion that treats all periods symmetrically and puts emphasis on the long run but at the same time is sensitive to a change in payoff in a single period. (Again we define the criterion in terms of the strict preference relation.)

- **Overtaking**: The sequence (v_i^t) is preferred to the sequence (w_i^t) if and only if $\liminf \sum_{t=1}^{T}(v_i^t - w_i^t) > 0$.

The following examples illustrate some of the differences between the three criteria. The sequence $(1, -1, 0, 0, \ldots)$ is preferred for any $\delta \in (0, 1)$ by the discounting criterion to the sequence $(0, 0, \ldots)$, but according to the other two criteria the two sequences are indifferent. The sequence $(-1, 2, 0, 0, \ldots)$ is preferred to the sequence $(0, 0, \ldots)$ according to the overtaking criterion, but the two sequences are indifferent according to the limit of means. The sequence $(0, \ldots, 0, 1, 1, \ldots)$ in which M zeros are followed by a constant sequence of 1's is preferred by the limit of means to $(1, 0, 0, \ldots)$ for every value of M, but for every δ there exists M^* large enough that for all $M > M^*$ the latter is preferred to the former according to the discounting criterion for that value of δ.

Let $G = \langle N, (A_i), (\succsim_i) \rangle$ be a strategic game and for each $i \in N$ let u_i be a payoff function that represents $\succsim_i$. We define the δ-**discounted infinitely repeated game of** $\langle N, (A_i), (u_i) \rangle$ to be the infinitely repeated game for which the constituent game is G and the preference ordering $\succsim_i^*$ of each player $i \in N$ is derived from the payoff function u_i using the discounting criterion with a discount factor of δ for each player. Similarly we define the **limit of means infinitely repeated game of** $\langle N, (A_i), (u_i) \rangle$ and the **overtaking infinitely repeated game of** $\langle N, (A_i), (u_i) \rangle$.

We denote by $u(a)$ the profile $(u_i(a))_{i \in N}$. Define a vector $v \in \mathbb{R}^N$ to be a **payoff profile** of $\langle N, (A_i), (u_i) \rangle$ if there is an outcome $a \in A$ for which $v = u(a)$. We refer to a vector $v \in \mathbb{R}^N$ as a **feasible payoff profile** of $\langle N, (A_i), (u_i) \rangle$ if it is a convex combination of payoff profiles of outcomes in A: that is, if $v = \sum_{a \in A} \alpha_a u(a)$ for some collection $(\alpha_a)_{a \in A}$ of nonnegative rational numbers α_a with $\sum_{a \in A} \alpha_a = 1$. (In the literature the coefficients α_a are allowed to be any real numbers, not necessarily rational, a generalization that complicates the argument while adding little substance.) Note that a feasible payoff profile of $\langle N, (A_i), (u_i) \rangle$ is not necessarily a payoff profile of $\langle N, (A_i), (u_i) \rangle$.

[?] EXERCISE 139.1 Consider an infinitely repeated game in which the players' preferences are derived from their payoffs in the constituent game using *different* discount factors. Show that a payoff profile in such a repeated game may not be a feasible payoff profile of the constituent game.

8.4   Strategies as Machines

As we discussed in the introduction to this chapter, the main achievement of the theory of repeated games is to give us insights into the
structure of behavior when individuals interact repeatedly. In this section we develop a language in which to describe conveniently the structure of the equilibria that we find. We begin by defining a machine,
which is intended as an abstraction of the process by which a player
implements a strategy in a repeated game. A **machine** (or *automa-
ton*) for player i in an infinitely repeated game of $\langle N, (A_i), (\succsim_i) \rangle$ has the
following components.

- A set Q_i (the set of **states**).
- An element $q_i^0 \in Q_i$ (the **initial state**).
- A function $f_i: Q_i \to A_i$ that assigns an action to every state (the
  output function).
- A function $\tau_i: Q_i \times A \to Q_i$ that assigns a state to every pair con-
  sisting of a state and an action profile (the **transition function**).

The set Q_i is unrestricted. The names of the states do not of course
have any significance (the fact that we call a state "cooperative", for
example, does not mean that the behavior associated with it matches
its name). In the first period the state of the machine is q_i^0 and the
machine chooses the action $f_i(q_i^0)$. Whenever the machine is in some
state q_i, it chooses the action $f_i(q_i)$ corresponding to that state. The
transition function τ_i specifies how the machine moves from one state to
another: if the machine is in state q_i and a is the action profile chosen
then its state changes to $\tau_i(q_i, a)$.

Note that the input of the transition function consists of the current
state and the profile of *all* the players' current actions. It is more natural
to take as the input the current state and the list of actions chosen by the
other players. This fits the natural description of a "rule of behavior" or
"strategy" as a plan of how to behave in all possible circumstances that
are consistent with one's plans. However, since the game-theoretic definition requires that a strategy specify an action for all possible histories,
including those that are inconsistent with the player's own strategy, we
have to include as an input into the transition function the action of the
player himself.

To illustrate the concept of a machine we now give four examples of
machines for a player in the repeated *Prisoner's Dilemma* (Figure 134.1).

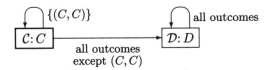

Figure 141.1 A machine that corresponds to the grim strategy in the *Prisoner's Dilemma*.

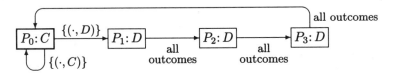

Figure 141.2 The machine M_1. This machine for player 1 in the *Prisoner's Dilemma* plays C as long as player 2 does so and punishes player 2 for choosing D by choosing D for three periods. (We use $\{(\cdot, X)\}$ to denote the set of all outcomes in which player 2's action is X.)

⋄ EXAMPLE 141.1 (*A machine for the "grim" strategy*) The machine $\langle Q_i, q_i^0, f_i, \tau_i \rangle$ defined as follows is the simplest one that carries out the ("grim") strategy that chooses C so long as both players have chosen C in every period in the past, and otherwise chooses D.

- $Q_i = \{\mathcal{C}, \mathcal{D}\}$.
- $q_i^0 = \mathcal{C}$.
- $f_i(\mathcal{C}) = C$ and $f_i(\mathcal{D}) = D$.
- $\tau_i(\mathcal{C}, (C, C)) = \mathcal{C}$ and $\tau_i(\mathcal{X}, (Y, Z)) = \mathcal{D}$ if $(\mathcal{X}, (Y, Z)) \neq (\mathcal{C}, (C, C))$.

This machine is illustrated in Figure 141.1. Each box corresponds to a state; inside each box is the name of the state followed (after the colon) by the action that the machine takes in that state. The box with the heavy boundary corresponds to the initial state. The arrows correspond to the transitions; adjacent to each arrow is the set of outcomes that induces the transition.

⋄ EXAMPLE 141.2 The machine M_1 of player 1 shown in Figure 141.2 plays C as long as player 2 plays C; it plays D for three periods, and then reverts back to C, if player 2 plays D when he should play C. (We can think of the other player being "punished" for three periods for playing D, and then "forgiven".) Notice that a machine must have at least four states in order to carry out this strategy.

⋄ EXAMPLE 141.3 The machine M_2 of player 2 shown in Figure 142.1 starts by playing C and continues to do so if the other player chooses D.

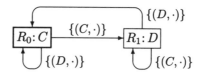

Figure 142.1 The machine M_2. This machine for player 2 in the *Prisoner's Dilemma* starts by playing C but switches to D if player 1 chooses C, returning to C only if player 1 chooses D.

period	state of M_1	state of M_2	outcome	payoffs
1	P_0	R_0	(C,C)	3, 3
2	P_0	R_1	(C,D)	0, 4
3	P_1	R_1	(D,D)	1, 1
4	P_2	R_0	(D,C)	4, 0
5	P_3	R_0	(D,C)	4, 0
6	P_0	R_0	(C,C)	3, 3

Figure 142.2 The outcomes in the first six periods of the repeated *Prisoner's Dilemma* when player 1 uses the machine M_1 in Figure 141.2 and player 2 uses the machine M_2 in Figure 142.1.

If the other player chooses C then it switches to D, which it continues to play until the other player again chooses D, when it reverts to playing C.

To illustrate the evolution of play in a repeated game when each player's strategy is carried out by a machine, suppose that player 1 uses the machine M_1 and player 2 uses the machine M_2 in the repeated *Prisoner's Dilemma*.

The machines start in the states P_0 and R_0 respectively. The outcome in the first period is (C,C) since the output function of M_1 assigns the action C to state P_0 and the output function of M_2 assigns the action C to state R_0. The states in the following period are determined by the transition functions. The transition function of M_1 leaves the machine in state P_0, since the outcome in the first period is (C,C), while the transition function of M_2 moves the machine from R_0 to R_1 in response to this input. Thus the pair of states in period 2 is (P_0, R_1). The output functions determine the outcome in period 2 to be (C,D), so that M_1 now moves from P_0 to P_1 while M_2 stays in R_1. Play continues through period 5 as in the table in Figure 142.2. In period 6 the pair of states is the same as it is in period 1; subsequently the states and outcomes cycle, following the pattern in the first five periods. The fact that cycles are generated is not peculiar to this example: whenever each player uses a machine with finitely many states a cycle is eventually reached, though not necessarily in period 1. (This follows from the fact that each machine takes as its input only the actions in the previous period (i.e. it is "Markovian").)

[?] EXERCISE 143.1 Show that not every strategy in an infinitely repeated game can be executed by a machine with a *finite* number of states.

8.5 Trigger Strategies: Nash Folk Theorems

We now study the set of Nash equilibrium outcomes of an infinitely repeated game. We show that this set includes outcomes that are not repetitions of Nash equilibria of the constituent game. To support such an outcome, each player must be deterred from deviating by being "punished". Such punishment may take many forms. One possibility is that each player uses a "trigger strategy": any deviation causes him to carry out a punitive action that lasts forever. In the equilibria that we study in this section each player uses such a strategy.

Let $\langle N, (A_i), (\succsim_i) \rangle$ be a strategic game and for each $i \in N$ let u_i be a payoff function that represents the preference ordering $\succsim_i$. Recall that we define a feasible payoff profile of $G = \langle N, (A_i), (u_i) \rangle$ to be a convex combination $\sum_{a \in A} \alpha_a u(a)$ for which the coefficients α_a are rational. Let $w = \sum_{a \in A} \alpha_a u(a)$ be such a profile and suppose that $\alpha_a = \beta_a / \gamma$ for each $a \in A$, where every β_a is an integer and $\gamma = \sum_{a \in A} \beta_a$. Then the sequence of outcomes in the repeated game that consists of an indefinite repetition of a cycle of length γ in which each $a \in A$ is played for β_a periods yields an average payoff profile over the cycle, and hence in the entire repeated game, of w.

Define player i's **minmax payoff in** G, henceforth denoted v_i, to be the lowest payoff that the other players can force upon player i:

$$v_i = \min_{a_{-i} \in A_{-i}} \max_{a_i \in A_i} u_i(a_{-i}, a_i). \tag{143.2}$$

A payoff profile w for which $w_i \geq v_i$ for all $i \in N$ is called **enforceable**; if $w_i > v_i$ for all $i \in N$ then w is **strictly enforceable**.[1] If $a \in A$ is an outcome of G for which $u(a)$ is (strictly) enforceable in G then we refer to a as a **(strictly) enforceable outcome of** G. Denote by $p_{-i} \in A_{-i}$ one of the solutions of the minimization problem on the right-hand side of (143.2). For each action profile $a \in A$ let

---

[1]In much of the literature the term *individually rational* is used instead of "enforceable".

$b_i(a_{-i}) \in A_i$ be an action of player i in G that is a best response to a_{-i} (i.e. $b_i(a_{-i}) \in B_i(a_{-i})$). (Notice that p_{-i} and the function b_i depend only on the players' preferences over A, not on the payoff representations of these preferences.) The collection of actions p_{-i} is the most severe "punishment" that the other players can inflict upon player i in G. (Note that we restrict punishments to be deterministic. In some of the literature punishers are allowed to randomize, possibly correlating their actions over time; this changes the set of feasible payoff profiles and the enforceable payoffs but not the structure of the set of equilibria of the repeated game.)

In the next two results we show that the set of Nash equilibrium payoff profiles of an infinitely repeated game in which the players evaluate streams of payoffs by the limit of means is the set of *all* feasible enforceable payoff profiles of the constituent game. The third result shows that the same is approximately true when the players discount future payoffs using a discount factor close to 1.

■ PROPOSITION 144.1 *Every Nash equilibrium payoff profile of the limit of means infinitely repeated game of $G = \langle N, (A_i), (u_i) \rangle$ is an enforceable payoff profile of G. The same is true, for any $\delta \in (0,1)$, of every Nash equilibrium payoff profile of the δ-discounted infinitely repeated game of G.*

Proof. Suppose that w is a payoff profile of the limit of means infinitely repeated game of G that is not enforceable; suppose that $w_i < v_i$. Then w is not a Nash equilibrium payoff profile of the repeated game, because for any strategy profile s the strategy s_i' of player i defined by $s_i'(h) = b_i(s_{-i}(h))$ for each history h gives player i a payoff of at least v_i in each period. The same argument applies to any δ-discounted infinitely repeated game of G.   □

The following exercise asks you to express the strategy s_i' of player i in this proof in the language of machines that we developed in Section 8.4.

[?] EXERCISE 144.2 Consider a two-player infinitely repeated game. For any given machine for player 2 construct a machine for player 1 that yields her a payoff of at least v_1.

■ PROPOSITION 144.3 (Nash folk theorem for the limit of means criterion) *Every feasible enforceable payoff profile of $G = \langle N, (A_i), (u_i) \rangle$ is a Nash equilibrium payoff profile of the limit of means infinitely repeated game of G.*

Proof. Let $w = \sum_{a \in A}(\beta_a/\gamma)u(a)$ be a feasible enforceable payoff profile, where β_a for each $a \in A$ is an integer and $\gamma = \sum_{a \in A} \beta_a$, and let (a^t) be the cycling sequence of action profiles for which the cycle (of length γ) contains β_a repetitions of a for each $a \in A$. Let s_i be the strategy of player i in the repeated game that chooses a_i^t in each period t unless there was a previous period t' in which a single player other than i deviated from $a^{t'}$, in which case it chooses $(p_{-j})_i$, where j is the deviant in the first such period t'. The strategy profile s is a Nash equilibrium of the repeated game since a player j who deviates receives at most his minmax payoff v_j in every subsequent period; the payoff profile generated by s is w. $\square$

[?] EXERCISE 145.1 Construct a machine that executes the equilibrium strategy s_i of player i in this proof.

The strategy s_i in this proof is a *trigger strategy*. Many other strategies can be used to prove the result (for example the strategy used in the proof of Proposition 146.2).

The following is an analog of Proposition 144.3 for an infinitely repeated game with discounting. The proof is similar to that of the previous result; we leave it to you.

■ PROPOSITION 145.2 (Nash folk theorem for the discounting criterion) *Let w be a strictly enforceable feasible payoff profile of $G = \langle N, (A_i), (u_i) \rangle$. For all $\epsilon > 0$ there exists $\underline{\delta} < 1$ such that if $\delta > \underline{\delta}$ then the δ-discounted infinitely repeated game of G has a Nash equilibrium whose payoff profile w' satisfies $|w' - w| < \epsilon$.*

To illuminate the character of equilibria in which each player uses a trigger strategy, consider two infinitely repeated games: one in which the constituent game is the *Prisoner's Dilemma*, which we denote G_1 (see Figure 134.1), and the other in which the constituent game is the game G_2 shown in Figure 146.1. In both G_1 and G_2 each player's minmax payoff is 1 and by playing D each player holds the other's payoff to this level ($p_{-1} = p_{-2} = D$).

In both games the trigger strategies used in the proof of Proposition 144.3 involve each player switching to D for good in response to any deviation from the equilibrium path. In G_1 the action D dominates the action C, so that it is a stable order for each player to choose D. Thus there is some rationale for a punisher who believes that a deviation signals the end of the current stable order to choose the action D in the future. By contrast, in G_2 a constant repetition of (D, D) is not a stable

$$
\begin{array}{cc}
 & A \qquad D \\
\begin{array}{c} A \\ \\ D \end{array} &
\begin{array}{|c|c|}
\hline
2,3 & 1,5 \\
\hline
0,1 & 0,1 \\
\hline
\end{array}
\end{array}
$$

Figure 146.1 The game G_2.

order since A strictly dominates D for player 1. Thus player 1 suffers from the punishment he inflicts on his opponent, making incredible his threat to punish a deviation and casting doubt on the plausibility of equilibria in which such trigger strategies are employed. We are led to study the notion of subgame perfect equilibrium, which rules out such strategies since it requires that each player's behavior after *every* history be optimal.

? EXERCISE 146.1 Consider the infinitely repeated game in which the players' preferences are represented by the discounting criterion, the common discount factor is $\frac{1}{2}$, and the constituent game is the game G_2 in Figure 146.1. Show that $((A,A),(A,A),\ldots)$ is not a subgame perfect equilibrium outcome path.

8.6   Punishing for a Limited Length of Time: A Perfect Folk Theorem for the Limit of Means Criterion

The strategies used in the proof of Proposition 144.3 to generate an arbitrary enforceable payoff profile punish a deviant indefinitely. Such punishment is unnecessarily harsh: a deviant's payoff needs to be held down to the minmax level only for enough periods to wipe out his (one-period) gain from the deviation. If the players' preferences satisfy the limit of means criterion then a strategy that returns to the equilibrium path after the punishment has the advantage that it yields the same payoff for the punishers as does the equilibrium path itself, so that the players have no reason not to adopt it. Hence under the limit of means criterion the social norm of punishing for only a finite number of periods is a subgame perfect equilibrium of the infinitely repeated game.

■ PROPOSITION 146.2 (Perfect folk theorem for the limit of means criterion) *Every feasible strictly enforceable payoff profile of $G = \langle N, (A_i), (u_i) \rangle$ is a subgame perfect equilibrium payoff profile of the limit of means infinitely repeated game of G.*

Proof. Let $w = \sum_{a \in A} (\beta_a/\gamma) u(a)$ be a feasible strictly enforceable payoff profile of G and let $(a^k)_{k=1}^{\gamma}$ be the sequence of action profiles that consists of β_a repetitions of a for each $a \in A$.

We now construct a strategy profile that generates a sequence of action profiles in G consisting of an indefinite repetition of the cycle $(a^k)_{k=1}^{\gamma}$. Each player punishes any deviation for only a limited number of periods. It is convenient to specify the strategy of each player so that any punishment begins only in the period that follows the completion of a cycle. If a single player deviates in some period in which nobody deserves to be punished then that player, say i, is declared to deserve punishment; beginning in the first period of the next cycle the other players punish i by choosing p_{-i} for enough periods to cancel out any possible gain for him. Subsequently the punishers return to the equilibrium, starting at the beginning of a cycle. (Simultaneous deviations by more than one player are ignored.) Given that the players' preferences satisfy the limit of means criterion, the payoff profile is w after every possible history.

To define the strategies precisely let g^* be the maximal amount that any player can gain by deviating from any action profile in G. That is, let g^* be the maximum of $u_i(a_{-i}, a_i') - u_i(a)$ over all $i \in N$, $a_i' \in A_i$ and $a \in A$. Since $w_i > v_i$ there exists an integer $m^* \geq 1$ that is an integral multiple of γ such that $\gamma g^* + m^* v_i \leq m^* w_i$ for all $i \in N$. The strategy of each player i punishes any deviant for m^* periods and is given by the following machine.

- Set of states: $\{(Norm^k, d) : \text{either } k = 1 \text{ and } d = 0 \text{ or } 2 \leq k \leq \gamma \text{ and } d \in \{0\} \cup N\} \cup \{P(j, t) : j \in N \text{ and } 1 \leq t \leq m^*\}$. (The state $(Norm^k, 0)$ means that we are in the kth period of the cycle and no player deserves punishment. The state $(Norm^k, j)$ means that we are in the kth period of the cycle and player j deserves punishment. The state $P(j, t)$ means that player j is being punished and there are t periods left in which he has to be punished.)

- Initial state: $(Norm^1, 0)$.

- Output function: In $(Norm^k, d)$ for any $d \in \{0\} \cup N$ choose a_i^k; in $P(j, t)$ choose $(p_{-j})_i$ if $i \neq j$ and $b_i(p_{-i})$ if $i = j$.

- Transition function:

  ∘ From $(Norm^k, d)$ move to[2] $(Norm^{k+1 \, (\text{mod } \gamma)}, d)$ unless:

---

[2]We define $m \, (\text{mod } \gamma)$ to be the integer q with $1 \leq q \leq \gamma$ satisfying $m = \ell\gamma + q$ for some integer ℓ (so that, in particular, $\gamma \, (\text{mod } \gamma) = \gamma$).

· $d = 0$ and player j alone deviated from a^k, in which case move to $(Norm^{k+1}, j)$ if $k \leq \gamma - 1$ and to $P(j, m^*)$ if $k = \gamma$.

· $d = j \neq 0$, in which case move to $(Norm^{k+1}, j)$ if $k \leq \gamma - 1$ and to $P(j, m^*)$ if $k = \gamma$.

◦ From $P(j, t)$ move to $P(j, t-1)$ if $2 \leq t \leq m^*$, and to $(Norm^1, 0)$ if $t = 1$.

We leave it to you to verify that the strategy profile thus defined is a subgame perfect equilibrium.   □

The strategies that we define in this proof do not initiate punishment immediately after a deviation, but wait until the end of a cycle before doing so. We define the strategies in this way in order to calculate easily the length of punishment necessary to deter a deviation: if punishment were to begin immediately after a deviation then we would have to take into account, when we calculated the length of the required punishment, the possibility that a deviant's payoffs in the remainder of the cycle are low, so that he has an additional gain from terminating the cycle.

? EXERCISE 148.1 (*A game with both long- and short-lived players*) Consider an infinite horizon extensive game in which the strategic game G is played between player 1 and an infinite sequence of players, each of whom lives for only one period and is informed of the actions taken in every previous period. Player 1 evaluates sequences of payoffs by the limit of means, and each of the other players is interested only in the payoff that he gets in the single period in which he lives.

a. Find the set of subgame perfect equilibria of the game when G is the *Prisoner's Dilemma* (see Figure 134.1).

b. Show that when G is the modification of the *Prisoner's Dilemma* in which the payoff to player 2 of (C, D) is 0 then for every rational number $x \in [1, 3]$ there is a subgame perfect equilibrium in which player 1's average payoff is x.

Consider the infinitely repeated game for which the constituent game is given in Figure 146.1. In this game $v_1 = v_2 = 1$. Consider the strategy profile defined in the proof of Proposition 146.2 to support the sequence (a^t) of outcomes in which $a^t = (A, A)$ for all t that takes the following form: each player chooses A in every period (the cycle is of length one) unless the other player deviated in the previous period, in which case he chooses D for $m^* = 2$ periods and then reverts to A.

This strategy profile is not a subgame perfect equilibrium of the infinitely repeated game when the players' preferences are represented by either the overtaking criterion or the discounting criterion. After a deviation by player 2, each player is supposed to choose D for two periods before reverting to A. Player 1 would be better off choosing A than punishing player 2, since the sequence of payoffs $(1, 1, 2, 2, \ldots)$ is preferred under both criteria to the sequence $(0, 0, 2, 2, \ldots)$. (The two sequences are indifferent under the limit of means criterion.) To support the path in which the outcome is (A, A) in every period as a subgame perfect equilibrium, player 2 has to punish player 1 if player 1 does not fulfill her obligations to punish player 2. Further, player 2 has to be punished if he does not punish player 1 for not punishing player 2, and so on. In the next two sections we use strategies with these features to prove perfect folk theorems when the players' preferences are represented by the overtaking and discounting criteria.

8.7 Punishing the Punisher: A Perfect Folk Theorem for the Overtaking Criterion

The next result is an analog of Proposition 146.2 for the overtaking criterion; it shows how strategies different from those used to prove the perfect folk theorem for the limit of means criterion can support desirable outcomes when the players' preferences are represented by the overtaking criterion. For simplicity we construct a strategy profile only for the case in which the equilibrium path consists of the repetition of a single (strictly enforceable) outcome.

■ PROPOSITION 149.1 (Perfect folk theorem for the overtaking criterion) *For any strictly enforceable outcome a^* of $G = \langle N, (A_i), (u_i) \rangle$ there is a subgame perfect equilibrium of the overtaking infinitely repeated game of G that generates the path (a^t) in which $a^t = a^*$ for all t.*

Proof. Let M be the maximum of $u_i(a)$ over all $i \in N$ and $a \in A$. Consider the strategy profile in which each player i uses the following machine.

- Set of states: $\{Norm\} \cup \{P(j, t) : j \in N \text{ and } t \text{ is a positive integer}\}$. (In the state $P(j, t)$ player j deserves to be punished for t periods more.)

- Initial state: *Norm*.

- Output function: In *Norm* choose a_i^*. In $P(j, t)$ choose $(p_{-j})_i$ if $i \neq j$ and $b_i(p_{-i})$ if $i = j$.

- Transitions in response to an outcome $a \in A$:

∘ From *Norm* stay in *Norm* unless for some player j we have
$a_{-j} = a^*_{-j}$ and $a_j \neq a^*_j$ (i.e. j is the only deviant from a^*), in
which case move to $P(j, t)$, where t is the smallest integer such
that $M + t v_j < (t + 1) u_j(a^*)$.

∘ From $P(j, t)$:

· If $a_{-j} = p_{-j}$ or $a_\ell \neq (p_{-j})_\ell$ for at least two players ℓ (i.e. all
punishers punish or at least two do not do so) then move to
$P(j, t - 1)$ if $t \geq 2$ and to *Norm* if $t = 1$.

· If $a_{-j} \neq p_{-j}$ and $a_\ell = (p_{-j})_\ell$ if $\ell \neq j^*$ (i.e. j^* is the only
punisher who does not punish) then move to $P(j^*, T(j, t))$,
where $T(j, t)$ is large enough that the sum of j^*'s payoff in
state $P(j, t)$ and his payoff in the subsequent $T(j, t)$ periods
if he does not deviate is greater than his payoff in the deviation plus $T(j, t) v_{j^*}$. (Such a number $T(j, t)$ exists since
after t periods the players were supposed to go back to the
equilibrium outcome a^* and $u_{j^*}(a^*) > v_{j^*}$.)

Under this strategy profile any attempt by a player to increase his
payoff by a unilateral deviation after any history, including one after
which punishment is supposed to occur, is offset by the other players'
subsequent punishment. Again we leave it to you to verify that the
strategy profile is a subgame perfect equilibrium.   □

8.8 Rewarding Players Who Punish: A Perfect Folk
Theorem for the Discounting Criterion

The strategy profile defined in the proof of Proposition 149.1, in which
players are punished for failing to mete out the punishment that they
are assigned, may fail to be a subgame perfect equilibrium when the
players' preferences are represented by the discounting criterion. The
reason is as follows. Under the strategy profile a player who fails to
participate in a punishment that was supposed to last, say, t periods is
himself punished for, say, t^* periods, where t^* may be much larger than
t. Further deviations may require even longer punishments, with the
result that the strategies should be designed to carry out punishments
that are unboundedly long. However slight the discounting, there may
thus be some punishment that results in losses that can never be recov-
ered. Consequently, the strategy profile may not be a subgame perfect
equilibrium if the players' preferences are represented by the discounting
criterion.

To establish an analog to Proposition 149.1 for the case that the players' preferences are represented by the discounting criterion, we construct a new strategy. In this strategy players who punish deviants as the strategy dictates are subsequently *rewarded*, making it worthwhile for them to complete their assignments. As in the previous section we construct a strategy profile only for the case in which the equilibrium path consists of the repetition of a single (strictly enforceable) outcome. The result requires a restriction on the set of games that is usually called *full dimensionality*.

■ PROPOSITION 151.1 (Perfect folk theorem for the discounting criterion) *Let a^* be a strictly enforceable outcome of $G = \langle N, (A_i), (u_i) \rangle$. Assume that there is a collection $(a(i))_{i \in N}$ of strictly enforceable outcomes of G such that for every player $i \in N$ we have $a^* \succ_i a(i)$ and $a(j) \succ_i a(i)$ for all $j \in N \setminus \{i\}$. Then there exists $\underline{\delta} < 1$ such that for all $\delta > \underline{\delta}$ there is a subgame perfect equilibrium of the δ-discounted infinitely repeated game of G that generates the path (a^t) in which $a^t = a^*$ for all t.*

Proof. The strategy profile in which each player uses the following machine is a subgame perfect equilibrium that supports the outcome a^* in every period. The machine has three types of states. In state $C(0)$ the action profile chosen by the players is a^*. For each $j \in N$ the state $C(j)$ is a state of "reconciliation" that is entered after any punishment of player j is complete; in this state the action profile that is chosen is $a(j)$. For each player j and period t between 1 and some number L that we specify later, the state $P(j,t)$ is one in which there remain t periods in which player j is supposed to be punished; in this state every player i other than j takes the action $(p_{-j})_i$, which holds j down to his minmax payoff. If any player i deviates in any state there is a transition to the state $P(i, L)$ (that is, the other players plan to punish player i for L periods). If in none of the L periods of punishment there is a deviation by a single punisher the state changes to $C(i)$. The set $\{C(i)\}$ of states serves as a system that punishes players who misbehave during a punishment phase: if player i does not punish player j as he is supposed to, then instead of the state becoming $C(j)$, in which the outcome is $a(j)$, player i is punished for L periods, after which the state becomes $C(i)$, in which the outcome is $a(i) \prec_i a(j)$.

To summarize, the machine of player i is defined as follows, where for convenience we write $a(0) = a^*$; we specify L later.

- Set of states: $\{C(j): j \in \{0\} \cup N\} \cup \{P(j,t): j \in N \text{ and } 1 \leq t \leq L\}$.
- Initial state: $C(0)$.
- Output function: In $C(j)$ choose $(a(j))_i$. In $P(j,t)$ choose $(p_{-j})_i$ if $i \neq j$ and $b_i(p_{-i})$ if $i = j$.
- Transitions in response to an outcome $a \in A$:
  - From $C(j)$ stay in $C(j)$ unless a single player k deviated from $a(j)$, in which case move to $P(k,L)$.
  - From $P(j,t)$:
    - If a single player $k \neq j$ deviated from p_{-j} then move to $P(k,L)$.
    - Otherwise move to $P(j, t-1)$ if $t \geq 2$ and to $C(j)$ if $t = 1$.

We now specify the values of $\underline{\delta}$ and L. As before, let M be the maximum of $u_i(a)$ over all $i \in N$ and $a \in A$. We choose L and $\underline{\delta}$ to be large enough that all possible deviations are deterred. To deter a deviation of any player in any state $C(j)$ we take L large enough that $M - u_i(a(j)) < L(u_i(a(j)) - v_i)$ for all $i \in N$ and all $j \in \{0\} \cup N$ and choose $\delta > \delta^*$ where δ^* is close enough to 1 that for all $\delta > \delta^*$ we have

$$M - u_i(a(j)) < \sum_{k=2}^{L+1} \delta^{k-1}(u_i(a(j)) - v_i).$$

(This condition is sufficient since $u_i(a(j)) > u_i(a(i))$ for $j \neq i$.) If a player i deviates from $P(j,t)$ for $j \neq i$ then he obtains at most M in the period that he deviates followed by L periods of $v_i < u_i(a(i))$ and $u_i(a(i))$ subsequently. If he does not deviate then he obtains $u_i(p_{-j}, b_j(p_{-j}))$ for between 1 and L periods and $u_i(a(j))$ subsequently. Thus to deter a deviation it is sufficient to choose $\underline{\delta} > \delta^*$ close enough to one that for all $\delta > \underline{\delta}$ we have

$$\sum_{k=1}^{L} \delta^{k-1}(M - u_i(p_{-j}, b_j(p_{-j}))) < \sum_{k=L+1}^{\infty} \delta^{k-1}(u_i(a(j)) - u_i(a(i))).$$

(Such a value of $\underline{\delta}$ exists because of our assumption that $u_i(a(j)) > u_i(a(i))$ if $i \neq j$.)   □

[?] EXERCISE 152.1 Consider the three-player symmetric infinitely repeated game in which each player's preferences are represented by the discounting criterion and the constituent game is $\langle \{1,2,3\}, (A_i), (u_i) \rangle$ where for $i = 1$, 2, 3 we have $A_i = [0,1]$ and $u_i(a_1, a_2, a_3) = a_1 a_2 a_3 + (1-a_1)(1-a_2)(1-a_3)$ for all $(a_1, a_2, a_3) \in A_1 \times A_2 \times A_3$.

a. Find the set of enforceable payoffs of the constituent game.

b. Show that for any discount factor $\delta \in (0, 1)$ the payoff of any player in any subgame perfect equilibrium of the repeated game is at least $\frac{1}{4}$.

c. Reconcile these results with Proposition 151.1.

8.9   The Structure of Subgame Perfect Equilibria Under the Discounting Criterion

The strategy of each player in the equilibrium constructed in the proof of Proposition 151.1, which concerns games in which the discount factor is close to 1, has the special feature that when any player deviates, the subsequent sequence of action profiles depends only on the identity of the deviant and not on the history that preceded the deviation. In this section we show that for *any* common discount factor a profile of such strategies can be found to support any subgame perfect equilibrium outcome.

We begin with two lemmas, the first of which extends the one deviation property proved for finite extensive games in Lemma 98.2 to infinitely repeated games with discounting.

■ LEMMA 153.1 *A strategy profile is a subgame perfect equilibrium of the δ-discounted infinitely repeated game of G if and only if no player can gain by deviating in a single period after any history.*

[?] EXERCISE 153.2 Prove this result.

The next result shows that under our assumptions the set of subgame perfect equilibrium payoff profiles of any δ-discounted infinitely repeated game is closed.

■ LEMMA 153.3 *Let $(w^k)_{k=1}^{\infty}$ be a sequence of subgame perfect equilibrium payoff profiles of the δ-discounted infinitely repeated game of G that converges to w^*. Then w^* is a subgame perfect equilibrium payoff profile of this repeated game.*

Proof. For each value of k let s^k be a subgame perfect equilibrium of the repeated game that generates the payoff profile w^k. We construct a strategy profile s that we show is a subgame perfect equilibrium and yields the payoff profile w^*. We define, by induction on the length of the history h, an action profile $s(h)$ of G and an auxiliary infinite subsequence (r^k) of the sequence (s^k) that has the property that the payoff profile generated by the members of the subsequence in the subgame following the history h has a limit and the action profile $r^k(h)$ converges to $s(h)$. Assume we have done so for all histories of length T or less, and consider a history (h, a) of length $T + 1$, where h is a history of length T. Let (r^k) be the sequence of strategy profiles that we chose for the history h and let $s(h)$ be the action profile we chose for that history. For $a = s(h)$ select for (h, a) a subsequence (r'^k) of (r^k) for which the sequence $(r'^k(h, a))$ converges, and let the action profile to which $r'^k(h, a)$ converges be $s(h, a)$. Obviously the limiting payoff profile of the subsequence that we have chosen is the same as that of (r^k). For $a \neq s(h)$ choose for (h, a) a subsequence (r''^k) of (r'^k) for which the sequence of payoff profiles and the sequence $(r''^k(h, a))$ both converge, and let the action profile to which $r''^k(h, a)$ converges be $s(h, a)$.

No player i can gain in deviating from s_i by changing his action after the history h and inducing some outcome a instead of $s(h)$ since if this were so then for large enough k he could profitably deviate from r_i^k, where (r^k) is the sequence that we chose for the history (h, a). Further, the payoff profile of s is w^*. $\square$

By this result the set of subgame perfect equilibrium payoffs of any player i in the repeated game is closed; since it is bounded it has a minimum, which we denote $m(i)$. Let $(a(i)^t)$ be the outcome of a subgame perfect equilibrium in which player i's payoff is $m(i)$.

■ PROPOSITION 154.1 *Let (a^t) be the outcome of a subgame perfect equilibrium of the δ-discounted infinitely repeated game of $G = \langle N, (A_i), (u_i) \rangle$. Then the strategy profile in which each player i uses the following machine is a subgame perfect equilibrium with the same outcome (a^t).*

- *Set of states:* {$Norm^t$: t is a positive integer} $\cup$ {$P(j, t)$: $j \in N$ and t is a positive integer}.

- *Initial state:* $Norm^1$.

- *Output function: In state $Norm^t$ play a_i^t. In state $P(j, t)$ play $a(j)_i^t$.*

- *Transition function:*

    ○ *In state $Norm^t$ move to $Norm^{t+1}$ unless exactly one player, say j, deviated from a^t, in which case move to $P(j, 1)$.*

    ○ *In state $P(j, t)$: Move to $P(j, t+1)$ unless exactly one player, say j', deviated from $a(j)^t$, in which case move to $P(j', 1)$.*

Proof. It is straightforward to verify, using Lemma 153.1, that this defines a subgame perfect equilibrium with the required property. $\square$

8.10   Finitely Repeated Games

8.10.1   Definition

We now turn to a study of finitely repeated games. The formal description of a finitely repeated game is very similar to that of an infinitely repeated game: for any positive integer T a T-period finitely repeated game of the strategic game $\langle N, (A_i), (\succsim_i) \rangle$ is an extensive game with perfect information that satisfies the conditions in Definition 137.1 when the symbol ∞ is replaced by T. We restrict attention to the case in which the preference relation $\succsim_i^*$ of each player i in the finitely repeated game is represented by the function $\Sigma_{t=1}^T u_i(a^t)/T$, where u_i is a payoff function that represents i's preferences in the constituent game. We refer to this game as the T-**period repeated game of** $\langle N, (A_i), (u_i) \rangle$.

8.10.2   Nash Equilibrium

The intuitive argument that drives the folk theorems for infinitely repeated games is that a mutually desirable outcome can be supported by a stable social arrangement in which a player is deterred from deviating by the threat that he will be "punished" if he does so. The same argument applies, with modifications, to a large class of finitely repeated games. The need for modification is rooted in the fact that the outcome in the last period of any Nash equilibrium of any finitely repeated game must be a Nash equilibrium of the constituent game, a fact that casts a shadow over the rest of the game. This shadow is longest in the special case in which every player's payoff in every Nash equilibrium of the constituent game is equal to his minmax payoff (as in the *Prisoner's Dilemma*). In this case the intuitive argument behind the folk theorems fails: the outcome in *every* period must be a Nash equilibrium of the constituent game, since if there were a period in which the outcome were not such an equilibrium then in the last such period some player could deviate with impunity. The following result formalizes this argument.

■ PROPOSITION 155.1 *If the payoff profile in every Nash equilibrium of the strategic game G is the profile (v_i) of minmax payoffs in G then for any value of T the outcome $(a^1, \ldots, a^T)$ of every Nash equilibrium of the T-period repeated game of G has the property that a^t is a Nash equilibrium of G for all $t = 1, \ldots, T$.*

Proof. Let $G = \langle N, (A_i), (u_i) \rangle$ and let $a = (a^1, \ldots, a^T)$ be the outcome of a Nash equilibrium s of the T-period repeated game of G. Suppose

that a^t is not a Nash equilibrium of G for some period t. Let $t \geq 1$ be
the last period for which a^t is not a Nash equilibrium of G; suppose that
$u_i(a^t_{-i}, a_i) > u_i(a^t)$. Consider a strategy $\hat{s}_i$ of player i that differs from
s_i only in that after the history $(a^1, \dots, a^{t-1})$ it chooses a_i, and after
any longer history h it chooses an action that, given the profile $s_{-i}(h)$ of
actions planned by the other players after the history h, yields at least
i's minmax payoff. The outcome of $(s_{-i}, \hat{s}_i)$ is a terminal history $\hat{a}$ that
is identical to a through period $t-1$; $u_i(\hat{a}^t) > u_i(a^t)$ and $u_i(\hat{a}^r) \geq u_i(a^r)$
for $r \geq t+1$. Thus player i prefers $\hat{a}$ to a, contradicting our assumption
that s is a Nash equilibrium of the repeated game. $\square$

This result applies to a very small set of games. If, contrary to the
assumptions of the result, the constituent game has a Nash equilibrium
a^* in which some player's payoff exceeds his minmax payoff then that
player can be punished for deviating in the penultimate period of the
game whenever the outcome in the final period is a^*. This punishment
may not be enough to deter the deviation if the difference between the
player's minmax payoff and his payoff in a^* is small. However, there is
always some integer L such that if the outcome is a^* in the last L periods then any deviation by the player in any period before this sequence
of L plays begins *is* deterred by the threat to impose upon the player
his minmax payoff in the remaining periods. Further, the value of L is
independent of the length T of the game, so that if for each player the
constituent game has a Nash equilibrium in which that player's payoff
exceeds his minmax payoff then for T large enough *any* feasible strictly
enforceable payoff profile can be approximately achieved as the average
payoff profile in a Nash equilibrium of the T-period repeated game. For
simplicity we state and prove this result only for the case in which the
constituent game has a *single* Nash equilibrium in which *every* player's
payoff exceeds his minmax payoff; we also restrict attention to equilibrium paths that are repetitions of a single outcome of the constituent
game.

■ PROPOSITION 156.1 (Nash folk theorem for finitely repeated games) *If
$G = \langle N, (A_i), (u_i) \rangle$ has a Nash equilibrium $\hat{a}$ in which the payoff of every
player i exceeds his minmax payoff v_i then for any strictly enforceable
outcome a^* of G and any $\epsilon > 0$ there exists an integer T^* such that
if $T > T^*$ the T-period repeated game of G has a Nash equilibrium in
which the payoff of each player i is within ϵ of $u_i(a^*)$.*

Proof. Consider the strategy of player i that is carried out by the following machine. The set of states consists of *Norm*^{t} for $t = 1, \dots, T - L$

(L is determined later), *Nash*, and $P(j)$ for each $j \in N$. Each player i chooses a_i^* in *Norm*^{t} for all values of t, $\hat{a}_i$ in *Nash*, and punishes player j by choosing $(p_{-j})_i$ in $P(j)$. If a single player j deviates in the state *Norm*^{t} then there is a transition to $P(j)$; otherwise there is a transition to *Norm*$^{t+1}$ if $t < T - L$ and to *Nash* if $t = T - L$. Once reached, the states $P(j)$ and *Nash* are never left. The outcome is that a^* is chosen in the first $T - L$ periods and $\hat{a}$ is chosen in the last L periods. To summarize, player i's machine is the following.

- Set of states: $\{Norm^t \colon 1 \leq t \leq T - L\} \cup \{P(j) \colon j \in N\} \cup \{Nash\}$.
- Initial state: *Norm*^{1}.
- Output function: In *Norm*^{t} choose a_i^*, in $P(j)$ choose $(p_{-j})_i$, and in *Nash* choose $\hat{a}_i$.
- Transition function:
  - From *Norm*^{t} move to *Norm*$^{t+1}$ unless either $t = T - L$, in which case move to *Nash*, or exactly one player, say j, deviated from a^*, in which case move to $P(j)$.
  - $P(j)$ for any $j \in N$ and *Nash* are absorbing.

It remains to specify L. A profitable deviation is possible only in one of the states *Norm*^{t}. To deter such a deviation we require L to be large enough that $\max_{a_i \in A_i} u_i(a_{-i}^*, a_i) - u_i(a^*) \leq L(u_i(\hat{a}) - v_i)$ for all $i \in N$. Finally, in order to obtain a payoff profile within ϵ of $u(a^*)$ we choose T^* so that $|[(T^* - L)u_i(a^*) + Lu_i(\hat{a})]/T^* - u_i(a^*)| < \epsilon$ for all $i \in N$. $\square$

[?] EXERCISE 157.1 Extend this result to the case in which the Nash equilibrium of G in which player i's payoff exceeds v_i may depend on i.

8.10.3 Subgame Perfect Equilibrium

In any subgame perfect equilibrium of a finitely repeated game the outcome in the last period after *any* history (not just after the history that occurs if every player adheres to his strategy) is a Nash equilibrium of the constituent game. Thus the punishment embedded in the strategies used to prove the Nash folk theorem (Proposition 156.1) is not consistent with a subgame perfect equilibrium; indeed, no punishment is possible if the constituent game has a unique Nash equilibrium payoff profile. Consequently we have the following result.

■ PROPOSITION 157.2 *If the strategic game G has a unique Nash equilibrium payoff profile then for any value of T the action profile chosen after*

$$
\begin{array}{c c}
 & \begin{array}{c c c} C & \quad D & \quad E \end{array} \\
\begin{array}{c} C \\[18pt] D \\[18pt] E \end{array} &
\begin{array}{|c|c|c|}
\hline
3,3 & 0,4 & 0,0 \\
\hline
4,0 & 1,1 & 0,0 \\
\hline
0,0 & 0,0 & \frac{1}{2},\frac{1}{2} \\
\hline
\end{array}
\end{array}
$$

Figure 158.1 A modified *Prisoner's Dilemma.*

any history in any subgame perfect equilibrium of the T-period repeated game of G is a Nash equilibrium of G.

Proof. The outcome in any subgame that starts in period T of any subgame perfect equilibrium of the repeated game is a Nash equilibrium of G. Thus each player's payoff in the last period of the game is independent of history. Consequently in any subgame that starts in period $T-1$ the action profile is a Nash equilibrium of G. An inductive argument completes the proof.   □

If the constituent game has more than one Nash equilibrium payoff profile then punishment *can* be embedded in a subgame perfect equilibrium strategy profile: the players' payoffs in the final periods of the game can depend on their behavior in previous periods. The following example illustrates the equilibria that can arise in this case. We argue that in the T-period repeated game of the strategic game in Figure 158.1 there is a subgame perfect equilibrium for which the outcome is (C, C) in every period but the last three, in which it is (D, D), so that if T is large the average payoff profile is close to $(3, 3)$. In the equilibrium each player uses the following strategy: choose C in every period through period $T-3$ unless one of the players chose D in some previous period, in which case choose E in every subsequent period, regardless of the subsequent outcomes; if the outcome is (C, C) in every period through $T-3$ then choose D in the last three periods. A player who deviates to D in any period up to $T-3$ after a history in which the outcome was (C, C) in every previous period gains one unit of payoff in that period, but then subsequently loses at least 1.5 units, since the other player chooses E in every subsequent period. That is, the threat to play E subsequently is enough to deter any deviation; this punishment is credible since (E, E) is a Nash equilibrium of the constituent game. (Note that the same strategy profile is a subgame perfect equilibrium also if in the constituent game the payoff profile $(\frac{1}{2}, \frac{1}{2})$ is replaced by $(0, 0)$, in which case the constituent game differs from the *Prisoner's Dilemma* only in that each player has an additional weakly dominated action.)

This example makes it clear that if there are two Nash equilibria of the constituent game G, one of which dominates the other, then any payoff profile in which every player obtains more than his payoff in the inferior Nash equilibrium of G can be achieved as the average payoff profile in a subgame perfect equilibrium of the T-period repeated game of G for T large enough. In fact a stronger result can be established: any strictly enforceable payoff profile can be achieved as the average payoff profile in a subgame perfect equilibrium of the repeated game. Such a payoff profile is supported by a strategy profile that, up until the final periods of the game, is akin to that constructed in the proof of the perfect folk theorem for the discounting criterion (Proposition 151.1).

The argument (which draws upon the ideas in the proofs of Propositions 151.1 and 156.1) is the following. Let a^* be a strictly enforceable outcome of G. A strategy profile in the T-period repeated game that generates a sequence of outcomes for which the average payoff profile is close to $u(a^*)$ when T is large has the following form. There are three stages. Throughout the first two stages each player i chooses a_i^* so long as no player deviates. In the third stage the players adhere, in the absence of a deviation, to a sequence of Nash equilibria of the constituent game for which each player's average payoff exceeds his lowest Nash equilibrium payoff in the constituent game. Deviations are punished as follows. A deviation that occurs during the first stage is punished by the other players' using an action that holds the deviant to his minmax payoff for long enough to wipe out his gain. After this punishment is complete, a state of "reconciliation" is entered for long enough to reward the players who took part in the punishment for completing their assignment (cf. the strategy in the proof of Proposition 151.1). A deviation by some player i that occurs during the second stage is ignored until the beginning of the third stage, during which the worst Nash equilibrium for player i is executed in every period. Deviations during the last stage do not need to be punished since the outcome in every period is a Nash equilibrium of the constituent game. The length of the second stage is chosen to be large enough that for a player who deviates in the last period of the first stage both the punishment and the subsequent reconciliation can be completed during the second stage. Given the length of the second stage, the length of the third stage is chosen to be large enough that a player who deviates in the first period of the second stage is worse off given his punishment, which begins in the first period of the third stage. The lower bounds on the lengths of the second and third stages are independent of T, so that for T large enough the average payoff profile induced by the strategy profile is close to $u(a^*)$.

In the following statement of the result we restrict attention to equilibrium paths that consist of the repetition of a single outcome of the constituent game (as we did in the discussion above). We omit a proof, which may be found in Krishna (1989), for example.

■ PROPOSITION 160.1 (Perfect folk theorem for finitely repeated games) *Let a^* be a strictly enforceable outcome of $G = \langle N, (A_i), (u_i) \rangle$. Assume that (i) for each $i \in N$ there are two Nash equilibria of G that differ in the payoff of player i and (ii) there is a collection $(a(i))_{i \in N}$ of strictly enforceable outcomes of G such that for every player $i \in N$ we have $a^* \succ_i a(i)$ and $a(j) \succ_i a(i)$ for all $j \in N \setminus \{i\}$. Then for any $\epsilon > 0$ there exists an integer T^* such that if $T > T^*$ the T-period repeated game of G has a subgame perfect equilibrium in which the payoff of each player i is within ϵ of $u_i(a^*)$.*

Notes

Early discussions of the notion of a repeated game and the ideas behind the Nash folk theorem (Section 8.5) appear in Luce and Raiffa (1957, pp. 97–105 (especially p. 102) and Appendix 8), Shubik (1959b, Ch. 10 (especially p. 226)), and Friedman (1971). Perfect folk theorems for the limit of means criterion were established by Aumann and Shapley and by Rubinstein in the mid 1970's; see Aumann and Shapley (1994) and Rubinstein (1994). The perfect folk theorem for the overtaking criterion (Proposition 149.1) is due to Rubinstein (1979). The perfect folk theorem for the discounting criterion (Proposition 151.1) is due to Fudenberg and Maskin (1986); the proof that we give is based on Abreu, Dutta, and Smith (1994). Section 8.9 is based on Abreu (1988). Proposition 155.1 and the Nash and perfect folk theorems for finitely repeated games (Propositions 156.1 and 160.1) are due to Benoît and Krishna (1985, 1987). (Luce and Raiffa (1957, Section 5.5) earlier argued that the conclusion of Proposition 155.1 holds for the *Prisoner's Dilemma*.)

For an early discussion of the difference between the models of finitely and infinitely repeated games (Section 8.2) see Aumann (1959, Section 6). For a detailed discussion of preference relations over streams of outcomes see, for example, Diamond (1965). For a presentation of some of the folk theorems in the language of machines see Ben-Porath and Peleg (1987). The example in Figure 158.1 is taken from Benoît and Krishna (1985); Friedman (1985) contains a similar example. Exercise 148.1 is due to Fudenberg and Levine (1989). Exercise 152.1 is taken from Fudenberg and Maskin (1986).

For a discussion of the issues that arise when the players use mixed strategies see Fudenberg and Maskin (1991). As we have seen, the equilibria of a repeated game are not all efficient; further, the outcome generated by an equilibrium after a deviation occurs may not be efficient even if the outcome in the absence of any deviation is efficient. Pearce (1992, Section 4) discusses models that examine the consequences of allowing the set of players, after any history, to switch from their current strategy profile to one that is Pareto superior (i.e. to "renegotiate"). If some or all of the players in a repeated game do not know the form of the constituent game then many new issues arise. Zamir (1992) and Forges (1992) are surveys of work in this area.

Krishna (1989), Sorin (1990, 1992), Fudenberg (1992), and Pearce (1992) are surveys that cover the material in this chapter and extensions of it.

9 Complexity Considerations in Repeated Games

In this chapter we investigate the structure of the equilibrium strategies in an infinitely repeated game in which each player is concerned about the complexity of his strategy.

9.1 Introduction

In the previous chapter we described representatives of a family of results known as "folk theorems", which establish, under a variety of assumptions about the players' preferences, that a very wide range of payoffs is compatible with Nash equilibrium and even subgame perfect equilibrium in an infinitely repeated game. A folk theorem entails the construction of *some* equilibria that generate the required outcomes. It does not demand that these equilibrium strategies be reasonable in any sense; our judgments about the nature of the equilibrium strategies used in the proofs are all informal. In this chapter we focus more closely on the structure of the equilibrium strategies rather than on the set of equilibrium payoffs, using the tool of a machine described in the previous chapter.

The basic assumption upon which the analysis is predicated is that players care about the complexity of their strategies. When choosing a strategy a player is confronted with a tradeoff: on the one hand he would like his strategy to serve his goals as well as possible, and on the other hand he would like it to be as simple as possible. There are many reasons why a player may value simplicity: a more complex plan of action is more likely to break down; it is more difficult to learn; it may require time to implement. We do not study these reasons here, but simply *assume* that complexity is costly and is under the control of the decision-maker.

We explore the effect of this assumption on the equilibrium outcomes of an infinitely repeated game, asking, in particular, how the introduction of a cost of complexity affects the predictions of the model. Although we limit attention to repeated games, complexity considerations may be studied in the context of any model of choice. A model that includes such "procedural" aspects of decision-making is known as a model of "bounded rationality".

9.2   Complexity and the Machine Game

In this chapter we restrict attention, for simplicity, to an infinitely repeated game in which the players' preferences are represented by the discounting criterion: we study the players' behavior in the two-player δ-discounted infinitely repeated game of $G = \langle \{1,2\}, (A_i), (u_i) \rangle$ (see Section 8.3). We study this behavior by analyzing a *machine game*, in which each player chooses a machine to play the infinitely repeated game. In this chapter we define a **machine** of player i to be a four-tuple $\langle Q_i, q_i^0, f_i, \tau_i \rangle$ in which

- Q_i is a set of **states**
- $q_i^0 \in Q_i$ is the **initial state**
- $f_i : Q_i \to A_i$ is the **output function**
- $\tau_i : Q_i \times A_j \to Q_i$ (where $j \neq i$) is the **transition function**.

This definition differs from that given in the previous chapter (Section 8.4) in that a player's transition function describes how the state changes in response to the *action of the other player*, not in response to an outcome of the strategic game (i.e. a pair of actions). As defined in the previous chapter, a machine corresponds to the notion of a strategy in an extensive game, which requires that a player's action be specified for *every* history, including those that are precluded by the strategy itself (see Section 6.4). Here we want a machine to correspond to a plan of action as it is usually understood, and thus take as an input to a player's transition function only the action of the *other* player.

Every pair (M_1, M_2) of machines induces a sequence $(a^t(M_1, M_2))_{t=1}^{\infty}$ of outcomes in G and a sequence $(q^t(M_1, M_2))_{t=1}^{\infty}$ of pairs of states defined as follows: for $i = 1, 2$ and $t \geq 1$ we have

- $q_i^1(M_1, M_2) = q_i^0$
- $a_i^t(M_1, M_2) = f_i(q_i^t(M_1, M_2))$
- $q_i^{t+1}(M_1, M_2) = \tau_i(q_i^t(M_1, M_2), a_j^t(M_1, M_2))$ (where $j \neq i$).

We restrict each player to choose a machine that has a finite number of states, and denote the set of all such machines for player i by $\mathcal{M}_i$. Thus the machine game is a two-player strategic game in which the set of actions of each player i is $\mathcal{M}_i$. To complete the description of this game we need to describe the players' preferences. If we assume that each player i cares only about his payoff $\mathcal{U}_i(M_1, M_2) = (1 - \delta) \sum_{t=1}^{\infty} \delta^{t-1} u_i(a^t(M_1, M_2))$ in the repeated game then we obtain the same conclusion as that of the Nash folk theorem (Proposition 145.2), since the trigger strategies that are used in the proof of this result can be implemented by finite machines. If, on the other hand, each player cares about both his payoff in the repeated game and the complexity of his strategy then, as we shall see, we obtain results that are very different from the folk theorem.

There are many ways of defining the complexity of a machine. We take a naïve approach: the *complexity* $c(M)$ of the machine $M = \langle Q, q^0, f, \tau \rangle$ is its number of states (i.e. the cardinality of Q). Our analysis is sensitive to the measure of complexity that we use. We view this measure as an additional piece of information about the strategic situation, which should reflect the relevant difficulties of the player in carrying out a strategy. From this perspective the sensitivity of the model to the complexity measure is desirable: in different circumstances different measures may be appropriate.

In the following definition we assume that each player's preferences in a machine game are positively sensitive to his payoff in the repeated game and negatively sensitive to the complexity of his machine.

▶ DEFINITION 165.1 A **machine game** of the δ-discounted infinitely repeated game of $\langle \{1,2\}, (A_i), (u_i) \rangle$ is a strategic game $\langle \{1,2\}, (\mathcal{M}_i), (\succsim_i) \rangle$ in which for each player i

- $\mathcal{M}_i$ is the set of all finite machines for player i in the infinitely repeated game

- $\succsim_i$ is a preference ordering that is increasing in player i's payoff in the repeated game and decreasing in the complexity of his machine: $(M_1, M_2) \succ_i (M_1', M_2')$ whenever either $\mathcal{U}_i(M_1, M_2) > \mathcal{U}_i(M_1', M_2')$ and $c(M_i) = c(M_i')$, or $\mathcal{U}_i(M_1, M_2) = \mathcal{U}_i(M_1', M_2')$ and $c(M_i) < c(M_i')$.

A special case is that in which each player's preferences are additive: $\succsim_i$ is represented by $\mathcal{U}_i(M_1, M_2) - \gamma c(M_i)$ for some $\gamma > 0$, in which case γ can be interpreted as the cost of each state of the machine. Another special case is that in which the preferences are **lexicographic**:

$$\begin{array}{c c}
 & \begin{array}{c c} C \qquad & D \end{array} \\
\begin{array}{c} C \\ \\ D \end{array} & \begin{array}{|c|c|}
\hline
3,3 & 0,5 \\
\hline
5,0 & 1,1 \\
\hline
\end{array}
\end{array}$$

Figure 166.1 The *Prisoner's Dilemma*.

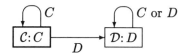

Figure 166.2 The machine M in Example 166.1 (a machine that carries out the grim strategy in the repeated *Prisoner's Dilemma*).

each player is concerned first with his payoff in the repeated game and second with the complexity of his machine. This case is especially interesting since lexicographic preferences are close to the preferences in the standard model of a repeated game in which complexity considerations are absent, a model that is the progenitor of the model of a machine game.

◇ EXAMPLE 166.1 Suppose that the game G is the *Prisoner's Dilemma*, with the payoffs given in Figure 166.1. Consider the two-state machine M that implements the grim strategy (see Figure 166.2). If the players' common discount factor δ is large enough then this machine is a best response to itself in the δ-discounted repeated game of G. Even by using a more complex machine, player 1 cannot achieve a higher payoff in the repeated game. However, while there is no machine of player 1 that achieves a *higher* payoff in the repeated game than M does, given that player 2 uses M, there *is* a machine of player 1 that achieves the *same* payoff and is less complex: that in which there is one state, in which C is chosen. The state $\mathcal{D}$ in the machine M is designed to allow a player to threaten his opponent, but in equilibrium this threat is redundant since each player always chooses C. Thus either player can drop the state $\mathcal{D}$ without affecting the outcome; hence (M, M) is not a Nash equilibrium of the machine game.

◇ EXAMPLE 166.2 For the *Prisoner's Dilemma* (as in the previous example) let M be the machine in Figure 167.1. The behavior that this machine generates can be interpreted as beginning with a display of the

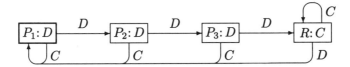

Figure 167.1 The machine M in Example 166.2.

ability to punish. After this display the player begins a cooperative phase in which he plays C, threatening to punish a deviant by moving back to the initial state. If both players use the machine M then the sequence of payoffs in the repeated game is $(1, 1, 1)$ followed by an infinite sequence of 3's.

We claim that if the players' common discount factor δ is large enough then (M, M) is a Nash equilibrium of the machine game if the players' preferences do not give too much weight to complexity (as is the case if their preferences are either lexicographic or additive with a small cost of complexity). The argument is as follows. To increase his payoff in the repeated game, a player must choose D at least sometimes when his opponent's machine is in state R. Any such choice of D causes the other machine to choose D for at least three periods, so that when δ is close enough to 1 a player does not gain by such a maneuver ($5 + \delta + \delta^2 + \delta^3 < 3 + 3\delta + 3\delta^2 + 3\delta^3$ for δ close enough to 1). Thus for δ large enough a player cannot increase his payoff in the repeated game by any machine, however complex.

We now show that a player cannot achieve the same payoff in the repeated game by using a less complex machine. To achieve the same payoff he must choose C at least once when the other player's machine is in state R. To do so his machine must have at least four states. To see this, consider the first period, say t, in which $q_j^t = R$ and $f_i(q_i^t) = C$. We must have $f_i(q_i^{t-3}) = f_i(q_i^{t-2}) = f_i(q_i^{t-1}) = D$ and hence, in particular, $q_i^t \neq q_i^{t-1}$. Further, $q_i^{t-2} \neq q_i^{t-1}$ since $\tau_i(q_i^{t-2}, D) = q_i^{t-1}$ while $\tau_i(q_i^{t-1}, D) = q_i^t$. Similarly, $q_i^{t-3} \neq q_i^{t-2}$ and $q_i^{t-3} \neq q_i^{t-1}$.

In a machine game a player has to solve a problem in which he balances his desires to achieve a high payoff and to employ a simple machine. In some sense this problem is more complicated than that of finding an optimal strategy in the repeated game, since the player must consider the complexity of his rule of behavior; we do not impose any constraint on the player's ability to solve this problem.

9.3   The Structure of the Equilibria of a Machine Game

We now characterize the structure of the Nash equilibria of a machine game. We first generalize an observation we made about Example 166.1: if the machine M_i of some player i has a state that is not used when M_1 and M_2 operate then (M_1, M_2) is not a Nash equilibrium, since the state can be eliminated without affecting the outcome, and player i prefers the machine in which the state is eliminated.

■ LEMMA 168.1 *If (M_1^*, M_2^*) is a Nash equilibrium of a machine game then for every state q_i of the machine M_i^* there exists a period t such that $q_i^t(M_1^*, M_2^*) = q_i$.*

Our next result shows that in a Nash equilibrium each machine has the same number of states and that any Nash equilibrium of a machine game corresponds to a Nash equilibrium of the repeated game.

■ LEMMA 168.2 *If (M_1^*, M_2^*) is a Nash equilibrium of a machine game then*

- $c(M_1^*) = c(M_2^*)$

- *the pair of strategies in the repeated game associated with (M_1^*, M_2^*) is a Nash equilibrium of the repeated game.*

Proof. For any strategy s_j of player j in the repeated game and any machine M_i of player i, denote by $\mathcal{U}_j(M_i, s_j)$ player j's payoff in the repeated game when he uses s_j and player i uses the strategy that corresponds to M_i. Since M_i^* is finite, player j's problem $\max_{s_j} \mathcal{U}_j(M_i^*, s_j)$ of finding a best response (ignoring complexity) to the machine M_i^* in the repeated game has a solution (see Derman (1970, Theorem 1 on p. 23)). Let $M_i^* = \langle Q_i, q_i^0, f_i, \tau_i \rangle$ and for each $q \in Q_i$ let $\mathcal{V}_j(q) = \max_{s_j} \mathcal{U}_j(M_i^*(q), s_j)$, where $M_i^*(q)$ is the machine that differs from M_i^* only in that the initial state is q. For each $q \in Q_i$ let $A_j(q)$ be the set of solutions to the problem

$$\max_{a_j \in A_j} \{u_j(f_i(q), a_j) + \delta \mathcal{V}_j(\tau_i(q, a_j))\}.$$

Then in the repeated game a strategy of player j is a best response to the strategy corresponding to M_i^* if and only if the action it plays when player i's machine is in state q is a member of $A_j(q)$. In particular, choosing $a_j^*(q) \in A_j(q)$ for each $q \in Q_i$, there is a best response that is implemented by the following machine, which has $c(M_i^*)$ states.

- The set of states is Q_i.

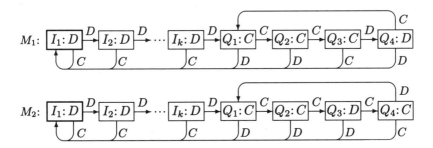

Figure 169.1 Machines M_1 for player 1 and M_2 for player 2 for the infinitely repeated *Prisoner's Dilemma*. The pair (M_1, M_2) generates the path in which (D, D) occurs for k periods and then the sequence (C, C), (C, C), (C, D), (D, C) is repeated indefinitely.

- The initial state is q_i^0.
- The output function f_j is defined by $f_j(q) = a_j^*(q)$.
- The transition function τ_j is defined by $\tau_j(q, x) = \tau_i(q, f_j(q))$ for any $x \in A_i$.

Since (M_1^*, M_2^*) is a Nash equilibrium of the machine game it follows that $c(M_j^*) \leq c(M_i^*)$ and hence $c(M_1^*) = c(M_2^*)$. Further, since player j can use a machine with $c(M_i^*)$ states to achieve a payoff in the repeated game equal to $\max_{s_j} U_j(M_i^*, s_j)$ it follows that the pair of strategies that is executed by (M_1^*, M_2^*) is a Nash equilibrium of the repeated game. $\square$

? EXERCISE 169.1 Give an example of a three-player game for which the associated machine game has a Nash equilibrium in which the numbers of states in the players' machines are not the same.

We now derive a result that has strong implications for the set of Nash equilibria of a machine game. To obtain some intuition for the result, consider the pair of machines for the infinitely repeated *Prisoner's Dilemma* that is shown in Figure 169.1. This pair of machines generates a path in which there are initially $k \geq 2$ periods in which the outcome is (D, D) (the players display their threats), after which a cycle of length four in which the outcomes are (C, C), (C, C), (C, D) and (D, C) is repeated indefinitely. Any deviation by a player from the prescribed behavior in the cycle causes his opponent's machine to go to its initial state, punishing the deviant for k periods. As you can check, the pair of machines is a Nash equilibrium of the repeated game when the discount factor δ is close enough to 1. However, it is not an equilibrium of the machine game. To see this, consider M_1. In each of the three states Q_1,

Q_2, and Q_3 player 1 takes the same action; she uses the fact that there are three states only to know when to choose the action D. However, she could obtain this information by observing player 2's action, as follows. Suppose that she adopts the machine M_1' in which the three states Q_1, Q_2, and Q_3 are replaced by a single state Q in which she chooses C, the state remains Q so long as player 2 chooses C, the state switches to Q_4 if player 2 chooses D, and the transition from the state I_k is to Q if player 2 chooses D. Then (M_1', M_2) generates the same sequence of outcomes of the *Prisoner's Dilemma* as does (M_1, M_2); thus in the machine game player 1 can profitably deviate to M_1' since it has fewer states than M_1.

Note that M_1' does not monitor player 2's behavior when player 2's machine is in Q_1 or Q_2: if player 2 chooses D in either of these states then M_1' does not return to the state I_1 but moves to Q_4. If player 1 uses the machine M_1' then player 2 can exploit this feature by choosing C in state Q_3.

The situation is similar to that in which a paratrooper has to jump after counting to 100 and another paratrooper has to jump after counting to 101. If the second paratrooper counts then he can monitor the first paratrooper, who is afraid of jumping. However, counting is costly in the tense environment of the plane, and the second paratrooper can avoid the burden of counting by simply watching his friend and jumping immediately after her. However, if the second paratrooper does not count then the first paratrooper can exploit this lack of monitoring and ... not jump.

In general we can show that if a Nash equilibrium pair of machines generates outcomes in which one of the players takes the same action in two different periods then the other player also takes the same action in these two periods (contrary to the behavior of the players in periods $k + 2$ and $k + 3$ of the example that we just discussed).

■ LEMMA 170.1 *If (M_1^*, M_2^*) is a Nash equilibrium of a machine game then there is a one-to-one correspondence between the actions of player 1 and player 2 prescribed by M_1^* and M_2^*: if $a_i^t(M_1^*, M_2^*) = a_i^s(M_1^*, M_2^*)$ for some $t \neq s$ then $a_j^t(M_1^*, M_2^*) = a_j^s(M_1^*, M_2^*)$.*

Proof. Let $M_i^* = \langle Q_i, q_i^0, f_i, \tau_i \rangle$ and for each $q_i \in Q_i$ define $A_j(q_i)$ as in the proof of Lemma 168.2. By the second part of Lemma 168.2 the machine M_j^* executes a strategy in the repeated game that is a solution of the problem $\max_{s_j} \mathcal{U}_j(M_i^*, s_j)$. Therefore $f_j(q_i^t(M_1^*, M_2^*)) \in A_j(q_i^t(M_1^*, M_2^*))$ for all t. Thus if there are two periods t and s in which $a_i^t(M_1^*, M_2^*) \neq a_i^s(M_1^*, M_2^*)$ and $a_j^t(M_1^*, M_2^*) = a_j^s(M_1^*, M_2^*)$ then there exists an optimal policy a_j' of player j for which $a_j'(q_i^t(M_1^*, M_2^*)) = a_j'(q_i^s(M_1^*, M_2^*))$. That is, player j uses the same action whenever player i's state is either $q_i^t(M_1^*, M_2^*)$ or $q_i^s(M_1^*, M_2^*)$. The following machine carries out the policy a_j' and has $c(M_i^*) - 1$ states, contradicting the first part of Lemma 168.2.

- The set of states is $Q_i \setminus \{q_i^s\}$.
- The initial state is q_i^0 if $q_i^s \neq q_i^0$ and is q_i^t otherwise.
- The output function is defined by $f_j(q) = a_j'(q)$.
- The transition function is defined as follows. If $\tau_i(q, f_j(q)) = q_i^s$ then $\tau_j(q, x) = q_i^t$ for all $x \in A_i$; otherwise $\tau_j(q, x) = \tau_i(q, f_j(q))$ for all $x \in A_i$ if $q \neq q_i^t$ and

$$\tau_j(q_i^t, a_i) = \begin{cases} \tau_i(q_i^s, f_j(q_i^s)) & \text{if } a_i = a_i^s(M_1^*, M_2^*) \\ \tau_i(q_i^t, f_j(q_i^t)) & \text{otherwise} \end{cases}$$

This completes the proof.   □

This result has a striking implication for the equilibrium outcome path in any game in which each player has two actions. For example, if in the repeated *Prisoner's Dilemma* two outcomes appear on the equilibrium path, then this pair of outcomes is either $\{(C, C), (D, D)\}$ or $\{(C, D), (D, C)\}$.

We now turn to an exploration of the structure of the equilibrium machines. Since each player's machine is finite there is a minimal number t' such that for some $t > t'$ we have $q_i^t = q_i^{t'}$ for both $i = 1$ and $i = 2$; let t^* be the minimal such t. The sequence of pairs of states starting in period t' consists of cycles of length $t^* - t'$. We refer to this stage as the **cycling phase**; the stage before period t' is the **introductory phase**.

We now show that the sets of states that a player uses in the cycling and introductory phases are disjoint. Further, in the introductory phase each state is entered only once and each of a player's states that is used in the cycling phase appears only once in each cycle. Thus in equilibrium there is a one-to-one correspondence between the states in the machines of players 1 and 2, a fact that may be interpreted to mean that in each period each machine "knows" the state that the other machine is in.

■ PROPOSITION 171.1 *If (M_1^*, M_2^*) is an equilibrium of a machine game then there exists a period t^* and an integer $\ell < t^*$ such that for $i = 1, 2$, the states in the sequence $(q_i^t(M_1^*, M_2^*))_{t=1}^{t^*-1}$ are distinct and $q_i^t(M_1^*, M_2^*) = q_i^{t-\ell}(M_1^*, M_2^*)$ for $t \geq t^*$.*

Proof. Let t^* be the first period in which one of the states of either of the two machines appears for the second time. That is, let t^* be the minimal time for which there is a player i and a period $t_i < t^*$ such that $q_i^{t^*} = q_i^{t_i}$. We have $a_i^{t^*} = a_i^{t_i}$ and hence, by Lemma 170.1, $a_j^{t^*} = a_j^{t_i}$. It follows that for all $k \geq 0$ we have $q_i^{t^*+k} = q_i^{t_i+k}$, and thus, using Lemma 168.1, $c(M_i^*) = t^* - 1$. By the selection of player i all states of M_j^* through time $t^* - 1$ are distinct, so that the first part of Lemma 168.2 implies that there exists $t_j < t^*$ such that $q_j^{t_j} = q_j^{t^*}$. It remains to show that $t_j = t_i$. Assume to the contrary that, say, $t_j > t_i$. Then player j can obtain the same path of outcomes with a machine in which $q_j^{t_i}$ is excluded by making a transition from $q_j^{t_i-1}$ to $q_j^{t^*}$, omitting $q_j^{t_i}$. But this contradicts the optimality of M_j^*. $\square$

A machine game is a strategic game, so that no considerations of the type modeled by the notion of subgame perfect equilibrium enter the analysis. To incorporate such considerations, we can modify the solution concept and require that after every history in the repeated game the pair of machines be an equilibrium of the machine game. Such a modification implies that the play of the machines does not have any introductory phase: a player who can change his machine in the course of play wants to omit any introductory states once the cycling phase is reached. It follows that the set of equilibrium paths is severely restricted by this modification of the solution, as Exercise 173.2 illustrates.

9.4   The Case of Lexicographic Preferences

The results of the previous section significantly limit the set of equilibria of a machine game. To limit the set of equilibria further we need to specify the tradeoff in each player's preferences between his payoff in the repeated game and the complexity of his machine. In this section we assume that the players' preferences are lexicographic (complexity being a secondary consideration, after the payoff in the repeated game); we restrict attention to the case in which the component game is the *Prisoner's Dilemma* (with payoffs as in Figure 166.1).

As we noted above, Lemma 170.1 implies that the set of outcomes that occurs on an equilibrium path is either a subset of $\{(C, C), (D, D)\}$ or a subset of $\{(C, D), (D, C)\}$. First consider equilibria of the former type. Let n_C and n_D be two nonnegative integers, at least one of which is positive. Then for δ close enough to 1 it can be shown that there is an equilibrium with a cycle of length $n_C + n_D$ in which (C, C) appears

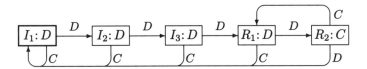

Figure 173.1 A machine M for either player in the infinitely repeated *Prisoner's Dilemma*.

n_C times and (D, D) appears n_D times. For the case $n_C = n_D = 1$ there is a symmetric equilibrium in which each player uses the machine M in Figure 173.1. (For preferences that are not lexicographic the pair (M, M) is an equilibrium only if the players's preferences do not put too much weight on complexity.)

? EXERCISE 173.1

    a. Show that if δ is close enough to 1 then the pair of machines (M, M) is a Nash equilibrium of the machine game.

    b. Show that if the machine M is modified so that in R_1 it plays C, in R_2 it plays D, and the transitions in R_1 and R_2 are reversed, then the new pair of machines is *not* a Nash equilibrium of the machine game.

In these equilibria the introductory phase is nonempty, and this is so for any equilibrium that supports a path in which (C, C) is an outcome.

? EXERCISE 173.2 Show that every equilibrium in which (C, C) is one of the outcomes has an introductory phase.

Now consider equilibria in which every outcome on the equilibrium path is either (C, D) or (D, C). Some such equilibria are cyclical, without any introductory phase. Precisely, for all positive integers n_1 and n_2 satisfying $5n_i/(n_1 + n_2) > 1$ for $i = 1$ and $i = 2$ there exists δ large enough that there is an equilibrium of the machine game in which the cycle consists of n_1 plays of (D, C) followed by n_2 plays of (C, D), without any introductory phase. (The condition on n_1 and n_2 ensures that each player's average payoff exceeds his minmax payoff of 1.)

An equilibrium for the case $n_1 = n_2 = 1$ is shown in Figure 174.1. One interpretation of this equilibrium is that the players alternate being generous towards each other. One can think of (C, D) as the event in which player 1 gives a gift to player 2 and (D, C) as the event in which player 2 gives a gift to player 1. In the equilibrium a player does not care if his opponent does not accept the gift (i.e. chooses C when he could have chosen D and received the gift), but he insists that his opponent

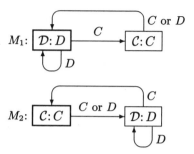

Figure 174.1 Machines M_1 for player 1 and M_2 for player 2 for the infinitely repeated *Prisoner's Dilemma.* For δ large enough the pair (M_1, M_2) is an equilibrium of the machine game; it generates the path consisting of repetitions of the cycle $((D, C), (C, D))$.

$$
\begin{array}{c|c|c|}
 & A & B \\
\hline
A & 3,1 & 1,3 \\
\hline
B & 2,0 & 2,0 \\
\hline
\end{array}
$$

Figure 174.2 The constituent game for the repeated game in Exercise 174.1.

give him a gift (play C) in periods in which he expects to get a gift: if he does not receive a gift then he does not move to the state in which he is generous.

In our analysis so far the constituent game is a strategic game. One can think also of the case in which the constituent game is an extensive game. While the analysis of the Nash equilibria of the repeated game is unchanged (though the set of subgame perfect equilibria may be somewhat different), the analysis of the Nash equilibria of the machine game is quite different in this case, as the following exercise demonstrates.

? EXERCISE 174.1 Consider the infinitely repeated game for which the constituent game is given in Figure 174.2.

    a. Show that the set of paths associated with the Nash equilibria of the machine game contains only the outcomes (A, A) and (B, B).

    b. Show that if the players' preferences in the machine game are lexicographic, then every finite sequence containing only the outcomes (A, A) and (B, B) is the cycling phase of the path associated with some Nash equilibrium of the machine game for δ large enough.

c. Notice that the game is the strategic form of an extensive game with perfect information. Assume that the players engage in the infinitely repeated game in which the constituent game is this extensive game, learning at the end of each round the terminal history that occurs. Show that the machine game for this repeated game has a unique Nash equilibrium, in which the payoff profile is $(2, 0)$ in every period. [Hint: When player 1 chooses B she cannot monitor whether player 2 plans to choose A or B if she chooses A.]

Notes

This chapter is based on Rubinstein (1986) and Abreu and Rubinstein (1988). The line of argument, and in particular the proof of Lemma 170.1, is a modification due to Piccione (1992) of the proof of Abreu and Rubinstein (1988). Exercise 174.1 is based on Piccione and Rubinstein (1993).

In a related strand of literature the complexity of the machines that a player can employ is taken to be exogenously bounded. The main aim of this line of research is to show that equilibrium outcomes that differ from repetitions of (D, D) can be supported in the finitely repeated *Prisoner's Dilemma*; see for example Neyman (1985) and Zemel (1989).

10 Implementation Theory

In this chapter we study the inverse of the problem considered in the previous chapters: rather than fix a game and look for the set of outcomes given by some solution concept, we fix a set of outcomes and look for a game that yields that set of outcomes as equilibria.

10.1 Introduction

The standard procedure in game theory is to formulate a model that captures a situation and to investigate the set of outcomes that are consistent with some solution concept. If we fix the structure of the game and vary the players' preferences then a solution concept induces a correspondence from preference profiles to the set of outcomes.

Our general approach in this book is that a game is not necessarily a description of some physical rules that exist: most strategic situations lack a clear structure, and even when one exists the players' perceptions of the situation do not necessarily coincide with an "objective" description of that situation. By contrast, in this chapter a planner is assumed to set the rules of the interaction, and the individuals, when confronted with these rules, are assumed to take them literally. The planner can design the structure of the game but cannot control the players' preferences or actions. She starts with a description of the outcomes she wishes to associate with each possible preference profile and looks for a game that "implements" this correspondence. On finding such a game she can realize her objective by having the individuals play the game, assuming of course that their behavior conforms with the solution concept.

An assumption underlying this interpretation of an implementation problem is that the planner can force the individuals to play the game but cannot enforce the desirable outcome directly, possibly because she lacks information about some parameters of the situation, information that is known to all participants but is either too costly or impossible for her to obtain.

To illustrate the nature of an implementation problem, consider a planner who wishes to assign an object to one of two individuals. Suppose that she wants to give it to the individual who values it most, but does not know who this is. Her problem then is to design a game form with the property that, for every possible pair of valuations, the outcome according to some solution concept is that the object is given to the individual who values it most. Whether this is possible depends on the outcomes that the planner can impose on the individuals. For example, she may be allowed only to transfer money from one individual to another, or she may be allowed also to impose fines on the individuals.

As in other chapters, we focus on the conceptual aspects of the theory and present only a sample of the main ideas. We restrict attention to implementation problems in which the individuals are fully informed about the parameters of the situation; we do not touch upon the large literature that considers the case in which there is asymmetric information.

10.2   The Implementation Problem

Let N be a set of individuals, C a set of feasible *outcomes*, and $\mathcal{P}$ a set of preference profiles over C. We denote individual i's preference relation by $\succsim_i$ and sometimes denote a preference profile $(\succsim_i)_{i \in N}$ simply by $\succsim$. A **choice rule** is a function that assigns a *subset* of C to each profile in $\mathcal{P}$. We refer to a choice rule that is singleton-valued as a **choice function**. The objective of the planner is to design a game form whose outcomes, for each preference profile $\succsim$ in $\mathcal{P}$, coincide with $f(\succsim)$, where f is the choice rule or the choice function. If f is not singleton-valued then the planner is concerned that each of the outcomes in $f(\succsim)$ be possible. For instance, in the example discussed in the previous section the planner may wish to assign the object to the individual whose valuation is highest, without discriminating between the individuals if their valuations are the same. In a more general problem, the planner may wish to implement the choice rule that associates with each preference profile the set of efficient outcomes.

The planner controls the rules of the game, formalized as a game form. A **strategic game form with consequences in** C is a triple $\langle N, (A_i), g \rangle$ where A_i, for each $i \in N$, is the set of actions available to player i, and $g: A \to C$ (where $A = \times_{i \in N} A_i$) is an *outcome function* that associates an outcome with every action profile. A strategic game form and a preference profile $(\succsim_i)$ induce a strategic game $\langle N, (A_i), (\succsim_i') \rangle$ where $\succsim_i'$ is defined by $a \succsim_i' b$ if and only if $g(a) \succsim_i g(b)$ for each $i \in N$. An **extensive game form (with perfect information) with consequences in** C is a tuple $\langle N, H, P, g \rangle$ where H is a set of histories, $P: H \setminus Z \to N$ is a player function, and $g: Z \to C$ is an outcome function ($Z \subseteq H$ being the set of terminal histories). (Cf. Definition 89.1 and the following definition of an extensive game form.) An extensive game form and a preference profile induce an extensive game.

The planner operates in an **environment** that consists of

- a finite set N of players, with $|N| \geq 2$
- a set C of outcomes
- a set $\mathcal{P}$ of preference profiles over C
- a set $\mathcal{G}$ of (either strategic or extensive) game forms with consequences in C.

When designing a game form to implement her objectives the planner must take into account how the individuals will play any possible game. A **solution concept** for the environment $\langle N, C, \mathcal{P}, \mathcal{G} \rangle$ is a set-valued function S with domain $\mathcal{G} \times \mathcal{P}$. If the members of $\mathcal{G}$ are strategic game forms then S takes values in the set of action profiles, while if the members of $\mathcal{G}$ are extensive game forms then S takes values in the set of terminal histories.

The following definition is one formulation of the planner's problem.

▷ DEFINITION 179.1 Let $\langle N, C, \mathcal{P}, \mathcal{G} \rangle$ be an environment and let S be a solution concept. The game form $G \in \mathcal{G}$ with outcome function g is said to S-**implement** the choice rule $f: \mathcal{P} \to C$ if for every preference profile $\succsim \in \mathcal{P}$ we have $g(S(G, \succsim)) = f(\succsim)$. In this case we say the choice rule f is S-**implementable** in $\langle N, C, \mathcal{P}, \mathcal{G} \rangle$.

In other notions of implementation via a strategic game form the set of actions of each player is required to be the set of possible preference profiles (each player must announce a preference relation for *every* player) and announcing the true profile is required to be consistent with the solution concept. One such notion is the following.

▷ DEFINITION 179.2 Let $\langle N, C, \mathcal{P}, \mathcal{G} \rangle$ be an environment in which $\mathcal{G}$ is a set of strategic game forms for which the set of actions of each player i is a set $\mathcal{P}$ of preference profiles, and let S be a solution concept. The strategic game form $G = \langle N, (A_i), g \rangle \in \mathcal{G}$ **truthfully S-implements** the choice rule $f: \mathcal{P} \to C$ if for every preference profile $\succsim \in \mathcal{P}$ we have

- $a^* \in \mathcal{S}(G, \succsim)$ where $a_i^* = \succsim$ for each $i \in N$ (every player reporting the true preference profile is a solution of the game)

- $g(a^*) \in f(\succsim)$ (the outcome if every player reports the true preference profile is a member of $f(\succsim)$).

In this case we say the choice rule f is **truthfully $\mathcal{S}$-implementable** in $\langle N, C, \mathcal{P}, \mathcal{G} \rangle$.

This notion of implementation differs from the previously defined notion in three respects. First, and most important, it requires the set of actions of each player to be the set of preference profiles and "truth telling" to be a solution to every game that may arise. Second, it allows (non truth-telling) solutions of the game to yield outcomes that are inconsistent with the choice rule. Third, it allows there to be preference profiles for which not every outcome prescribed by the choice rule corresponds to a solution of the game.

Our discussion is organized according to the set of game forms under consideration and the solution concept used. We begin with strategic game forms and the solution of dominant strategy equilibrium; we then consider strategic game forms and the solution of Nash equilibrium; finally we consider extensive game forms and the solution of subgame perfect equilibrium.

We establish two types of results, one negative and one positive. The negative results give conditions under which only degenerate choice rules can be implemented. The positive ones give conditions under which every rule in a very large set can be (at least approximately) implemented. Results of the latter type are reminiscent of the "folk theorems" of Chapter 8. Like the folk theorems their main interest is not the fact that "anything is possible"; rather, the *structures* of the mechanisms that we use to prove the results are the most interesting aspects of the investigation. Some features of these structures sometimes correspond to mechanisms that we observe, giving us insight into the rationale for these mechanisms.

10.3   Implementation in Dominant Strategies

In this section we assume that the planner is restricted to use a strategic game form. We assume also that, desiring to avoid strategic complications, she aims to achieve her goals by designing the game so that the outcomes that she wishes to implement are consistent with the solution concept of dominant strategy equilibrium (DSE), defined as follows.

▶ DEFINITION 181.1 A **dominant strategy equilibrium of a strategic game** $\langle N, (A_i), (\succsim_i) \rangle$ is a profile $a^* \in A$ of actions with the property that for every player $i \in N$ we have $(a_{-i}, a_i^*) \succsim_i (a_{-i}, a_i)$ for all $a \in A$.

Thus the action of every player in a dominant strategy equilibrium is a best response to *every* collection of actions for the other players, not just the *equilibrium* actions of the other players as in a Nash equilibrium (Definition 14.1). (Note that the fact that a^* is a dominant strategy equilibrium does not imply that for any player i the action a_i^* dominates (even weakly) all the other actions of player i: it could be that for some $a_i \neq a_i^*$ we have $(a_{-i}, a_i^*) \sim_i (a_{-i}, a_i)$ for all $a_{-i} \in A_{-i}$.) The notion of DSE-implementation is strong, since a dominant strategy is optimal no matter what the other players do. We now see that it is hard to DSE-implement a choice rule.

We say that a choice rule $f: \mathcal{P} \to C$ is **dictatorial** if there is a player $j \in N$ such that for any preference profile $\succsim \in \mathcal{P}$ and outcome $a \in f(\succsim)$ we have $a \succsim_j b$ for all $b \in C$. The following result, known as the Gibbard–Satterthwaite theorem, is a milestone in implementation theory.

■ PROPOSITION 181.2 *Let* $\langle N, C, \mathcal{P}, \mathcal{G} \rangle$ *be an environment in which* C *contains at least three members,* $\mathcal{P}$ *is the set of all possible preference profiles, and* $\mathcal{G}$ *is the set of strategic game forms. Let* $f: \mathcal{P} \to C$ *be a choice rule that is DSE-implementable and satisfies the condition*

for every $a \in C$ *there exists* $\succsim \in \mathcal{P}$ *such that* $f(\succsim) = \{a\}$. (181.3)

Then f *is dictatorial.*

A proof of this result uses the following result.

■ LEMMA 181.4 (Revelation principle for DSE-implementation) *Let* $\langle N, C, \mathcal{P}, \mathcal{G} \rangle$ *be an environment in which* $\mathcal{G}$ *is the set of strategic game forms. If a choice rule* $f: \mathcal{P} \to C$ *is DSE-implementable then*

a. *f is truthfully DSE-implementable*

b. *there is a strategic game form* $G^* = \langle N, (A_i), g^* \rangle \in \mathcal{G}$ *in which* A_i *is the set of all preference relations (rather than profiles) such that for all* $\succsim \in \mathcal{P}$ *the action profile* $\succsim$ *is a dominant strategy equilibrium of the strategic game* $\langle G^*, \succsim \rangle$ *and* $g^*(\succsim) \in f(\succsim)$.

Proof. Let $G = \langle N, (A_i), g \rangle$ be a game form that DSE-implements f. We first prove (b). The set of dominant actions for any player j depends only on $\succsim_j$, so that we can define $a_j(\succsim_j)$ to be a dominant action for player j in any game $\langle G, (\succsim_{-j}, \succsim_j) \rangle$. Define the outcome function g^* of G^* by

$g^*(\succsim) = g((a_i(\succsim_i)))$. Since G DSE-implements f we have $g^*(\succsim) \in f(\succsim)$. Now suppose that there is a preference profile $\succsim$ for which $\succsim_j$ is not a dominant strategy for player j in G^*. Then there is a preference profile $\succsim'$ such that $g^*(\succsim'_{-j}, \succsim'_j) \succ_j g^*(\succsim'_{-j}, \succsim_j)$, so that $a_j(\succsim_j)$ is not a best response in $\langle G, \succsim \rangle$ to the collection of actions $(a_i(\succsim'_i))_{i \in N \setminus \{j\}}$, a contradiction. Thus $\succsim$ is a dominant strategy equilibrium of $\langle G^*, \succsim \rangle$.

It is immediate that $\succsim$ is a dominant strategy for every player in the game $\langle G', \succsim \rangle$ in which the set of actions of each player is $\mathcal{P}$ and the outcome function is given by $g'((\succsim(i))) = g^*((\succsim_i(i)))$ (where $\succsim(i)$ is a preference profile for each $i \in N$), so that f is truthfully DSE-implementable, proving (a).   □

From this lemma it follows that if a choice rule cannot be truthfully DSE-implemented then it cannot be DSE-implemented. Thus for example if $\mathcal{P}$ contains only strict preference relations then the choice function that chooses the second ranked outcome in player 1's preferences is not DSE-implementable: if it were, then by the lemma a dominant strategy in G^* for player 1 would be to announce her true preference relation, but in fact in G^* it is better for her to announce a preference relation in which her most preferred action is ranked second.

Notice that the game G^* in the lemma does not necessarily DSE-implement the choice rule since, as we noted earlier, the notion of truthful DSE-implementation does not exclude the possibility that there are non-truthful dominant strategy profiles for which the outcome is different from any given by the choice rule. In brief, it does *not* follow from Lemma 181.4 that DSE-implementation is equivalent to truthful DSE-implementation.

[?] EXERCISE 182.1 Show that if the set $\mathcal{P}$ of preference profiles contains only strict preferences then a choice function is truthfully DSE-implementable if and only if it is DSE-implementable.

The main part of the proof of the Gibbard–Satterthwaite theorem (181.2) is the proof of the following result in social choice theory, which we omit. (The standard proof of this result relies on Arrow's impossibility theorem (for a proof of which see, for example, Sen (1986)).)

■ LEMMA 182.2 *Let C be a set that contains at least three members and let $\mathcal{P}$ be the set of all possible preference profiles. If a choice function $f: \mathcal{P} \to C$ satisfies (181.3) and for every preference profile $\succsim \in \mathcal{P}$ we have $f(\succsim_{-j}, \succsim_j) \succsim_j f(\succsim_{-j}, \succsim'_j)$ for every preference relation $\succsim'_j$ then f is dictatorial.*

Proof of Proposition 181.2. It follows from the proof of Lemma 181.4 that if the choice rule f is DSE-implementable, say by the game form G, then any selection g^* of f (i.e. $g^*(\succsim) \in f(\succsim)$ for all $\succsim \in \mathcal{P}$) has the property that for every preference profile $\succsim$ we have $g^*(\succsim_{-j}, \succsim_j) \succsim_j g^*(\succsim_{-j}, \succsim'_j)$ for every preference relation $\succsim'_j$. Since f satisfies (181.3), g^* does also. Consequently by Lemma 182.2 g^* is dictatorial, so that f is also. $\qquad\qquad\qquad\qquad\qquad\qquad\qquad\qquad\qquad\qquad\qquad\qquad\square$

? EXERCISE 183.1 Explain, without making reference to the Gibbard–Satterthwaite theorem (181.2), why the following choice function is not DSE-implementable in an environment $\langle N, C, \mathcal{P}, \mathcal{G} \rangle$ in which C contains at least three members, $\mathcal{P}$ is the set of all possible preference profiles, and $\mathcal{G}$ is the set of strategic game forms:

$$f(\succsim) = \begin{cases} a & \text{if for all } i \in N \text{ we have } a \succ_i b \text{ for all } b \neq a \\ a^* & \text{otherwise,} \end{cases}$$

where a^* is an arbitrary member of C.

The Gibbard–Satterthwaite theorem (181.2) applies to an environment $\langle N, C, \mathcal{P}, \mathcal{G} \rangle$ in which $\mathcal{P}$ is the set of *all* possible preference profiles. There are environments in which $\mathcal{P}$ does not contain all possible preference profiles for which we can construct game forms that DSE-implement nondegenerate choice rules. The most well-known such game forms are those studied by Clarke (1971) and Groves (1973), which are designed for a situation in which a set of individuals has to decide whether or not to pursue some costly joint project and, if they decide to go ahead, how to assign the expenses. Clarke and Groves take the set C of outcomes to consist of all pairs $(x, (m_i))$ where $x = 1$ or 0 according to whether or not the project is undertaken and m_i is a payment by individual i. They impose the condition that the preference relation over C of each player i be represented by a utility function of the form $\theta_i x - m_i$ for some $\theta_i \in \mathbb{R}$; under this assumption we can identify the set $\mathcal{P}$ of preference profiles with the set $\mathbb{R}^N$ of profiles of real numbers. The aim of the game forms that Clarke and Groves construct is to implement choice functions $f : \mathbb{R}^N \to C$ with the property that the project is undertaken if and only if $\sum_{i \in N} \theta_i \geq \gamma$, where $\gamma \geq 0$ is the cost of the project.

Not all such choice functions are DSE-implementable. The next proposition and exercise establish that such a choice function f is truthfully DSE-implementable if and only if for each $j \in N$ there is a function h_j such that $m_j(\theta) = x(\theta)(\gamma - \sum_{i \in N \setminus \{j\}} \theta_i) + h_j(\theta_{-j})$ for all $\theta \in \mathbb{R}^N$, where $f(\theta) = (x(\theta), m(\theta))$. In the strategic game form used to implement f, each player j announces a number a_j, interpreted as a declaration of his value of the project, and the project is executed if and only if the sum of these declarations is at least γ; the payment made by player j is equal to $h_j(a_{-j})$ (which is independent of his announcement), plus, if the project is carried out, an amount equal to the difference between the cost of the project and the sum of the announcements made by the other players. Formally, in this strategic game form $\langle N, (A_i), g \rangle$ we have $A_i = \mathbb{R}$ and $g(a) = (x(a), m(a))$ for each $a \in A$ where

$$\begin{cases} x(a) = 1 \text{ if and only if } \sum_{i \in N} a_i \geq \gamma \\ m_j(a) = x(a)(\gamma - \sum_{i \in N \setminus \{j\}} a_i) + h_j(a_{-j}) \text{ for each } j \in N. \end{cases} \quad (184.1)$$

Such a game form is called a *Groves mechanism*.

■ **PROPOSITION 184.2** *Let* $\langle N, C, \mathcal{P}, \mathcal{G} \rangle$ *be an environment in which* $C = \{(x, m): x \in \{0, 1\} \text{ and } m \in \mathbb{R}^N\}$, $\mathcal{P}$ *is the set of profiles* $(\succsim_i)$ *in which each* $\succsim_i$ *is represented by a utility function of the form* $\theta_i x - m_i$ *for some* $\theta_i \in \mathbb{R}$, *and* $\mathcal{G}$ *is the set of strategic game forms; identify* $\mathcal{P}$ *with* $\mathbb{R}^N$. *A choice function* $f : \mathbb{R}^N \to C$ *with* $f(\theta) = (x(\theta), m(\theta))$ *for which*

- $x(\theta) = 1$ *if and only if* $\sum_{i \in N} \theta_i \geq \gamma$

- *for each* $j \in N$ *there is a function* h_j *such that* $m_j(\theta) = x(\theta)(\gamma - \sum_{i \in N \setminus \{j\}} \theta_i) + h_j(\theta_{-j})$ *for all* $\theta \in \mathbb{R}^N$

is truthfully DSE-implemented by the Groves mechanism $\langle N, (A_i), g \rangle$ *defined in (184.1).*

Proof. Let $j \in N$ and let a_{-j} be an arbitrary vector of actions of the players other than j. We argue that when the players other than j choose a_{-j}, j's payoff when he chooses $a_j = \theta_j$ is at least as high as his payoff when he chooses any other action in A_j. There are three cases.

- If $x(a_{-j}, \theta_j) = x(a_{-j}, a_j')$ then $m_j(a_{-j}, a_j') = m_j(a_{-j}, \theta_j)$ and hence $g(a_{-j}, a_j') = g(a_{-j}, \theta_j)$.

- If $x(a_{-j}, \theta_j) = 0$ and $x(a_{-j}, a_j') = 1$ then j's payoff under (a_{-j}, θ_j) is $-m_j(a_{-j}, \theta_j) = -h_j(a_{-j})$, while his payoff under (a_{-j}, a_j') is $\theta_j - m_j(a_{-j}, a_j') = \theta_j - (\gamma - \sum_{i \in N \setminus \{j\}} a_i) - h_j(a_{-j}) < -h_j(a_{-j})$, since $x(a_{-j}, \theta_j) = 0$ implies that $\sum_{i \in N \setminus \{j\}} a_i + \theta_j < \gamma$.

- If $x(a_{-j}, \theta_j) = 1$ and $x(a_{-j}, a_j') = 0$ then j's payoff under (a_{-j}, θ_j) is $\theta_j - m_j(a_{-j}, \theta_j) = \theta_j - (\gamma - \sum_{i \in N \setminus \{j\}} a_i) - h_j(a_{-j})$, while his payoff under (a_{-j}, a_j') is $-m_j(a_{-j}, a_j') = -h_j(a_{-j}) \leq \theta_j - (\gamma - \sum_{i \in N \setminus \{j\}} a_i) - h_j(a_{-j})$, since $x(a_{-j}, \theta_j) = 1$ implies that $\sum_{i \in N \setminus \{j\}} a_i + \theta_j \geq \gamma$.

Hence a dominant action for each player j is to choose $a_j = \theta_j$. The outcome $g(\theta)$ is equal to $f(\theta)$, so that $\langle N, (A_i), g \rangle$ truthfully DSE-implements f. $\square$

Note that the Groves mechanism (184.1) does *not* Nash-implement a choice function f satisfying the conditions of the proposition: for example, if $\gamma = 2$, $|N| = 2$, and $\theta_i = 1$ for both players then the associated game has, in addition to $(1,1)$, inefficient equilibria (e.g. $(0,0)$).

⁇ EXERCISE 185.1 In an environment like that in the previous proposition, show that if a choice function f with $f(\theta) = (x(\theta), m(\theta))$ and $x(\theta) = 1$ if and only if $\sum_{i \in N} \theta_i \geq \gamma$ is truthfully DSE-implementable then for each $j \in N$ there is a function h_j such that $m_j(\theta) = x(\theta)(\gamma - \sum_{i \in N \setminus \{j\}} \theta_i) - h_j(\theta_{-j})$ for all $\theta \in \mathbb{R}^N$. [You need to show that whenever $x(\theta_{-j}, \theta_j) = 1$ and $x(\theta_{-j}, \theta'_j) = 0$ then $m_j(\theta_{-j}, \theta_j) - m_j(\theta_{-j}, \theta'_j) = \gamma - \sum_{i \in N \setminus \{j\}} \theta_i$.]

10.4 Nash Implementation

We now turn to the case in which the planner, as in the previous section, uses strategic game forms, but assumes that for any game form she designs and for any preference profile the outcome of the game may be any of its Nash equilibria.

The first result is a version of the revelation principle (see also Lemma 181.4). It shows that any Nash-implementable choice rule is also truthfully Nash-implementable: there is a game form in which (*i*) each player has to announce a preference profile and (*ii*) for any preference profile truth-telling is a Nash equilibrium. This result serves two purposes. First, it helps to determine the boundaries of the set of Nash-implementable choice rules. Second, it shows that a simple game can be used to achieve the objective of a planner who considers truthful Nash equilibrium to be natural and is not concerned about the outcome so long as it is in the set given by the choice rule.

■ LEMMA 185.2 (Revelation principle for Nash implementation) *Let* $\langle N, C, \mathcal{P}, \mathcal{G} \rangle$ *be an environment in which* $\mathcal{G}$ *is the set of strategic game forms. If a choice rule is Nash-implementable then it is truthfully Nash-implementable.*

Proof. Let $G = \langle N, (A_i), g \rangle$ be a game form that Nash-implements the choice rule $f \colon \mathcal{P} \to C$ and for each $\succsim \in \mathcal{P}$ let $(a_i(\succsim))$ be a Nash equilibrium of the game $\langle G, \succsim \rangle$. Define a new game form $G^* = \langle N, (A_i^*), g^* \rangle$ in which $A_i^* = \mathcal{P}$ for each $i \in N$ and $g^*(p) = g((a_i(p_i)))$ for each

$p \in \times_{i \in N} A_i^*$. (Note that each p_i is a preference profile and p is a profile of preference profiles.) Clearly the profile p^* in which $p_i^* = \succsim$ for each $i \in N$ is a Nash equilibrium of $\langle G^*, \succsim \rangle$ and $g^*(p^*) \in f(\succsim)$.   □

Note that it does *not* follow from this result that in an analysis of Nash implementation we can restrict attention to games in which each player announces a preference profile, since the game that truthfully Nash-implements the choice rule may have non-truthful Nash equilibria that generate outcomes different from that dictated by the choice rule. Note also that it is essential that the set of actions of each player be the set of preference *profiles*, not the (smaller) set of preference relations, as in part (b) of the revelation principle for DSE-implementation (Lemma 181.4).

We now define a key condition in the analysis of Nash implementation.

▶ DEFINITION 186.1 A choice rule $f: \mathcal{P} \to C$ is **monotonic** if whenever $c \in f(\succsim)$ and $c \notin f(\succsim')$ there is some player $i \in N$ and some outcome $b \in C$ such that $c \succsim_i b$ and $b \succ_i' c$.

That is, in order for an outcome c to be selected by a monotonic choice rule when the preference profile is $\succsim$ but not when it is $\succsim'$ the ranking of c relative to some other alternative must be worse under $\succsim'$ than under $\succsim$ for at least one individual.

An example of a monotonic choice rule f is that in which $f(\succsim)$ is the set of weakly Pareto efficient outcomes: $f(\succsim) = \{c \in C$: there is no $b \in C$ such that $b \succ_i c$ for all $i \in N\}$. Another example is the rule f in which $f(\succsim)$ consists of every outcome that is a favorite of at least one player: $f(\succsim) = \{c \in C$: there exists $i \in N$ such that $c \succsim_i b$ for all $b \in C\}$.

■ PROPOSITION 186.2 *Let $\langle N, C, \mathcal{P}, \mathcal{G} \rangle$ be an environment in which $\mathcal{G}$ is the set of strategic game forms. If a choice rule is Nash-implementable then it is monotonic.*

Proof. Suppose that the choice rule $f: \mathcal{P} \to C$ is Nash-implemented by a game form $G = \langle N, (A_i), g \rangle$, $c \in f(\succsim)$, and $c \notin f(\succsim')$. Then there is an action profile a for which $g(a) = c$ that is a Nash equilibrium of the game $\langle G, \succsim \rangle$ but not of $\langle G, \succsim' \rangle$. That is, there is a player j and action $a_j' \in A_j$ such that $g(a_{-j}, a_j') \succ_j' g(a)$ and $g(a) \succsim_j g(a_{-j}, a_j')$. Hence f is monotonic.   □

◇ EXAMPLE 186.3 (*Solomon's predicament*) The biblical story of the Judgment of Solomon illustrates some of the main ideas of implementation theory. Each of two women, 1 and 2, claims a baby; each of them knows who is the true mother, but neither can prove her motherhood. Solomon tries to educe the truth by threatening to cut the baby in two, relying on the fact that the false mother prefers this outcome to that in which the true mother obtains the baby while the true mother prefers to give the baby away than to see it cut in two. Solomon can give the baby to either of the mothers or order its execution.

Formally, let a be the outcome in which the baby is given to mother 1, b that in which the baby is given to mother 2, and d that in which the baby is cut in two. Two preference profiles are possible:

θ (1 is the real mother): $a \succ_1 b \succ_1 d$ and $b \succ_2 d \succ_2 a$

θ' (2 is the real mother): $a \succ_1' d \succ_1' b$ and $b \succ_2' a \succ_2' d$.

Despite Solomon's alleged wisdom, the choice rule f defined by $f(\theta) = \{a\}$ and $f(\theta') = \{b\}$ is not Nash-implementable, since it is not monotonic: $a \in f(\theta)$ and $a \notin f(\theta')$ but there is no outcome y and player $i \in N$ such that $a \succsim_i y$ and $y \succ_i' a$. (In the biblical story Solomon succeeds in assigning the baby to the true mother: he gives it to the only woman to announce that she prefers that it be given to the other woman than be cut in two. Probably the women did not perceive Solomon's instructions as a strategic game form.)

The next result provides sufficient conditions for a choice rule to be Nash-implementable.

▸ DEFINITION 187.1 A choice rule $f : \mathcal{P} \to C$ has **no veto power** if $c \in f(\succsim)$ whenever for at least $|N| - 1$ players we have $c \succsim_i y$ for all $y \in C$.

■ PROPOSITION 187.2 *Let $\langle N, C, \mathcal{P}, \mathcal{G} \rangle$ be an environment in which $\mathcal{G}$ is the set of strategic game forms. If $|N| \geq 3$ then any choice rule that is monotonic and has no veto power is Nash-implementable.*

Proof. Let $f : \mathcal{P} \to C$ be a monotonic choice rule that has no veto power. We construct a game form $G = \langle N, (A_i), g \rangle$ that Nash-implements f as follows. The set of actions A_i of each player i is the set of all triples (p_i, c_i, m_i), where $p_i \in \mathcal{P}$, $c_i \in C$, and m_i is a nonnegative integer. The values $g((p_i, c_i, m_i)_{i \in N})$ of the outcome function are defined as follows.

• If for some $j \in N$ and some $(\succsim, c, m)$ with $c \in f(\succsim)$ we have $(p_i, c_i, m_i) = (\succsim, c, m)$ for all $i \in N \setminus \{j\}$ then

$$g((p_i, c_i, m_i)) = \begin{cases} c_j & \text{if } c \succsim_j c_j \\ c & \text{if } c \prec_j c_j. \end{cases}$$

- Otherwise $g((p_i, c_i, m_i)) = c_k$ where k is such that $m_k \geq m_j$ for all $j \in N$ (in the case of a tie the identity of k is immaterial).

This game form has three components. First, if all the players agree about the preference profile $\succsim$ and the outcome $c \in f(\succsim)$ to be implemented then the outcome is indeed c. Second, if there is almost agreement—all players but one agree—then the majority prevails unless the exceptional player announces an outcome that, under the preference relation announced by the majority, is not better for him than the outcome announced by the majority (which persuades the planner that the preference relation announced for him by the others is incorrect). Third, if there is significant disagreement then the law of the jungle applies: the player who "shouts loudest" chooses the outcome.

We now show that this game form Nash-implements f. Let $c \in f(\succsim)$ for some $\succsim \in \mathcal{P}$. Let $a_i = (\succsim, c, 0)$ for each $i \in N$. Then (a_i) is a Nash equilibrium of the game $\langle G, \succsim \rangle$ with the outcome c: any deviation by any player j, say to $(\succsim', c', m')$, that affects the outcome has the property that the outcome is $c' \prec_j c$.

Now let (a_i^*) be a Nash equilibrium of the game $\langle G, \succsim \rangle$ with the outcome c^*. We show that $c^* \in f(\succsim)$.

There are three cases to consider. First suppose that $a_i^* = (\succsim', c^*, m')$ for all $i \in N$ and $c^* \in f(\succsim')$. If $c^* \notin f(\succsim)$ then the monotonicity of f implies that there is a player $i \in N$ and $b \in C$ such that $c^* \succsim_i' b$ and $b \succ_i c^*$. But then the deviation by player i to the action $(\succsim, b, 0)$ changes the action profile to one that yields his preferable outcome b. Hence $c^* \in f(\succsim)$.

Second suppose that $a_i^* = (\succsim', c^*, m')$ for all $i \in N$ and $c^* \notin f(\succsim')$. If there is some $i \in N$ and outcome $b \in C$ such that $b \succ_i c^*$ then player i can deviate to $(\succsim', b, m'')$ for some $m'' > m'$, yielding the preferred outcome b. Thus c^* is a favorite outcome of every player; since f has no veto power we have $c^* \in f(\succsim)$.

Third suppose that $a_i^* \neq a_j^*$ for some players i and j. We show that for at least $|N| - 1$ players c^* is a favorite outcome, so that since f has no veto power we have $c^* \in f(\succsim)$. Since $|N| \geq 3$ there exists $h \in N \setminus \{i, j\}$; a_h^* is different from either a_i^* or a_j^*, say $a_h^* \neq a_i^*$. If there is an outcome b such that $b \succ_k c^*$ for some $k \in N \setminus \{i\}$ then k can profitably deviate by choosing $(\succsim', b, m'')$ for some $m'' > m_\ell$ for all $\ell \neq k$. Thus for all $k \in N \setminus \{i\}$ we have $c^* \succsim_k b$ for all $b \in C$. (Note that player i, unlike the other players, may not be able to achieve his favorite outcome by deviating since all the other players might be in agreement.)   □

The interest of a result of this type, like that of the folk theorems in Chapter 8, depends on the reasonableness of the game form constructed in the proof. A natural component of the game form constructed here is that a complaint against a consensus is accepted only if the suggested alternative is worse for the complainant under the preference profile claimed by the other players. A less natural component is the "shouting" part of the game form, especially since shouting bears no cost here.

The strength of the result depends on the size of the set of choice rules that are monotonic and have no veto power. If there are at least three alternatives and $\mathcal{P}$ is the set of all preference profiles then *no* monotonic choice *function* has no veto power. (This follows from Muller and Satterthwaite (1977, Corollary on p. 417); note that a monotonic choice function satisfies Muller and Satterthwaite's condition SPA.) Thus the proposition is of interest only for either a nondegenerate choice *rule* or a choice function with a limited domain.

The game form in the proof of the proposition is designed to cover all possible choice rules. A specific choice rule may be implemented by a game form that is much simpler. Two examples follow.

◇ EXAMPLE 189.1 Suppose that an object is to be assigned to a player in the set $\{1, \ldots, n\}$. Assume first that for all possible preference profiles there is a single player who prefers to have the object than not to have it. The choice function that assigns the object to this player can be implemented by the game form in which the set of actions of each player is $\{Yes, No\}$ and the outcome function assigns the object to the player with the lowest index who announces *Yes* if there is such a player, and to player n otherwise. It is easy to check that if player i is the one who prefers to have the object than not to have it then the only equilibrium outcome is that i gets the object.

Now assume that in each preference profile there are two ("privileged") players who prefer to have the object than to not have it, and that we want to implement the choice rule that assigns to each preference profile the two outcomes in which the object is assigned to one of these players. The game form just described does not work since, for example, for the profile in which these players are 1 and 2 there is no equilibrium in which player 2 gets the object. The following game form does implement the rule. Each player announces a name of a player and a number. If $n-1$ players announce the same name, say i, then i obtains the object unless he names a different player, say j, in which case j obtains the object. In any other case the player who names the largest number gets the

	Mine	Hers	Mine+
Mine	$(0, \epsilon, \epsilon)$	$(1, 0, 0)$	$(2, \epsilon, M)$
His	$(2, 0, 0)$	$(0, \epsilon, \epsilon)$	$(0, 0, 0)$
Mine+	$(1, M, \epsilon)$	$(0, 0, 0)$	$(0, 2\epsilon, 2\epsilon)$

Figure 190.1 A game form that implements the choice function considered in Example 190.1 in which the legitimate owner obtains the object. (Note that the entries in the boxes are outcomes, not payoffs.)

object. Any action profile in which all players announce the name of the same privileged player is an equilibrium. Any other action profile is not an equilibrium, since if at least $n-1$ players agree on a player who is not privileged then that player can deviate profitably by announcing somebody else; if there is no set of $n-1$ players who agree then there is at least one privileged player who can deviate profitably by announcing a larger number than anyone else.

◇ EXAMPLE 190.1 (*Solomon's predicament*) Consider again Solomon's predicament, described in Example 186.3. Assume that the object of dispute has monetary value to the two players and that Solomon may assign the object to one of the players, or to neither of them, and may also impose fines on them. The set of outcomes is then the set of triples (x, m_1, m_2) where either $x = 0$ (the object is not given to either player) or $x \in \{1, 2\}$ (the object is given to player x) and m_i is a fine imposed on player i. Player i's payoff if he gets the object is $v_H - m_i$ if he is the legitimate owner of the object and $v_L - m_i$ if he is not, where $v_H > v_L > 0$; it is $-m_i$ if he does not get the object. There are two possible preference profiles, $\succsim$ in which player 1 is the legitimate owner and $\succsim'$ in which player 2 is.

King Solomon wishes to implement the choice function f for which $f(\succsim) = (1, 0, 0)$ and $f(\succsim') = (2, 0, 0)$. This function is monotonic: for example $(1, 0, 0) \succ_2 (2, 0, (v_H + v_L)/2)$ and $(2, 0, (v_H + v_L)/2) \succ'_2 (1, 0, 0)$. Proposition 187.2 does not apply since there are only two players. However, the following game form (which is simpler than that in the proof of the proposition) implements f: each player has three actions, and the outcome function is that given in Figure 190.1, where $M = (v_H + v_L)/2$ and $\epsilon > 0$ is small enough. (The action "Mine+" can be interpreted

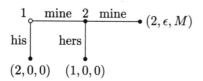

Figure 191.1 An extensive game form that implements the choice function given in Example 190.1. The vector near each terminal history is the outcome associated with that history.

as an impudent demand for the object, which is penalized if the other player does not dispute the ownership.)

Given our interest in the structure of the game forms that we construct, the fact that the game form in this example is simple and lacks a "shouting" component is attractive. In the next section (see Example 191.2) we show that the choice function in the example can be implemented by an even simpler scheme.

☐ EXERCISE 191.1 Consider the case in which there are two individuals. Let $N = \{1,2\}$ and $C = \{a,b,c\}$, and suppose that there are two possible preference profiles, $\succsim$ with $a \succ_1 c \succ_1 b$ and $c \succ_2 b \succ_2 a$ and $\succsim'$ with $c \succ'_1 a \succ'_1 b$ and $b \succ'_2 c \succ'_2 a$. Show that the choice function f defined by $f(\succsim) = a$ and $f(\succsim') = b$ is monotonic but not Nash-implementable.

10.5 Subgame Perfect Equilibrium Implementation

Finally, we turn to the case in which the planner uses extensive game forms with perfect information and assumes that for any preference profile the outcome of the game may be any subgame perfect equilibrium (SPE). To motivate the possibilities for implementing choice rules in this case, consider Solomon's quandary once again.

◇ EXAMPLE 191.2 (*Solomon's predicament*) The choice function f given in the previous example (190.1) is SPE-implemented by the following game form. First player 1 is asked whether the object is hers. If she says "no" then the object is given to player 2. If she says "yes" then player 2 is asked if he is the owner. If he says "no" then the object is given to player 1, while if he says "yes" then he obtains the object and must pay a fine of M satisfying $v_L < M < v_H$ while player 1 has to pay a small fine $\epsilon > 0$. This game form is illustrated in Figure 191.1 (in which outcomes, not payoffs, are shown near the terminal histories).

If player 1 is the legitimate owner (i.e. the preference profile is $\succsim$) then the game has a unique subgame perfect equilibrium, in which player 2 chooses "hers" and player 1 chooses "mine", achieving the desirable outcome $(1, 0, 0)$. If player 2 is the real owner then the game has a unique subgame perfect equilibrium, in which he chooses "mine" and player 1 chooses "his", yielding the outcome $(2, 0, 0)$. Thus the game SPE-implements the choice function given in Example 190.1.

The key idea in the game form described in this example is that player 2 is confronted with a choice that leads him to choose truthfully. If he does so then player 1 is faced with a choice that leads her to choose truthfully also. The tricks used in the literature to construct game forms to SPE-implement choice functions in other contexts are in the same spirit. In the remainder of the chapter we present a result that demonstrates the richness of the possibilities for SPE-implementation.

Let C^* be a set of deterministic consequences. We study the case in which the set C of outcomes has the form

$$C = \{(L, m): L \text{ is a lottery over } C^* \text{ and } m \in \mathbb{R}^N\}. \qquad (192.1)$$

If $(L, m) \in C$ then we interpret m_i as a fine paid by player i. (Note that m_i is *not* transferred to another player.)

We assume that for each player i there is a payoff function $u_i: C^* \to \mathbb{R}$ such that player i's preference relation over C is represented by the function $E_L(u_i(c^*)) - m_i$; we identify a preference profile with a profile $(u_i)_{i \in N}$ of such payoff functions and denote $E_L u_i(c^*)$ simply by $u_i(L)$. We assume further that $\mathcal{P} = U^N$, where U is a finite set that excludes the constant function. The set $\mathcal{G}$ of game forms that we consider is the set of extensive game forms with perfect information with consequences in C.

The notion of implementation that we explore is weaker than those studied previously: we construct a game form $\Gamma \in \mathcal{G}$ with the property that for any preference profile $u \in \mathcal{P}$ the game $\langle \Gamma, u \rangle$ has a unique subgame perfect equilibrium in which the desired alternative is realized with very high probability, though not necessarily with certainty. More precisely, we say that a choice function $f: \mathcal{P} \to C^*$ is **virtually SPE-implementable** if for any $\epsilon > 0$ there is an extensive game form $\Gamma \in \mathcal{G}$ such that for any preference profile $u \in \mathcal{P}$ the extensive game $\langle \Gamma, u \rangle$ has a unique subgame perfect equilibrium, in which the outcome is $f(u)$ with probability at least $1 - \epsilon$.

■ PROPOSITION 193.1 *Let C^* be a set of deterministic consequences. Let $\langle N, C, \mathcal{P}, \mathcal{G} \rangle$ be an environment in which $|N| \geq 3$, C is given by (192.1), $\mathcal{P} = U^N$, where U is the (finite) set of payoff functions described above, and $\mathcal{G}$ is the set of extensive game forms with perfect information and consequences in C. Then every choice function $f: \mathcal{P} \to C^*$ is virtually SPE-implementable.*

Proof. First note that since for no payoff function in U are all outcomes indifferent, for any pair (v, v') of distinct payoff functions there is a pair $(L(v, v'), L'(v, v'))$ of lotteries over C^* such that $v(L(v, v')) > v(L'(v, v'))$ and $v'(L'(v, v')) > v'(L(v, v'))$. (A player's choice between the lotteries $L(v, v')$ and $L'(v, v')$ thus indicates whether his payoff function is v or v'.) For any triple (u, v, v') of payoff functions let $L^*(u, v, v')$ be the member of the set $\{L(v, v'), L'(v, v')\}$ that is preferred by u. Then for any pair (v, v') of payoff functions we have $u(L^*(u, v, v')) = \max\{u(L(v, v')), u(L'(v, v'))\}$, so that $u(L^*(u, v, v')) \geq u(L^*(u', v, v'))$ for any payoff function u'. Further, $u(L^*(u, u, u')) > u(L^*(u', u, u'))$.

Now suppose that for some pair (v, v') a player who announces the payoff function u is given the lottery $L^*(u, v, v')$. Let B be the minimum, over all pairs (u, u') of distinct payoff functions, of the average gain, over all pairs (v, v'), of any player with payoff function u from announcing u rather than u':

$$B = \min_{(u,u') \in \mathcal{W}} \left(\frac{1}{M} \sum_{(v,v') \in \mathcal{W}} \{u(L^*(u, v, v')) - u(L^*(u', v, v'))\} \right),$$

where $\mathcal{W}$ is the set of all pairs of distinct payoff functions and $M = |U|(|U| - 1)$ (the number of members of $\mathcal{W}$). By the argument above we have $B > 0$.

For every $\epsilon > 0$ we construct a game form that has $K + 1$ stages (K being defined below). Each stage consists of $|N|$ substages. Let $N = \{1, \ldots, n\}$. In substage i of each of the first K stages player i announces a preference *profile* (a member of U^N); in substage i of stage $K + 1$ player i announces a payoff *function* (a member of U).

For any terminal history the outcome, which consists of a lottery and a profile of fines, is defined as follows. Each stage k for $k = 1, \ldots, K$ contributes to the lottery a consequence with probability $(1 - \epsilon)/K$. If in stage k all the players except possibly one announce the same preference profile, say (u_i), then this consequence is $f((u_i))$; otherwise it is some fixed consequence $c^* \in C^*$.

Each substage of the last stage contributes the probability $\epsilon/|N|$ to the lottery. This probability is split into M equal parts, each corresponding to a pair of payoff functions. The probability $\epsilon/|N|M$ that corresponds to $(i, (v, v')) \in N \times \mathcal{W}$ is assigned to the lottery $L^*(u_i', v, v')$, where u_i' is the payoff function that player i announces in stage $K + 1$.

As for the fines, a player has to pay $\delta > 0$ if he is the last player in the first K stages to announce a preference profile different from the profile of announcements in stage $K + 1$. In addition, a player has to pay a fine of δ/K for each stage in the first K in which all the other players announce the same profile, different from the one that he announces. (In order for the odd player out to be well-defined we need at least three players.)

Finally, we choose δ so that $\epsilon B/|N| > \delta$ and K so that $(1 - \epsilon)D/K + \delta/K < \delta$, where

$$D = \max_{v,c,c'}\{v(c) - v(c'): v \in U, \ c \in C^*, \text{ and } c' \in C^*\}.$$

We now show that for any $(u_i) \in U^N$ the game $\langle \Gamma, (u_i) \rangle$, where Γ is the game form described above, has a unique subgame perfect equilibrium, in which the outcome is $f((u_i))$ with probability at least $1 - \epsilon$. We first show that after every history in every subgame perfect equilibrium each player i announces his true payoff function in stage $K + 1$. If player i announces a false payoff function at this stage then, relative to the case in which he announces his true payoff function, there are two changes in the outcome. First, the lotteries contributed to the outcome by substage i of stage $K + 1$ change, reducing player i's expected payoff by at least $\epsilon B/|N|$. Second, player i may avoid a fine of δ if by changing his announcement in the last period he avoids being the last player to announce a preference profile in one of the first K stages that is different from the profile of announcements in the final stage. Since $\epsilon B/|N| > \delta$ the net effect is that the best action for any player is to announce his true payoff function in the final period, whatever history precedes his decision.

We now show that in any subgame perfect equilibrium all players announce the true preference profile (u_i) in each of the first K stages. Suppose to the contrary that some player does not do so; let player i in stage k be the last player not to do so. We argue that player i can increase his payoff by deviating and announcing the true preference profile (u_i). There are two cases to consider.

- If no other player announces a profile different from (u_i) in stage k then player i's deviation has no effect on the outcome; it reduces the fine he has to pay by δ/K, since he no longer announces a profile different from that announced by the other players, and may further reduce his fine by δ (if he is no longer the last player to announce a profile different from (u_i)).

- If some other player announces a profile different from (u_i) in stage k then the component of the final lottery attributable to stage k may change, reducing player i's payoff by at most $(1 - \epsilon)D/K$. In addition he may become the odd player out at stage k and be fined δ/K. At the same time he avoids the fine δ (since he is definitely not the last player to announce a profile different from (u_i)). Since $(1 - \epsilon)D/K + \delta/K < \delta$, the net effect is that the deviation is profitable.

We conclude that in every subgame perfect equilibrium every player, after every history at which he has to announce a preference profile, announces the true preference profile, so that the outcome of the game assigns probability of at least $1 - \epsilon$ to $f((u_i))$.   □

The game form constructed in this proof is based on two ideas. Stage $K + 1$ is designed so that it is dominant for every player to announce his true payoff function. In the earlier stages a player may wish to announce a preference profile different from the true one, since by doing so he may affect the final outcome to his advantage; but no player wants to be the *last* to do so, with the consequence that no player ever does so.

Notes

The Gibbard–Satterthwaite theorem (181.2) appears in Gibbard (1973) and Satterthwaite (1975). For alternative proofs see Schmeidler and Sonnenschein (1978) and Barberá (1983). Proposition 184.2 is due to Groves and Loeb (1975); the result in Exercise 185.1 is due to Green and Laffont (1977). Maskin first proved Proposition 187.2 (see Maskin (1985)); the proof that we give is due to Repullo (1987). The discussion in Section 10.5 is based on Abreu and Matsushima (1992), who prove a result equivalent to Proposition 193.1 for implementation via iterated elimination of strictly dominated strategies in strategic game forms; the variant that we present is that of Glazer and Perry (1996). The analysis of Solomon's predicament in Examples 186.3, 190.1, and 191.2 first appeared in Glazer and Ma (1989).

For a characterization of choice functions that are SPE-implementable see Moore and Repullo (1988).

In writing this chapter we benefited from Moore (1992) (a survey of the literature) and from unpublished lecture notes by Repullo.

III Extensive Games with Imperfect Information

The model of an extensive game with imperfect information allows a player, when taking an action, to have only partial information about the actions taken previously. The model is rich; it encompasses not only situations in which a player is imperfectly informed about the other players' previous actions, but also, for example, situations in which during the course of the game a player forgets an action that he previously took and situations in which a player is uncertain about whether another player has acted.

We devote Chapter 11 to an exploration of the concept of an extensive game with imperfect information, leaving until Chapter 12 a study of the main solution concept for such games, namely the notion of sequential equilibrium.

11 Extensive Games with Imperfect Information

In this chapter we explore the concept of an extensive game with imperfect information, in which each player, when taking an action, may have only partial information about the actions taken previously.

11.1 Extensive Games with Imperfect Information

11.1.1 Introduction

In each of the models we studied previously there is a sense in which the players are not perfectly informed when making their choices. In a strategic game a player, when taking an action, does not know the actions that the other players take. In a Bayesian game a player knows neither the other players' private information nor the actions that they take. In an extensive game with perfect information a player does not know the future moves planned by the other players.

The model that we study here—an extensive game with imperfect information—differs in that the players may in addition be imperfectly informed about some (or all) of the choices that have *already* been made. We analyze the model by assuming, as we did previously, that each player, when choosing an action, forms an expectation about the unknowns. However, these expectations differ from those we considered before. Unlike those in strategic games, they are not derived solely from the players' equilibrium behavior, since the players may face situations inconsistent with that behavior. Unlike those in Bayesian games, they are not deduced solely from the equilibrium behavior and the exogenous information about the moves of chance. Finally, unlike those in extensive games with perfect information, they relate not only to the other players' future behavior but also to events that happened in the past.

11.1.2 Definitions

The following definition generalizes that of an extensive game with perfect information (89.1) to allow players to be imperfectly informed about past events when taking actions. It also allows for exogenous uncertainty: some moves may be made by "chance" (see Section 6.3.1). It does not incorporate the other generalization of the definition of an extensive game with perfect information that we discussed in Section 6.3, in which more than one player may move after any history (see however the discussion after Example 202.1).

▸ DEFINITION 200.1 An **extensive game** has the following components.

- A finite set N (the set of **players**).

- A set H of sequences (finite or infinite) that satisfies the following three properties.

  - The empty sequence $\varnothing$ is a member of H.

  - If $(a^k)_{k=1,...,K} \in H$ (where K may be infinite) and $L < K$ then $(a^k)_{k=1,...,L} \in H$.

  - If an infinite sequence $(a^k)_{k=1}^{\infty}$ satisfies $(a^k)_{k=1,...,L} \in H$ for every positive integer L then $(a^k)_{k=1}^{\infty} \in H$.

  (Each member of H is a **history**; each component of a history is an **action** taken by a player.) A history $(a^k)_{k=1,...,K} \in H$ is **terminal** if it is infinite or if there is no a^{K+1} such that $(a^k)_{k=1,...,K+1} \in H$. The set of actions available after the nonterminal history h is denoted $A(h) = \{a : (h, a) \in H\}$ and the set of terminal histories is denoted Z.

- A function P that assigns to each nonterminal history (each member of $H \setminus Z$) a member of $N \cup \{c\}$. (P is the **player function**, $P(h)$ being the player who takes an action after the history h. If $P(h) = c$ then chance determines the action taken after the history h.)

- A function f_c that associates with every history h for which $P(h) = c$ a probability measure $f_c(\cdot|h)$ on $A(h)$, where each such probability measure is independent of every other such measure. ($f_c(a|h)$ is the probability that a occurs after the history h.)

- For each player $i \in N$ a partition $\mathcal{I}_i$ of $\{h \in H : P(h) = i\}$ with the property that $A(h) = A(h')$ whenever h and h' are in the same member of the partition. For $I_i \in \mathcal{I}_i$ we denote by $A(I_i)$ the set $A(h)$ and by $P(I_i)$ the player $P(h)$ for any $h \in I_i$. ($\mathcal{I}_i$ is the **information partition** of player i; a set $I_i \in \mathcal{I}_i$ is an **information set** of player i.)

- For each player $i \in N$ a preference relation $\succsim_i$ on lotteries over Z (the **preference relation** of player i) that can be represented as the expected value of a payoff function defined on Z.

We refer to a tuple $\langle N, H, P, f_c, (\mathcal{I}_i)_{i \in N} \rangle$ (which excludes the players' preferences) whose components satisfy the conditions in the definition as an **extensive game form**.

Relative to the definition of an extensive game with perfect information and chance moves (see Section 6.3.1), the new element is the collection $(\mathcal{I}_i)_{i \in N}$ of information partitions. We interpret the histories in any given member of $\mathcal{I}_i$ to be indistinguishable to player i. Thus the game models a situation in which after any history $h \in I_i \in \mathcal{I}_i$ player i is informed that some history in I_i has occurred but is not informed that the history h has occurred. The condition that $A(h) = A(h')$ whenever h and h' are in the same member of $\mathcal{I}_i$ captures the idea that if $A(h) \neq A(h')$ then player i could deduce, when he faced $A(h)$, that the history was not h', contrary to our interpretation of $\mathcal{I}_i$. (Note that Definition 200.1, unlike the standard definition of an extensive game, does not rule out the possibility that an information set contains two histories h and h' where $h' = (h, a^1, \ldots, a^K)$ for some sequence of actions $(a^1, \ldots, a^K)$.)

Each player's information partition is a primitive of the game; a player can distinguish between histories in different members of his partition without having to make any inferences from the actions that he observes. As the game is played, a participant may be able, given his conjectures about the other players' behavior, to make inferences that refine this information. Suppose, for example, that the first move of a game is made by player 1, who chooses between a and b, and the second move is made by player 2, one of whose information sets is $\{a, b\}$. We interpret this game to model a situation in which player 2 does not observe the choice of player 1: when making his move, he is not informed whether player 1 chose a or b. Nevertheless, when making his move player 2 may infer (from his knowledge of a steady state or from introspection about player 1) that the history is a, even though he does not observe the action chosen by player 1.

Each player's preference relation is defined over *lotteries* on the set of terminal histories, since even if the players' actions are deterministic the chance moves that the model allows induce such lotteries.

Note that Definition 200.1 extends our definition of an extensive game with perfect information and chance moves (see Section 6.3.1) in the

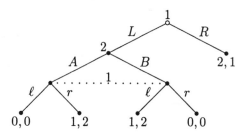

Figure 202.1 An extensive game with imperfect information.

sense that the extensive game with perfect information and chance moves $\langle N, H, P, f_c, (\succsim_i)_{i \in N} \rangle$ may naturally be identified with the extensive game $\langle N, H, P, f_c, (\mathcal{I}_i)_{i \in N}, (\succsim_i)_{i \in N} \rangle$ in which every member of the information partition of every player is a singleton.

◇ EXAMPLE 202.1 Figure 202.1 shows an example of an extensive game with imperfect information. In this game player 1 moves first, choosing between L and R. If she chooses R, the game ends. If she chooses L, player 2 moves; he is informed that player 1 chose L, and chooses A or B. In either case it is player 1's turn to move, and when doing so she is not informed whether player 2 chose A or B, a fact indicated in the figure by the dotted line connecting the ends of the histories after which player 1 has to move for the second time, choosing an action from the set $\{\ell, r\}$. Formally, we have $P(\varnothing) = P(L, A) = P(L, B) = 1$, $P(L) = 2$, $\mathcal{I}_1 = \{\{\varnothing\}, \{(L, A), (L, B)\}\}$, and $\mathcal{I}_2 = \{\{L\}\}$ (player 1 has two information sets and player 2 has one). The numbers under the terminal histories are the players' payoffs. (The first number in each pair is player 1's payoff and the second is player 2's payoff.)

In Definition 200.1 we do not allow more than one player to move after any history. However, there is a sense in which an extensive game as we have defined it can model such a situation. To see this, consider the example above. After player 1 chooses L, the situation in which players 1 and 2 are involved is essentially the same as that captured by a game with perfect information in which they choose actions simultaneously. (This is the reason that in much of the literature the definition of an extensive game with perfect information does not include the possibility of simultaneous moves.)

A player's strategy in an extensive game with perfect information is a function that specifies an action for every history after which the player chooses an action (Definition 92.1). The following definition is

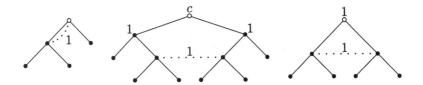

Figure 203.1 Three one-player extensive games with imperfect recall.

an extension to a general extensive game; since we later consider the possibility that the players may randomize, we add the qualifier "pure".

▶ DEFINITION 203.1 A **pure strategy of player** $i \in N$ in an extensive game $\langle N, H, P, f_c, (\mathcal{I}_i), (\succsim_i) \rangle$ is a function that assigns an action in $A(I_i)$ to each information set $I_i \in \mathcal{I}_i$.

As for an extensive game with perfect information, we can associate with any extensive game a strategic game; see the definitions of the strategic form (94.1) and reduced strategic form (95.1). (Note that the outcome of a strategy profile here may be a lottery over the terminal histories, since we allow moves of chance.)

11.1.3 Perfect and Imperfect Recall

The model of an extensive game is capable of capturing a wide range of informational environments. In particular, it can capture situations in which at some points players forget what they knew earlier. We refer to games in which at every point every player remembers whatever he knew in the past as *games with perfect recall*. To define such games formally, let $\langle N, H, P, f_c, (\mathcal{I}_i) \rangle$ be an extensive game form and let $X_i(h)$ be the record of player i's experience along the history h: $X_i(h)$ is the sequence consisting of the information sets that the player encounters in the history h and the actions that he takes at them, in the order that these events occur. In the game in Figure 202.1, for example, $X_1((L, A)) = (\varnothing, L, \{(L, A), (L, B)\})$.

? EXERCISE 203.2 Give a formal definition of $X_i(h)$.

▶ DEFINITION 203.3 An extensive game form has **perfect recall** if for each player i we have $X_i(h) = X_i(h')$ whenever the histories h and h' are in the same information set of player i.

The game in Figure 202.1 has perfect recall, while the three (one-player) game forms in Figure 203.1 do not. In the left-hand game a player does not know if she has made a choice or not: when choosing an

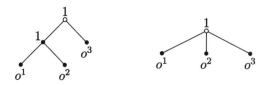

Figure 204.1 Two distinct one-player extensive games that appear to model the same situation.

action she does not know whether she is at the beginning of the game or has already chosen her left-hand action. In the middle game the player forgets something that she previously knew: when making a choice at her last information set she is not informed of the action of chance, though she was so informed when she made her previous choice. In the right-hand game she does not remember the action she took in the past.

The literature on games with imperfect recall is very small. An example of a game theoretic treatment of a situation with imperfect recall is that of the machine games in Chapter 9. In the underlying repeated game that a machine game models, each player, when taking an action, is not informed of past events, including his own previous actions. The size of his memory depends on the structure of his machine. More memory requires more states; since states are costly, even in equilibrium a player still may imperfectly recall his own past actions.

11.2 Principles for the Equivalence of Extensive Games

Some extensive games appear to represent the same strategic situation as others. Consider, for example, the two one-player games in Figure 204.1. (In these games, as in the others in this section, we associate letters with terminal histories. If two terminal histories are assigned the same letter then the two histories represent the same event; in particular, all the players are indifferent between them.) Formally, the two games are different: in the left-hand game player 1 makes two decisions, while in the right-hand game she makes only one. However, principles of rationality suggest that the two games model the same situation.

We now give further examples of pairs of games that arguably represent the same situation and discuss some principles that generalize these examples. We do not argue that these principles should be taken as axioms; we simply believe that studying them illuminates the meaning of an extensive game, especially one with imperfect information.

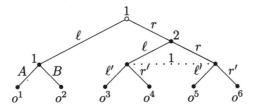

Figure 205.1 The game Γ_1.

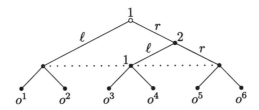

Figure 205.2 The game Γ_2, equivalent to Γ_1 according to the inflation–deflation principle.

The four principles that we consider all preserve the reduced strategic form of the game: if one extensive game is equivalent to another according to the principles then the reduced strategic forms of the two games are the same. Thus a solution concept that does not depend solely on the reduced strategic form may assign different outcomes to games that are equivalent according to the principles; to justify such a solution concept one has to argue that at least one of the principles is inappropriate.

Let Γ_1 be the game in Figure 205.1. The principles that we discuss claim that this game is equivalent to four other extensive games, as follows.

Inflation–Deflation According to this principle Γ_1 is equivalent to the game Γ_2 in Figure 205.2. In Γ_2 player 1 has imperfect recall: at her second information set she is not informed whether she chose r or ℓ at the start of the game. That is, the three histories ℓ, (r, ℓ), and (r, r) are all in the same information set in Γ_2, while in Γ_1 the history ℓ lies in one information set and the histories (r, ℓ) and (r, r) lie in another. The interpretation that we have given to a game like Γ_2 is that player 1, when acting at the end of the game, has forgotten the action she took at the beginning of the game. However, another interpretation of an information set is that it represents the information about history that is inherent in the structure of the game, information that may be refined

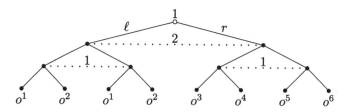

Figure 206.1 The game Γ_3, equivalent to Γ_1 according to the principle of addition of a superfluous move.

by inferences that the players may make. Under this interpretation a player always remembers what he knew and did in the past and may obtain information by making inferences from this knowledge. Indeed, the argument that Γ_1 and Γ_2 are equivalent relies on the assumption that player 1 is capable of making such inferences. The fact that she is informed whether the history was ℓ or a member of $\{(r, \ell), (r, r)\}$ is irrelevant to her strategic calculations, according to the argument, since in any case she can infer this information from her knowledge of her action at the start of the game. Under this interpretation it is inappropriate to refer to a game like that in Figure 205.2 as having "imperfect recall": the information sets reflect imperfections in the information inherent in the situation that can be overridden by the players' abilities to remember their past experience.

Formally, according to the *inflation–deflation principle* the extensive game Γ is equivalent to the extensive game Γ' if Γ' differs from Γ only in that there is an information set of some player i in Γ that is a union of information sets of player i in Γ' with the following property: any two histories h and h' in different members of the union have subhistories that are in the same information set of player i and player i's action at this information set is different in h and h'. (To relate this to the examples above, let $\Gamma = \Gamma_2$, $\Gamma' = \Gamma_1$, and $i = 1$.)

Addition of a Superfluous Move   According to this principle Γ_1 is equivalent to the game Γ_3 in Figure 206.1. The argument is as follows. If in the game Γ_3 player 1 chooses ℓ at the start of the game then the action of player 2 is irrelevant, since it has no effect on the outcome (note the outcomes in the bottom left-hand part of the game). Thus in Γ_3 whether player 2 is informed of player 1's choice at the start of the game should make no difference to his choice.

Formally the *principle of addition of a superfluous move* is the following. Let Γ be an extensive game, let $P(h) = i$, and let $a \in A(h)$. Suppose

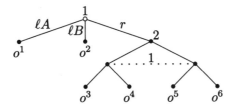

Figure 207.1 The game Γ_4, equivalent to Γ_1 according to the principle of coalescing of moves.

that for any sequence h' of actions (including the empty sequence) that follows the history (h, a) and for any $b \in A(h)$ we have

- $(h, a, h') \in H$ if and only if $(h, b, h') \in H$, and (h, a, h') is terminal if and only if (h, b, h') is terminal
- if both (h, a, h') and (h, b, h') are terminal then $(h, a, h') \sim_i (h, b, h')$ for all $i \in N$
- if both (h, a, h') and (h, b, h') are nonterminal then they are in the same information set.

Then Γ is equivalent to the game Γ' that differs from Γ only in that (i) all histories of the form (h, c, h') for $c \in A(h)$ are replaced by the single history (h, h'), (ii) if the information set I_i that contains the history h in Γ is not a singleton then h is excluded from I_i in Γ', (iii) the player who is assigned to the history (h, h') in Γ' is the one who is assigned to (h, a, h') in Γ, (iv) (h, h') and (h, h'') are in the same information set of Γ' if and only if (h, a, h') and (h, a, h'') are in the same information set of Γ, and (v) the players' preferences are modified accordingly. (Note that Γ is the game that has the superfluous move, which is removed to create Γ'. To relate the definition to Γ_1 and Γ_3, let $\Gamma = \Gamma_3$, $\Gamma' = \Gamma_1$, $i = 2$, and $h = \ell$, and let a be one of the actions of player 2.)

Coalescing of moves According to this principle, Γ_1 is equivalent to the game Γ_4 in Figure 207.1. In Γ_1 player 1 first chooses between ℓ and r, then chooses between A and B in the event that she chooses ℓ. The idea is that this decision problem is equivalent to that of deciding between ℓA, ℓB, and r, as in Γ_4. The argument is that if player 1 is rational then her choice at the start of Γ_1 requires her to compare the outcomes of choosing ℓ and r; to determine the outcome of choosing ℓ requires her to plan at the start of the game whether to choose A or B.

Formally the *principle of coalescing of moves* is the following. Let Γ be an extensive game and let $P(h) = i$, with $h \in I_i$. Let $a \in A(I_i)$ and

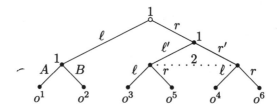

Figure 208.1 The game Γ_5, equivalent to Γ_1 according to the principle of inter-change of moves.

suppose that $\{(h', a): h' \in I_i\} = I_i'$ is an information set of player i. Let Γ' be the game that differs from Γ only in that the information set I_i' is deleted, for all $h' \in I_i$ the history (h', a) is deleted and every history (h', a, b, h'') where $b \in A(h', a)$ is replaced by the history (h', ab, h'') where ab is a new action (that is not a member of $A(h')$), and the information sets, player function, and players' preferences are changed accordingly. Then Γ and Γ' are equivalent. (In the example let $\Gamma = \Gamma_1$, $\Gamma' = \Gamma_4$, $h = \varnothing$, $i = 1$, and $a = \ell$.)

Interchange of moves   According to this principle Γ_1 is equivalent to the game Γ_5 in Figure 208.1. The idea is that the order of play is immaterial if one player does not have any information about the other player's action when making his choice.

  Formally the *principle of interchange of moves* is the following (which allows transformations more general than that from Γ_1 to Γ_5). Let Γ be an extensive game and let $h \in I_i$, an information set of player i. Suppose that for all histories h' in some subset H' of I_i the player who takes an action after i has done so is j, who is not informed of the action that i takes at h'. That is, suppose that $(h', a) \in I_j$ for all $h' \in H'$ and all $a \in A(h')$, where I_j is an information set of player j. The information set I_j may contain other histories; let H'' be the subset of I_j consisting of histories of the form (h', a) for some $h' \in H'$. Then Γ is equivalent to the extensive game in which every history of the type (h', a, b) for $h' \in H'$ is replaced by (h', b, a), the information set I_i of player i is replaced by the union of $I_i \setminus H'$ and all histories of the form (h', b) for $h' \in H'$ and $b \in A(h', a)$, and the information set I_j of player j is replaced by $(I_j \setminus H'') \cup H'$. (In the example we have $\Gamma = \Gamma_1$, $\Gamma' = \Gamma_5$, $h = r$, $i = 2$, $j = 1$, $H' = I_2 = \{r\}$, and $H'' = I_1 = \{(r, \ell), (r, r)\}$.)

[?] EXERCISE 208.1 Formulate the principles of coalescing of moves and inflation–deflation for one-player extensive games and show that every one-player extensive game with imperfect information and no chance moves (but possibly with imperfect recall) in which no information set contains both a history h and a subhistory of h is equivalent to a decision problem with a single nonterminal history. (The result holds even for games with chance moves, which are excluded only for simplicity.)

Thompson (1952) shows that these four transformations preserve the reduced strategic form. He restricts attention to finite extensive games in which no information set contains both a history h and some subhistory of h and shows that if any two such games have the same reduced strategic form then one can be obtained from the other by a sequence of the four transformations. We are not aware of any elegant proof of this result. We simply give an example to illustrate the procedure: starting with the game at the top of Figure 210.1 the series of transformations shown in Figures 210.1 and 211.1 leads to the extensive game with perfect information and simultaneous moves at the bottom of Figure 211.1.

11.3 Framing Effects and the Equivalence of Extensive Games

The principles of equivalence between extensive games discussed in the previous section are based on a conception of rationality that ignores framing effects. This conception is inconsistent with the findings of psychologists that even minor variations in the framing of a problem may dramatically affect the participants' behavior (see for example Tversky and Kahneman (1986)).

To illustrate that games that are equivalent according to these principles may differ in their framing and lead to different behavior, consider the strictly competitive games in Figure 212.1. The middle game is obtained from the top one by adding a superfluous move; the bottom game is the strategic form of each extensive game.

A reasonable principle for behavior in these games is that of maxminimizing. However, this principle yields different outcomes in the games. In the bottom game player 1's maxminimizer is the pure strategy r while in the top game the logic of maxminimizing directs her towards using the mixed strategy $(\frac{1}{2}, \frac{1}{2})$ (since she is informed that chance played right).

This example was originally proposed as a demonstration of the difficulties with the principle of maxminimizing, but we view it as a part of a deeper problem: how to analyze game theoretic situations taking into account framing effects, an intriguing issue of current research.

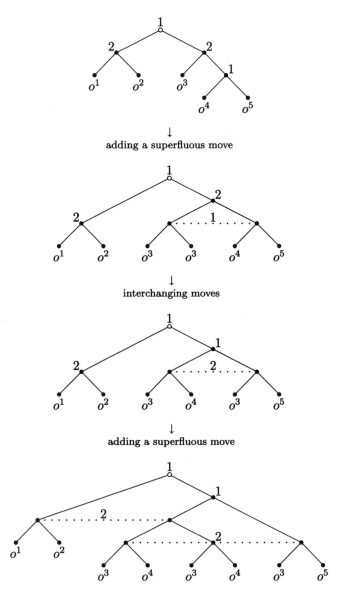

Figure 210.1 The first three transformations in a series that converts the top game into an extensive game with perfect information and simultaneous moves. The transformations continue in Figure 211.1.

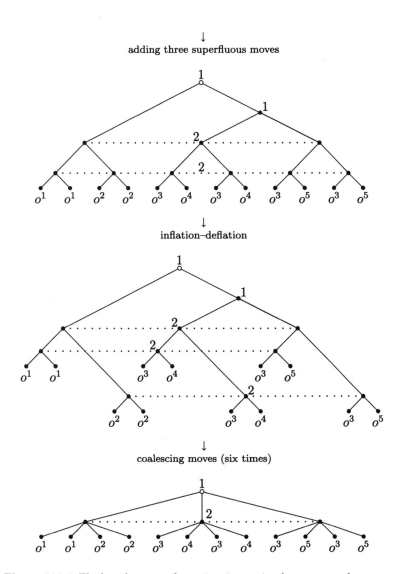

Figure 211.1 The last three transformations in a series that converts the top game in Figure 210.1 into the extensive game with perfect information and simultaneous moves in the bottom of this figure.

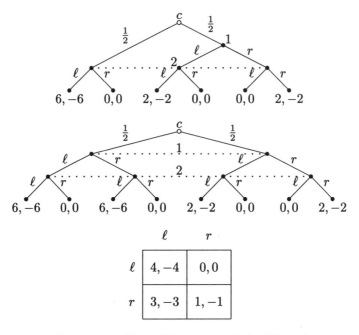

Figure 212.1 Three games. The middle game is obtained from the top game by adding a superfluous move. The bottom game is the strategic form of both the top and middle games.

11.4   Mixed and Behavioral Strategies

In Definition 203.1 we defined the notion of a pure strategy in an extensive game. There are two ways to model the possibility that a player's actions in such a game depend upon random factors.

▶ DEFINITION 212.1   A **mixed strategy of player** i in an extensive game $\langle N, H, P, f_c, (\mathcal{I}_i), (\succsim_i) \rangle$ is a probability measure over the set of player i's pure strategies. A **behavioral strategy of player** i is a collection $(\beta_i(I_i))_{I_i \in \mathcal{I}_i}$ of independent probability measures, where $\beta_i(I_i)$ is a probability measure over $A(I_i)$.

For any history $h \in I_i \in \mathcal{I}_i$ and action $a \in A(h)$ we denote by $\beta_i(h)(a)$ the probability $\beta_i(I_i)(a)$ assigned by $\beta_i(I_i)$ to the action a.

Thus, as in a strategic game, a mixed strategy of player i is a probability measure over player i's set of pure strategies. By contrast, a behavioral strategy specifies a probability measure over the actions available to player i at each of his information sets. The two notions reflect two different ways in which a player might randomize: he might randomly select a pure strategy, or he might plan a collection of randomizations, one for each of the points at which he has to take an action. The difference between the two notions can be appreciated by examining the game in Figure 202.1. In this game player 1 has two information sets, at each of which she has two possible actions. Thus she has four pure strategies, which assign to the information sets $\{\varnothing\}$ and $\{(L, A), (L, B)\}$ respectively the actions L and ℓ, L and r, R and ℓ, and R and r. (If you are puzzled by the last two strategies, read (or reread) Section 6.1.2.) A mixed strategy of player 1 is a probability distribution over these four pure strategies. By contrast, a behavioral strategy of player 1 is a pair of probability distributions, one for each information set; the first is a distribution over $\{L, R\}$ and the second is a distribution over $\{\ell, r\}$.

In describing a mixed or behavioral strategy we have used the language of the naïve interpretation of actions that depend on random factors, according to which a player consciously chooses a random device (see Section 3.2). When discussing mixed strategies in Chapter 3, we describe some other interpretations, which have analogs here. For example, we may think of the mixed and behavioral strategies of player i as two ways of describing the other players' beliefs about player i's behavior. The other players can organize their beliefs in two ways: they can form conjectures about player i's pure strategy in the entire game (a mixed strategy), or they can form a collection of independent beliefs about player i's actions for each history after which he has to act (a behavioral strategy).

For any profile $\sigma = (\sigma_i)_{i \in N}$ of either mixed or behavioral strategies in an extensive game, we define the **outcome** $O(\sigma)$ **of** σ to be the probability distribution over the terminal histories that results when each player i follows the precepts of σ_i. For a finite game this outcome is defined precisely as follows. For any history $h = (a^1, \dots, a^k)$ define a pure strategy s_i of player i to be *consistent* with h if for every subhistory $(a^1, \dots, a^\ell)$ of h for which $P(a^1, \dots, a^\ell) = i$ we have $s_i(a^1, \dots, a^\ell) = a^{\ell+1}$. For any history h let $\pi_i(h)$ be the sum of the probabilities according to σ_i of all the pure strategies of player i that are consistent with h. (Thus for example if h is a history in which player i never moves then $\pi_i(h) = 1$.) Then for any profile σ of mixed strategies the probability that $O(\sigma)$ assigns to any terminal history h is $\Pi_{i \in N \cup \{c\}} \pi_i(h)$. For any profile β of behavioral strategies the probability that $O(\beta)$ assigns to the terminal history $h = (a^1, \dots, a^K)$ is $\Pi_{k=0}^{K-1} \beta_{P(a^1, \dots, a^k)}(a^1, \dots, a^k)(a^{k+1})$ (where for $k = 0$ the history $(a^1, \dots, a^k)$ is the initial history).

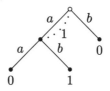

Figure 214.1 A one-player extensive game in which there is a behavioral strategy that is not outcome-equivalent to any mixed strategy.

Two (mixed or behavioral) strategies of any player are **outcome-equivalent** if for every collection of pure strategies of the other players the two strategies induce the same outcome. In the remainder of this section we examine the conditions under which for any mixed strategy there is an outcome-equivalent behavioral strategy and *vice versa*; we show, in particular, that this is so in any game with perfect recall.

We first argue that, in a set of games that includes all those with perfect recall, for any behavioral strategy there is an outcome-equivalent mixed strategy. Consider an extensive game in which no information set contains both some history h and a history of the form (h, h') for some $h' \neq \varnothing$. (Note that this condition is satisfied by any game with perfect recall; it is often included as part of the definition of an extensive game.) For every behavioral strategy β_i of any player i in such a game, the mixed strategy defined as follows is outcome-equivalent: the probability assigned to any pure strategy s_i (which specifies an action $s_i(I_i)$ for every information set $I_i \in \mathcal{I}_i$) is $\Pi_{I_i \in \mathcal{I}_i} \beta_i(I_i)(s_i(I_i))$. (Note that the derivation of this mixed strategy relies on the assumption that the collection $(\beta_i(I_i))_{I_i \in \mathcal{I}_i}$ is independent. Note also that in a game in which some information set contains histories of the form h and (h, h') with $h' \neq \varnothing$ there may be a behavioral strategy for which there is no equivalent mixed strategy: in the game in Figure 214.1, for example, the behavioral strategy that assigns probability $p \in (0, 1)$ to a generates the outcomes (a, a), (a, b), and b with probabilities p^2, $p(1 - p)$, and $1 - p$ respectively, a distribution that cannot be duplicated by any mixed strategy.)

We now show that, in a game with perfect recall, for every mixed strategy there is an outcome-equivalent behavioral strategy.

■ PROPOSITION 214.1 *For any mixed strategy of a player in a finite extensive game with perfect recall there is an outcome-equivalent behavioral strategy.*

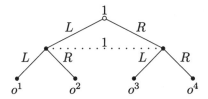

Figure 215.1 An extensive game in which mixed and behavioral strategies are not equivalent.

Proof. Let σ_i be a mixed strategy of player i. As above, for any history h let $\pi_i(h)$ be the sum of the probabilities according to σ_i of all the pure strategies of player i that are consistent with h. Let h and h' be two histories in the same information set I_i of player i, and let $a \in A(h)$. Since the game has perfect recall, the sets of actions of player i in h and h' are the same. Thus $\pi_i(h) = \pi_i(h')$. Since in any pure strategy of player i the action a is taken after h if and only if it is taken after h', we also have $\pi_i(h, a) = \pi_i(h', a)$. Thus we can define a behavioral strategy β_i of player i by $\beta_i(I_i)(a) = \pi_i(h, a)/\pi_i(h)$ for any $h \in I_i$ for which $\pi_i(h) > 0$ (clearly $\sum_{a \in A(h)} \beta_i(I_i)(a) = 1$); how we define $\beta_i(I_i)(a)$ if $\pi_i(h) = 0$ is immaterial.

We claim that β_i is outcome-equivalent to σ_i. Let s_{-i} be a collection of pure strategies for the players other than i. Let h be a terminal history. If h includes moves that are inconsistent with s_{-i} then the probability of h is zero under both σ_i and β_i. Now assume that all the moves of players other than i in h are consistent with s_{-i}. If h includes a move after a subhistory $h' \in I_i$ of h that is inconsistent with σ_i then $\beta_i(I_i)$ assigns probability zero to this move, and thus the probability of h according to β_i is zero. Finally, if h is consistent with σ_i then $\pi_i(h') > 0$ for all subhistories h' of h and the probability of h according to β_i is the product of $\pi_i(h', a)/\pi_i(h')$ over all (h', a) that are subhistories of h; this product is $\pi_i(h)$, the probability of h according to σ_i. $\square$

In a game with imperfect recall there may be a mixed strategy for which there is no outcome-equivalent behavioral strategy, as the one-player game with imperfect recall in Figure 215.1 shows. Consider the mixed strategy in which player 1 chooses LL with probability $\frac{1}{2}$ and RR with probability $\frac{1}{2}$. The outcome of this strategy is the probability distribution $(\frac{1}{2}, 0, 0, \frac{1}{2})$ over the terminal histories. This outcome cannot be achieved by any behavioral strategy: the behavioral strategy $((p, 1 - p), (q, 1 - q))$ induces a distribution over the terminal histories in which

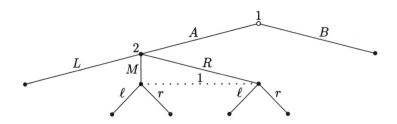

Figure 216.1 The extensive game form for Exercise 216.1.

LR has zero probability only if either $p = 0$ or $q = 1$, in which case the probability of either LL or RR is zero.

? EXERCISE 216.1 Consider the game form in Figure 216.1. Find the behavioral strategy of player 1 that is equivalent to her mixed strategy in which she plays (B, r) with probability 0.4, (B, ℓ) with probability 0.1, and (A, ℓ) with probability 0.5.

11.5 Nash Equilibrium

A **Nash equilibrium in mixed strategies** of an extensive game is (as before) a profile σ^* of mixed strategies with the property that for every player $i \in N$ we have

$$O(\sigma^*_{-i}, \sigma^*_i) \succsim_i O(\sigma^*_{-i}, \sigma_i) \text{ for every mixed strategy } \sigma_i \text{ of player } i.$$

For finite games an equivalent definition of a mixed strategy equilibrium is that every pure strategy in the support of each player's mixed strategy is a best response to the strategies of the other players (cf. Definition 44.1). A **Nash equilibrium in behavioral strategies** is defined analogously.

Given Proposition 214.1, the two definitions are equivalent for games with perfect recall. For games with imperfect recall they are not equivalent, as the game in Figure 214.1 shows. In this game the player is indifferent among all her mixed strategies, which yield her a payoff of 0, while the behavioral strategy that assigns probability p to a yields her a payoff of $p \cdot 0 + p \cdot (1 - p) \cdot 1 + (1 - p)^2 \cdot 0 = p(1 - p)$, so that the best behavioral strategy has $p = \frac{1}{2}$, and yields her a payoff of $\frac{1}{4}$.

In Chapter 6 we argue that the notion of Nash equilibrium is often unsatisfactory in extensive games with perfect information and we introduce the notion of subgame perfect equilibrium to deal with the problems. To extend the ideas behind this notion to general extensive games

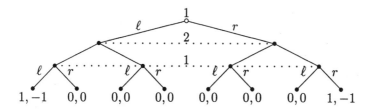

Figure 217.1 The extensive game with imperfect information in Exercise 217.1.

is challenging, mainly because when making a choice at a non-singleton information set a player has to form an expectation about the history that occurred, an expectation that may not be uniquely determined by the equilibrium strategies. The next chapter is devoted to a discussion of this issue.

? EXERCISE 217.1 Consider the strictly competitive extensive game with imperfect recall in Figure 217.1. Show that player 1's best behavioral strategy assures her a payoff of 1 with probability $\frac{1}{4}$, while there is a mixed strategy that assures her the payoff 1 with probability $\frac{1}{2}$.

? EXERCISE 217.2 Let Γ_2 be an extensive game with imperfect information in which there are no chance moves, and assume that the game Γ_1 differs from Γ_2 only in that one of the information sets of player 1 in Γ_2 is split into two information sets in Γ_1. Show that all Nash equilibria in pure strategies of Γ_2 correspond to Nash equilibria of Γ_1. Show that the requirement that there be no chance moves is essential for this result.

? EXERCISE 217.3 Formulate the following parlor game as an extensive game with imperfect information. First player 1 receives a card that is either H or L with equal probabilities. Player 2 does not see the card. Player 1 may announce that her card is L, in which case she must pay \$1 to player 2, or may claim that her card is H, in which case player 2 may choose to concede or to insist on seeing player 1's card. If player 2 concedes then he must pay \$1 to player 1. If he insists on seeing player 1's card then player 1 must pay him \$4 if her card is L and he must pay her \$4 if her card is H. Find the Nash equilibria of this game.

Notes

The model of an extensive game with imperfect information studied in this chapter is due to Kuhn (1950, 1953), as are the notions of perfect and imperfect recall. Section 11.2 is based on Thompson (1952); the example at the end of the section is based on one in Elmes and Reny (1994). The example in Section 11.3 is based on Aumann and Maschler (1972). Proposition 214.1 is due to Kuhn (1950, 1953). The game in Figure 214.1 is a variant of one due to Isbell (1957, p. 85).

12 Sequential Equilibrium

In this chapter we extend the notion of subgame perfect equilibrium to extensive games with imperfect information. We focus on the concept of sequential equilibrium and briefly discuss some of its refinements.

12.1 Strategies and Beliefs

Recall that a subgame perfect equilibrium of an extensive game with perfect information is a strategy profile for which every player's strategy is optimal (given the other players' strategies) at any history after which it is his turn to take an action, whether or not the history occurs if the players follow their strategies. The natural application of this idea to extensive games with imperfect information leads to the requirement that each player's strategy be optimal at each of his information sets.

For the game in Figure 219.1 this requirement is substantial. The pair of strategies (L, R) is a Nash equilibrium of this game. If player 1 adheres to this equilibrium then player 2's information set is not reached.

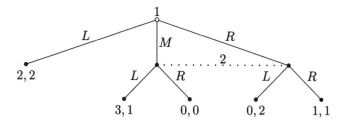

Figure 219.1 An extensive game with imperfect information in which the requirement that each player's strategy be optimal at every information set eliminates a Nash equilibrium.

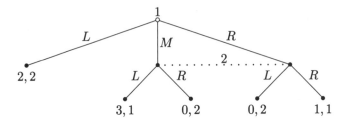

Figure 220.1 An extensive game with imperfect information that has a Nash equilibrium that is not ruled out by an implementation of the idea behind the notion of subgame perfect equilibrium.

However, if for some reason player 2's information set is reached then his action R is inferior to his action L *whatever* he thinks caused him to have to act (i.e. whether player 1, contrary to her plan, chose M or R). Thus for this game the natural extension of the idea of subgame perfect equilibrium is unproblematic: the equilibrium (L, R) does not satisfy the conditions of this extension (while the equilibrium (M, L) does). The games for which this is so are rare; a more common situation is that which arises in the game in Figure 220.1. In this game too the strategy profile (L, R) is a Nash equilibrium in which player 2's information set is not reached. But in this case player 2's optimal action in the event that his information set *is* reached depends on his belief about the history that has occurred. The action R is optimal if he assigns probability of at least $\frac{1}{2}$ to the history M, while L is optimal if he assigns probability of at most $\frac{1}{2}$ to this history. Thus his optimal action depends on his explanation of the cause of his having to act. His belief cannot be derived from the equilibrium strategy, since this strategy assigns probability zero to his information set being reached.

The solutions for extensive games that we have studied so far have a single component: a strategy profile. We now study a solution—sequential equilibrium—that consists of both a strategy profile and a belief system, where a belief system specifies, for each information set, the beliefs held by the players who have to move at that information set about the history that occurred. It is natural to include a belief system as part of the equilibrium, given our interpretation of the notion of subgame perfect equilibrium (see Section 6.4). When discussing this notion of equilibrium we argue that to describe fully the players' reasoning about a game we have to specify their expectations about the actions that will be taken after histories that will not occur if the players adhere to their plans, and that these expectations should be consistent with rationality. In particular, we interpret the components of a strategy that specify actions after histories that are not consistent with the strategy as beliefs about what will happen in these unexpected events. In games with imperfect information, beliefs about unexpected events must include beliefs not only about the future but also about the past.

To summarize, the basic idea behind the notion of sequential equilibrium is that an equilibrium should specify not only the players' strategies but also their beliefs at each information set about the history that occurred. We refer to such a pair as an *assessment*. That is, an assessment consists of (i) a profile of behavioral strategies and (ii) a belief system consisting of a collection of probability measures, one for each information set. (Note that the notion of an assessment coincides with that of a strategy profile for an extensive game with perfect information since in such a game all information sets are singletons and hence there is only one possible (degenerate) belief system.)

The extension of the requirement in a subgame perfect equilibrium that each player's strategy be optimal after any history is the following, which we refer to as *sequential rationality*: for each information set of each player i the (behavioral) strategy of player i is a best response to the other players' strategies, given player i's beliefs at that information set.

So far we have imposed no restriction on the players' beliefs. Several classes of additional constraints are discussed in the literature, including the following.

Consistency with Strategies   In the spirit of Nash equilibrium we should require that the belief system be consistent with the strategy profile, in the sense that at any information set consistent with the players' strategies the belief about the history that has occurred should be derived from the strategies using Bayes' rule. For example in the game in Figure 220.1 we do not want (M, L) to be a solution supported by the belief of player 2 that the history that led to his information set is R. If player 1's strategy is consistent with her choosing either M or R (that is, her strategy assigns positive probability to at least one of these choices), then we want to require that player 2's belief that the history M has occurred be derived from player 1's strategy using Bayes' rule. That is, player 2 should assign probability $\beta_1(\varnothing)(M)/(\beta_1(\varnothing)(M) + \beta_1(\varnothing)(R))$ (where β_1 is player 1's behavioral strategy) to this event.

Structural Consistency   Even at an information set that is not reached if all players adhere to their strategies we may wish to require that a player's belief be derived from *some* (alternative) strategy profile using Bayes' rule. (This constraint on the beliefs is referred to as "structural" since it does not depend on the players' payoffs or on the equilibrium strategy.)

Common Beliefs   Game theoretic solution concepts require that all asymmetries be included in the description of the game; every player is assumed to analyze the situation in the same way. In the context of subgame perfect equilibrium this leads to the (implicit) requirement that all the players' beliefs about the plans of some player i in case an unexpected event occurs are the same. In the current context it leads to the requirement that all players share the *same* belief about the cause of any unexpected event.

For some families of games the formal expression of these three restrictions is not problematic, though the reasonableness of the restrictions is in dispute. One example is the family of games in which the first move, about which the players may be asymmetrically informed, is made by chance, and subsequently every player is informed of every other player's moves. However, for arbitrary games even the formalization of the restrictions presents difficulties, as we shall see. The most widely-used formulation is that of sequential equilibrium, which we define in the next section. This notion usually leaves many degrees of freedom and is frequently consistent with a large set of outcomes, a fact that has motivated game theorists to impose additional restrictions on beliefs. In later sections we briefly discuss some of these restrictions.

12.2   Sequential Equilibrium

We restrict attention throughout to games with perfect recall (see Definition 203.3) in which every information set contains a finite number of histories. As we discuss above, a candidate for a sequential equilibrium of such a game is an assessment, defined formally as follows.

▶ DEFINITION 222.1 An **assessment** in an extensive game is a pair (β, μ), where β is a profile of behavioral strategies and μ is a function that assigns to every information set a probability measure on the set of histories in the information set.

Let (β, μ) be an assessment in $\Gamma = \langle N, H, P, f_c, (\mathcal{I}_i), (\succsim_i) \rangle$. The interpretation of μ, which we refer to as a **belief system**, is that $\mu(I)(h)$ is the probability that player $P(I)$ assigns to the history $h \in I$, conditional on I being reached.

An assessment is *sequentially rational* if for every player i and every information set $I_i \in \mathcal{I}_i$ the strategy of player i is a best response to the other players' strategies given i's beliefs at I_i. To state this condition more formally, define the **outcome** $O(\beta, \mu | I)$ **of** (β, μ) **conditional on** I to be the distribution over terminal histories determined by β and μ conditional on I being reached, as follows. Let $h^* = (a^1, \ldots, a^K)$ be a terminal history. Then

- $O(\beta, \mu | I)(h^*) = 0$ if there is no subhistory of h^* in I (i.e. the information that the game has reached I rules out h^*)

- $O(\beta, \mu | I)(h^*) = \mu(I)(h) \cdot \Pi_{k=L}^{K-1} \beta_{P(a^1, \ldots, a^k)}(a^1, \ldots, a^k)(a^{k+1})$ if the subhistory $h = (a^1, \ldots, a^L)$ of h^* is in I, where $L < K$.

(If I is the information set consisting of the initial history then $O(\beta, \mu | I)$ is just the outcome $O(\beta)$ defined in Section 11.4.) Note that the assumption of perfect recall implies that there is at most one subhistory of h^* in I. Note also that the rationale for taking the product in the second case is that by perfect recall the histories $(a^1, \ldots, a^k)$ for $k = L, \ldots, K-1$ lie in different information sets, and thus for $k = L, \ldots, K-1$ the events $\{a^{k+1}$ follows $(a^1, \ldots, a^k)$ conditional on $(a^1, \ldots, a^k)$ occurring$\}$ are independent.

While at first sight this definition of $O(\beta, \mu | I)$ is natural, it has undesirable features in a game in which there are two information sets I and I' and histories $h \in I$ and $h' \in I'$ with the property that a subhistory of h is in I' and a subhistory of h' is in I. The following example demonstrates this point.

◇ EXAMPLE 223.1 Consider the game form shown in Figure 224.1. (Sometimes, as here, we represent the initial history by several small circles rather than a single circle. In this example the number adjacent to each such circle is the probability assigned by chance to one of its actions at the initial history.) In an assessment (β, μ) in which $\beta_1 = \beta_3 = Out$ player 2's information set is not reached; if he is called upon to move then an unexpected event must have occurred. Suppose that his belief at his information set, I, satisfies $\mu(I)(A, C) > 0$ and $\mu(I)(B, C) > 0$. In deciding the action to take in the event that I is reached, he must calculate $O(\beta, \mu | I)$. The definition of this distribution given above assumes

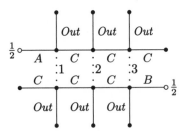

Figure 224.1 The game form in Example 223.1. (The game begins with a move of chance in which A and B are each selected with probability $\frac{1}{2}$.)

that he continues to hold expectations about the moves of players 1 and 3 that are derived from β. However, any strategy profile that generates the belief $\mu(I)$ must differ from β: it must assign positive probability to both player 1 and player 3 choosing C. That is, if his belief is derived from an alternative strategy profile, then his explanation of the past is inconsistent with his expectation of the future.

This example illustrates the complexity of defining a reasonable extension of the notion of subgame perfect equilibrium for games in which one information set can occur both before and after another. The definition of sequential equilibrium that we now present covers such games but, as the example indicates, can lack appeal in them. We begin with a formal definition of sequential rationality.

▶ DEFINITION 224.1 Let $\Gamma = \langle N, H, P, f_c, (\mathcal{I}_i), (\succsim_i) \rangle$ be an extensive game with perfect recall. The assessment (β, μ) is **sequentially rational** if for every player $i \in N$ and every information set $I_i \in \mathcal{I}_i$ we have

$$O(\beta, \mu | I_i) \succsim_i O((\beta_{-i}, \beta_i'), \mu | I_i) \text{ for every strategy } \beta_i' \text{ of player } i.$$

The following definition aims to capture some of the restrictions on beliefs discussed in the previous section. Define a behavioral strategy profile to be *completely mixed* if it assigns positive probability to every action at every information set.

▶ DEFINITION 224.2 Let $\Gamma = \langle N, H, P, f_c, (\mathcal{I}_i), (\succsim_i) \rangle$ be a finite extensive game with perfect recall. An assessment (β, μ) is **consistent** if there is a sequence $((\beta^n, \mu^n))_{n=1}^{\infty}$ of assessments that converges to (β, μ) in Euclidian space and has the properties that each strategy profile β^n is completely mixed and that each belief system μ^n is derived from β^n using Bayes' rule.

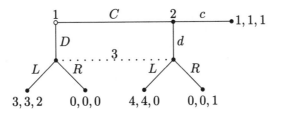

Figure 225.1 The game in Example 225.2 (*Selten's horse*).

The idea behind this requirement is that the probability of events conditional on zero-probability events must approximate probabilities that are derived from strategies that assign positive probability to every action. We do not find this consistency requirement to be natural, since it is stated in terms of limits; it appears to be a rather opaque technical assumption. To quote Kreps (1990a, p. 430), "[r]ather a lot of bodies are buried in this definition". The assumptions embodied in the definition are unclear to us, though we shall see that the definition does capture some appealing requirements that we may wish to impose on assessments.

▶ DEFINITION 225.1 An assessment is a **sequential equilibrium** of a finite extensive game with perfect recall if it is sequentially rational and consistent.

We show later (Proposition 249.1) that every finite extensive game with perfect recall has a sequential equilibrium. It is clear that if (β, μ) is a sequential equilibrium then β is a Nash equilibrium. Further, in an extensive game with perfect information (β, μ) is a sequential equilibrium if and only if β is a subgame perfect equilibrium.

Consider again the game in Figure 220.1. The assessment (β, μ) in which $\beta_1 = L$, $\beta_2 = R$, and $\mu(\{M, R\})(M) = \alpha$ for any $\alpha \in [0, 1]$ is consistent since it is the limit as $\epsilon \to 0$ of assessments $(\beta^\epsilon, \mu^\epsilon)$ where $\beta_1^\epsilon = (1 - \epsilon, \alpha\epsilon, (1 - \alpha)\epsilon)$, $\beta_2^\epsilon = (\epsilon, 1 - \epsilon)$, and $\mu^\epsilon(\{M, R\})(M) = \alpha$ for every ϵ. For $\alpha \geq \frac{1}{2}$ this assessment is also sequentially rational, so that it is a sequential equilibrium.

◇ EXAMPLE 225.2 (*Selten's horse*) The game in Figure 225.1 has two types of Nash equilibria: one in which $\beta_1(\varnothing)(D) = 1$, $\frac{1}{3} \leq \beta_2(C)(c) \leq 1$, and $\beta_3(I)(L) = 1$, and one in which $\beta_1(\varnothing)(C) = 1$, $\beta_2(C)(c) = 1$, and $\frac{3}{4} \leq \beta_3(I)(R) \leq 1$ (where $I = \{(D), (C, d)\}$, player 3's information set). A Nash equilibrium of the first type is not part of any sequential

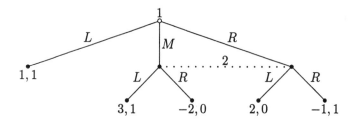

Figure 226.1 The game in Exercise 226.1.

equilibrium since the associated assessment violates sequential rationality at player 2's (singleton) information set. For every Nash equilibrium
β of the second type there is a sequential equilibrium (β, μ) in which
$\mu(I)(D) = \frac{1}{3}$. (To verify consistency, consider the sequence (β^ϵ) of
strategy profiles in which $\beta_1^\epsilon(\varnothing)(C) = 1 - \epsilon$, $\beta_2^\epsilon(C)(c) = 2\epsilon/(1 - \epsilon)$, and
$\beta_3^\epsilon(I)(R) = \beta_3(I)(R) - \epsilon$.)

⟨?⟩ EXERCISE 226.1 Find the set of sequential equilibria of the game in
Figure 226.1.

The following example shows that the notion of sequential equilibrium
is not invariant to the principle of coalescing of moves considered in
Section 11.2, a principle that seems very reasonable.

◇ EXAMPLE 226.2 Consider the games in Figure 227.1. The top game
(Γ_1) is obtained from the bottom one (Γ_2) by coalescing the moves
of player 1. In Γ_1 the assessment (β, μ) in which $\beta_1 = L$, $\beta_2 = L$,
and $\mu(\{M, R\})(R) = 0$ is a sequential equilibrium. (To verify consis-
tency, consider the sequence (β^ϵ) in which $\beta_1^\epsilon(\varnothing) = (1 - \epsilon - \epsilon^2, \epsilon, \epsilon^2)$
and $\beta_2^\epsilon(\{M, R\}) = (1 - \epsilon, \epsilon)$.) This equilibrium yields the payoff profile
$(3, 3)$. On the other hand, in any sequential equilibrium of Γ_2 player 1's
action at her second information set is R, by sequential rationality (since
R dominates L). Thus in any consistent assessment player 2's belief
$\mu(\{(C, L), (C, R)\})$ assigns probability 1 to (C, R), so that player 2 must
choose R. Hence $\beta_1 = (C, R)$ is the only equilibrium strategy of player 1.
Thus the only sequential equilibrium payoff profile in Γ_2 is $(5, 1)$.

The significant difference between the two games is that player 2's
belief in Γ_2 is based on the assumption that the relative likelihood of
the actions L and R at player 1's second information set is the outcome
of a rational choice by player 1, whereas player 2's belief in Γ_1 about
the relative probabilities of M and R is not constrained by any choice
of player 1. We argued in Section 6.4 that in some games a player's

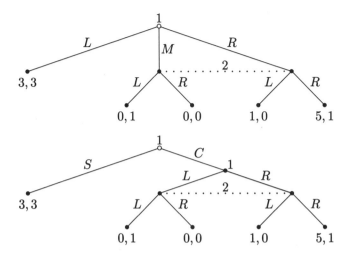

Figure 227.1 Two similar games. The top game (Γ_1) is obtained from the bottom game (Γ_2) by coalescing the moves of player 1.

strategy is not only a plan of action but also a specification of the other players' beliefs about his actions in the event that he does not follow this plan. In Γ_2 player 1's strategy has this role—it specifies player 2's belief about the relative probabilities of player 1 choosing L and R when she begins by choosing C—while in Γ_1 player 1's strategy describes only her move, without giving any relative probabilities to the other choices.

The following exercise gives an extension of the one deviation property for subgame perfect equilibria of extensive games with perfect information (Lemma 98.2) to sequential equilibria of extensive games with imperfect information.

▣ EXERCISE 227.1 (*The one deviation property for sequential equilibrium*) Let (β, μ) be a consistent assessment in a finite extensive game with perfect recall and let β_i' be a strategy of player i; denote $\beta' = (\beta_{-i}, \beta_i')$. Show that if I_i and I_i' are information sets of player i with the property that I_i' contains histories that have subhistories in I_i then $O(\beta', \mu | I_i)(h) = O(\beta', \mu | I_i')(h) \cdot \Pr(\beta', \mu | I_i)(I_i')$ for any terminal history h that has a subhistory in I_i', where $\Pr(\beta', \mu | I_i)(I_i')$ is the probability (according to (β', μ)) that I_i' is reached given that I_i is reached. Use this fact to show that (β, μ) is sequentially rational if and only if no player i has an information set I_i at which a change in $\beta_i(I_i)$ (holding the remainder of β_i fixed) increases his expected payoff conditional on reaching I_i.

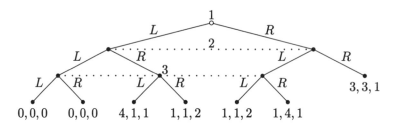

Figure 228.1 An extensive game in which there is a sequential equilibrium for which the belief system is not structurally consistent.

In Section 12.1 we discuss three conditions that relate the beliefs and strategies in an assessment. One of these is structural consistency, which may be defined formally as follows.

▸ DEFINITION 228.1 The belief system μ in an extensive game with perfect recall is **structurally consistent** if for each information set I there is a strategy profile β with the properties that I is reached with positive probability under β and $\mu(I)$ is derived from β using Bayes' rule.

(Note that *different* strategy profiles may justify the beliefs at different information sets.)

In many games, for any assessment (β, μ) that is consistent (in the sense of Definition 224.2) the belief system μ is structurally consistent. However, the following example shows that in some games there are consistent assessments (β, μ) (in fact, even sequential equilibria) in which μ is not structurally consistent: the beliefs cannot be derived from any alternative strategy profile.

◇ EXAMPLE 228.2 The game in Figure 228.1 has a unique Nash equilibrium outcome, in which players 1 and 2 choose R. To see this, suppose to the contrary that player 3's information set is reached with positive probability. Let the strategy profile used be β and let $\beta_i(I_i)(R) = \alpha_i$ for $i = 1, 2, 3$, where I_i is the single information set of player i.

  a. If $\alpha_3 \leq \frac{1}{2}$ then L yields player 2 a payoff of $\alpha_1(1 + 3\alpha_3) \leq \frac{5}{2}\alpha_1 < 1 + 2\alpha_1$, his payoff to R. Thus player 2 chooses R. But then $\mu(I_3)((L, R)) = 1$ and hence player 3 chooses R with probability 1, contradicting $\alpha_3 \leq \frac{1}{2}$.

  b. If $\alpha_3 \geq \frac{1}{2}$ then L yields player 1 a payoff of $\alpha_2(4 - 3\alpha_3) \leq \frac{5}{2}\alpha_2 < 1 + 2\alpha_2$, her payoff to R. Thus player 1 chooses R. Now if player 2 chooses L with positive probability then $\mu(I_3)((R, L)) = 1$, and hence player 3 chooses L with probability 1, contradicting $\alpha_3 \geq \frac{1}{2}$. Thus

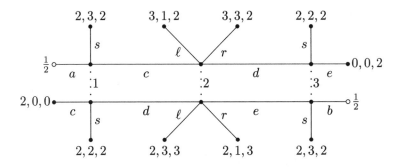

Figure 229.1 The game in Exercise 229.1.

player 2 chooses R with probability 1, contradicting our assumption that player 3's information set is reached with positive probability.

Thus in any Nash equilibrium $\alpha_1 = \alpha_2 = 1$; in addition we need $\alpha_3 \in [\frac{1}{3}, \frac{2}{3}]$, otherwise either player 1 or player 2 can profitably deviate. Let β be such an equilibrium. For an assessment (β, μ) to be sequentially rational, player 3's belief $\mu(I_3)$ must assign equal probabilities to the histories (L, R) and (R, L), and thus must take the form $(1 - 2\gamma, \gamma, \gamma)$; an assessment in which $\mu(I_3)$ takes this form is consistent if and only if $\gamma = \frac{1}{2}$. (A sequence of strategy profiles that demonstrates consistency for $\gamma = \frac{1}{2}$ is (β^ϵ) in which $\beta_1^\epsilon(I_1)(R) = \beta_2^\epsilon(I_2)(R) = 1 - \epsilon$ and $\beta_3^\epsilon(I_3)(R) = \alpha_3$.) However, the belief $(0, \frac{1}{2}, \frac{1}{2})$ of player 3 violates structural consistency since any strategy profile that yields (L, L) with probability zero also yields either (L, R) or (R, L) with probability zero.

[?] EXERCISE 229.1 Consider the game in Figure 229.1. As in the game in Figure 224.1 the first move is made by chance, and the information sets are not ordered (player 1's information set comes either before or after player 3's information set, depending on the move of chance). Show that the game has three sequential equilibria in pure strategies, in one of which players 1 and 3 both choose S. Discuss the reasonableness of these sequential equilibria.

The next example further illustrates the relationship between consistency and structural consistency. It shows that a sequentially rational assessment (β, μ) in which μ is structurally consistent may not be consistent (and hence may not be a sequential equilibrium).

◇ EXAMPLE 229.2 In the game in Figure 230.1 the assessment (β, μ) in which β is the pure strategy profile (R, S, R), player 2's belief assigns probability 1 to the history R, and player 3's belief assigns probability 1

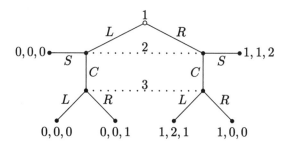

Figure 230.1 An extensive game in which there is a sequentially rational assessment with structurally consistent beliefs that is not a sequential equilibrium.

to the history (L, C), is sequentially rational. Further, the belief μ is structurally consistent: in particular, player 3's belief is supported by the alternative pure strategy profile in which player 1 chooses L and player 2 chooses C. That is, if player 3 has to move then she believes that player 1, as well as player 2, deviated from her equilibrium strategy. To rationalize her having to move it is sufficient for her to believe that player 2 deviated, and in fact she *knows* that he did so. Thus structural consistency allows player 3 to revise her belief about player 1 even though the only evidence she has is that player 2 deviated, and this is enough to explain what has happened.

These structurally consistent beliefs are not consistent: every sequence of assessments that involves strategies that are completely mixed and converge to β generates beliefs of player 3 that converge to the belief that assigns probability 1 to the history (R, C) (while μ assigns probability 1 to (L, C)). Thus (β, μ) is not a sequential equilibrium. (In the only sequential equilibrium of the game the strategy profile is (R, C, L), the belief of player 2 assigns probability 1 to R, and the belief of player 3 assigns probability 1 to (R, C).)

The next example illustrates how subtle the consistency requirement can be.

◇ EXAMPLE 230.1 Consider a two-stage three-player game in which in the first stage players 1 and 2 simultaneously choose from the set $\{L, M, R\}$, and in the second stage player 3 finds out how many players chose R and how many chose L. In this game player 3 has six information sets (for example $\{(M, M)\}$—the case in which she is informed that two players chose M—and $\{(R, L), (L, R)\}$—the case in which she is informed that one player chose R and one chose L).

If the strategies of players 1 and 2 call for them to choose the same action, but if in fact they choose different actions, then player 3 has to form a belief about the action chosen by each of them. At first sight it may seem that the notion of sequential equilibrium does not constrain these beliefs. Consider for example an assessment (β, μ) in which all three beliefs $\mu(\{(M, L), (L, M)\})(M, L)$, $\mu(\{(L, R), (R, L)\})(L, R)$, and $\mu(\{(M, R), (R, M)\})(R, M)$ are equal to $\frac{2}{3}$. These beliefs are clearly structurally consistent (recall that a different strategy profile can support the belief at each information set). However, the fact that consistency requires the belief at each information set to be justified by the *same* sequence of strategy profiles implies that (β, μ) is not consistent, as the following argument shows.

For (β, μ) to be consistent there must be a sequence (β^ϵ) for which the limits of $\beta_1^\epsilon(M)\beta_2^\epsilon(L)/\beta_1^\epsilon(L)\beta_2^\epsilon(M)$, $\beta_1^\epsilon(L)\beta_2^\epsilon(R)/\beta_1^\epsilon(R)\beta_2^\epsilon(L)$, and $\beta_1^\epsilon(R)\beta_2^\epsilon(M)/\beta_1^\epsilon(M)\beta_2^\epsilon(R)$ as $\epsilon \to 0$ are all 2 (where $\beta_i^\epsilon(a)$ is an abbreviation for $\beta_i^\epsilon(\varnothing)(a)$). But the product of these three ratios is 1, independent of ϵ, while the product of their limits is 8. Thus consistency rules out the belief system μ (regardless of β).

12.3   Games with Observable Actions: Perfect Bayesian Equilibrium

We now examine a family of games in which we can define a notion of equilibrium that is closely related to sequential equilibrium but is simpler. A *Bayesian extensive game with observable actions* models a situation in which every player observes the action of every other player; the only uncertainty is about an initial move of chance that distributes payoff-relevant personal information among the players in such a way that the information received by each player does not reveal any information about any of the other players. We say that chance selects *types* for the players and refer to player i after he receives the information θ_i as *type* θ_i. The formal definition follows.

▶ DEFINITION 231.1 A **Bayesian extensive game with observable actions** is a tuple $\langle \Gamma, (\Theta_i), (p_i), (u_i) \rangle$ where

- $\Gamma = \langle N, H, P \rangle$ is an extensive game form with perfect information and simultaneous moves

and for each player $i \in N$

- Θ_i is a finite set (the set of possible **types** of player i); we write $\Theta = \times_{i \in N} \Theta_i$

- p_i is a probability measure on Θ_i for which $p_i(\theta_i) > 0$ for all $\theta_i \in \Theta_i$, and the measures p_i are stochastically independent ($p_i(\theta_i)$ is the probability that player i is selected to be of type θ_i)

- $u_i\colon \Theta \times Z \to \mathbb{R}$ is a von Neumann–Morgenstern utility function ($u_i(\theta, h)$ is player i's payoff when the profile of types is θ and the terminal history of Γ is h).

The situation that such a game models is one in which chance selects the types of the players, who are subsequently fully cognizant at all points of all moves taken previously. We can associate with any such game an extensive game (with imperfect information and simultaneous moves) in which the set of histories is $\{\varnothing\} \cup (\Theta \times H)$ and each information set of each player i takes the form $I(\theta_i, h) = \{((\theta_i, \theta'_{-i}), h)\colon \theta'_{-i} \in \Theta_{-i}\}$ for $i \in P(h)$ and $\theta_i \in \Theta_i$ (so that the number of histories in $I(\theta_i, h)$ is the number of members of Θ_{-i}).

A candidate for an equilibrium of such a game is a pair $((\sigma_i), (\mu_i)) = ((\sigma_i(\theta_i))_{i \in N, \theta_i \in \Theta_i}, (\mu_i(h))_{i \in N, h \in H \setminus Z})$, where each $\sigma_i(\theta_i)$ is a behavioral strategy of player i in Γ (the strategy used by type θ_i of player i) and each $\mu_i(h)$ is a probability measure on Θ_i (the common belief, after the history h, of all players other than i about player i's type). Such a pair is closely related to an assessment. The profile (σ_i) rephrases the information in a profile of behavioral strategies in the associated extensive game; the profile (μ_i) summarizes the players' beliefs and is tailored to the assumption that each player is perfectly informed about the other players' previous moves and may be uncertain only about the other players' types.

Let s be a profile of behavioral strategies in Γ. Define $O_h(s)$ to be the probability measure on the set of terminal histories of Γ generated by s given that the history h has occurred (see Section 6.2). Define $O(\sigma_{-i}, s_i, \mu_{-i} | h)$ to be the probability measure on the set of terminal histories of Γ given that player i uses the strategy s_i in Γ, each type θ_j of each player j uses the strategy $\sigma_j(\theta_j)$, the game has reached h, and the probability that i assigns to θ_{-i} is derived from $\mu_{-i}(h)$. That is, $O(\sigma_{-i}, s_i, \mu_{-i} | h)$ is the compound lottery in which the probability of the lottery $O_h((\sigma_j(\theta_j))_{j \in N \setminus \{i\}}, s_i)$ is $\Pi_{j \in N \setminus \{i\}} \mu_j(h)(\theta_j)$ for each $\theta_{-i} \in \Theta_{-i}$.

The solution concept that we define is the following.

▶ DEFINITION 232.1 Let $\langle \Gamma, (\Theta_i), (p_i), (u_i) \rangle$ be a Bayesian extensive game with observable actions, where $\Gamma = \langle N, H, P \rangle$. A pair $((\sigma_i), (\mu_i)) = ((\sigma_i(\theta_i))_{i \in N, \theta_i \in \Theta_i}, (\mu_i(h))_{i \in N, h \in H \setminus Z})$, where $\sigma_i(\theta_i)$ is a behavioral strategy of player i in Γ and $\mu_i(h)$ is a probability measure on Θ_i, is a

perfect Bayesian equilibrium of the game if the following conditions are satisfied.

Sequential rationality For every nonterminal history $h \in H \setminus Z$, every player $i \in P(h)$, and every $\theta_i \in \Theta_i$ the probability measure $O(\sigma_{-i}, \sigma_i(\theta_i), \mu_{-i}|h)$ is at least good for type θ_i as $O(\sigma_{-i}, s_i, \mu_{-i}|h)$ for any strategy s_i of player i in Γ.

Correct initial beliefs $\mu_i(\varnothing) = p_i$ for each $i \in N$.

Action-determined beliefs If $i \notin P(h)$ and $a \in A(h)$ then $\mu_i(h, a) = \mu_i(h)$; if $i \in P(h)$, $a \in A(h)$, $a' \in A(h)$, and $a_i = a_i'$ then $\mu_i(h, a) = \mu_i(h, a')$.

Bayesian updating If $i \in P(h)$ and a_i is in the support of $\sigma_i(\theta_i)(h)$ for some θ_i in the support of $\mu_i(h)$ then for any $\theta_i' \in \Theta_i$ we have

$$\mu_i(h, a)(\theta_i') = \frac{\sigma_i(\theta_i')(h)(a_i) \cdot \mu_i(h)(\theta_i')}{\sum_{\theta_i \in \Theta_i} \sigma_i(\theta_i)(h)(a_i) \cdot \mu_i(h)(\theta_i)}.$$

The conditions in this definition are easy to interpret. The first requires that the strategy $\sigma_i(\theta_i)$ of each type θ_i of each player i be optimal for type θ_i after every sequence of events. The second requires that initially the other players' beliefs about the type of each player i be given by p_i.

The condition of action-determined beliefs requires that only a player's actions influence the other players' beliefs about his type: (i) if player i does not have to move at the history h then the actions taken at h do not affect the other players' beliefs about player i's type and (ii) if player i is one of the players who takes an action at h then the other players' beliefs about player i's type depend only on h and the action taken by player i, not on the other players' actions. This condition excludes the possibility that, for example, player j's updating of his belief about player i is affected by a move made by some player $k \neq i$. Thus the condition is consonant with the general approach that assumes independence between the players' strategies.

The condition of Bayesian updating relates to a case in which player i's action at the history h is consistent with the other players' beliefs about player i at h, given σ_i. In such a case the condition requires not only that the new belief depend only on player i's action (as required by the condition of action-determined beliefs) but also that the players' beliefs be derived via Bayes' rule from their observation of player i's actions. Thus the players update their beliefs about player i using Bayes' rule until his behavior contradicts his strategy σ_i, at which point they form a new conjecture about player i's type that is the basis for future Bayesian updating until there is another conflict with σ.

We now show that every sequential equilibrium of the extensive game associated with a finite Bayesian extensive game with observable actions is equivalent to a perfect Bayesian equilibrium of the Bayesian extensive game, in the sense that it induces the same behavior and beliefs.

■ PROPOSITION 234.1 *Let (β, μ) be a sequential equilibrium of the extensive game associated with the finite Bayesian extensive game with observable actions $\langle\langle N, H, P\rangle, (\Theta_i), (p_i), (u_i)\rangle$. For every $h \in H$, $i \in P(h)$, and $\theta_i \in \Theta_i$, let $\sigma_i(\theta_i)(h) = \beta_i(I(\theta_i, h))$. Then there is a collection $(\mu_i(h))_{i \in N, h \in H}$, where $\mu_i(h)$ is a probability measure on Θ_i, such that*

$$\mu(I(\theta_i, h))(\theta, h) = \Pi_{j \in N \setminus \{i\}} \mu_j(h)(\theta_j) \text{ for all } \theta \in \Theta \text{ and } h \in H$$

and $((\sigma_i), (\mu_i))$ is a perfect Bayesian equilibrium of the Bayesian extensive game.

? EXERCISE 234.2 Prove the proposition. (The main difficulty is to confirm that the beliefs in the sequential equilibrium can be reproduced by a collection of common independent beliefs about the players' types.)

The concept of perfect Bayesian equilibrium is easier to work with than that of sequential equilibrium (since there is no need to mess with consistency) but applies to a significantly smaller set of situations. The following example shows that even in this restricted domain the two notions are not equivalent.

◇ EXAMPLE 234.3 Consider a Bayesian extensive game with observable actions with the structure given in Figure 235.1. Player 1 has three equally likely possible types, x, y, and z, and player 2 has a single type. Consider a perfect Bayesian equilibrium $((\sigma_i), (\mu_i))$ in which $\sigma_1(x) = (Out, L)$, $\sigma_1(y) = (Out, M)$, $\sigma_1(z) = (C, R)$, $\mu_1(C, L)(y) = 1$, $\mu_1(C, M)(x) = 1$, and $\mu_1(C, R)(z) = 1$. That is, player 2 believes that player 1 is certainly of type y if he observes the history (C, L), certainly of type x if he observes the history (C, M), and certainly of type z if he observes the history (C, R) (the only history that is consistent with σ_1).

We claim that $((\sigma_i), (\mu_i))$ may (depending on the payoffs) be a perfect Bayesian equilibrium of such a game, since it satisfies the conditions of action-determined beliefs and Bayesian updating. (Note that $\mu_1(C, L)$ and $\mu_1(C, M)$ are not constrained by the condition of Bayesian updating since the probabilities of the histories (C, L) and (C, M) are both zero, given σ_1.) However, the associated assessment (β, μ) is not

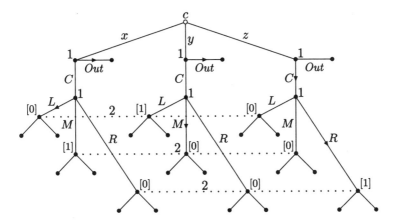

Figure 235.1 The structure of the Bayesian extensive games with observable actions in Example 234.3. Such a game may have a perfect Bayesian equilibrium (in which player 1's actions are those indicated by the arrows and player 2's beliefs are indicated by [1]'s and [0]'s) that is not a sequential equilibrium of the associated extensive game.

consistent, and hence is not a sequential equilibrium of any associated extensive game, whatever the payoffs. To see this, let (β^n, μ^n) be a sequence of assessments that converges to (β, μ), with the properties that each β^n assigns positive probability to each choice at every information set and each μ^n is derived from β^n using Bayes' rule. Denote by c_θ^n the probability, according to β_1^n, that player 1 chooses C after the history θ, and denote by ℓ_θ^n and m_θ^n the probabilities, according to β_1^n, that she chooses L and M respectively after the history (θ, C). Let $I_2^K = \{(x, C, K), (y, C, K), (z, C, K)\}$ for $K = L, M, R$, be the information set of player 2 that is reached if player 1 chooses C and then K. Then by Bayes' rule we have $\mu^n(I_2^L)(y, C, L) = c_y^n \ell_y^n / (c_x^n \ell_x^n + c_y^n \ell_y^n + c_z^n \ell_z^n)$ (using the fact that the three types of player 1 are equally likely), which converges (by assumption) to $\mu(I_2^L)(y, C, L) = 1$. Since $\ell_y^n \to \beta_1(y, C)(L) = 0$ and $\ell_x^n \to \beta_1(x, C)(L) = 1$ we conclude, dividing the numerator and denominator of $\mu^n(I_2^L)(y, C, L)$ by c_y^n, that $c_x^n / c_y^n \to 0$. Performing a similar calculation for the belief at I_2^M, we reach the contradictory conclusion that $c_y^n / c_x^n \to 0$. Thus (β, μ) is not consistent.

This example reflects the fact that the notion of sequential equilibrium requires that the beliefs of player 2 at two information sets not reached in the equilibrium not be independent: they must be derived from the *same* sequence of perturbed strategies of player 1. The notion of perfect Bayesian equilibrium imposes no such restriction on beliefs.

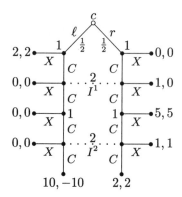

Figure 236.1 The extensive game discussed in Example 236.1. In all sequential equilibria of this game player 2 believes at his first information set that chance chose r with certainty, but believes at his second information set that there is a positive probability that chance chose ℓ.

The notion of perfect Bayesian equilibrium may be refined by imposing additional constraints on players' beliefs after unexpected events. For example, one can require that if at some point the players other than i conclude that i is certainly not of type θ_i then they never subsequently reverse this conclusion. This requirement has been used in some of the literature (see for example Osborne and Rubinstein (1990, pp. 96–97)). The following example shows, however, that there are games in which no perfect Bayesian equilibrium satisfies it: in all perfect Bayesian equilibria of the game we describe, a player who at some point assigns probability zero to some history later assigns positive probability to this history.

◇ EXAMPLE 236.1 In any sequential equilibrium of the game in Figure 236.1

- player 1 chooses C after the history r
- player 1 chooses X after the history (r, C, C)
- player 2 chooses C at his information set I^1
- player 2 chooses X with probability at least $\frac{4}{5}$ at his information set I^2 (otherwise player 1 chooses C after the histories ℓ and (ℓ, C, C), so that player 2 assigns probability 1 to the history (ℓ, C, C, C) at his information set I^2, making C inferior to X)
- player 1 chooses X after the history ℓ.

Thus player 2's belief at I^1 assigns probability 1 to the history r while his belief at I^2 assigns positive probability to chance having chosen ℓ (otherwise C is better than X).

[?] EXERCISE 237.1 Players 1 and 2 bargain over an item whose value for player 1 is 0 or 3, with equal probabilities. Player 1 knows the value of the object, while player 2 is informed of this value only after he purchases it. The value of the object to player 2 is its value to player 1 plus 2. The bargaining procedure is: player 1 makes an offer, which player 2 either accepts or rejects; in the event of rejection player 1 makes another offer, which player 2 either accepts or rejects. If no offer is accepted, player 1 is left with the object and obtains a payoff equal to its value; player 2's payoff is 0. Take the set of possible offers to be finite, including 2 and 5. Show that there is a sequential equilibrium in pure strategies in which there is no deal when player 1's valuation is 3, while the object is sold at the price of two in the first period when player 1's valuation is 0.

12.3.1 Signaling Games

A signaling game is a Bayesian extensive game with observable actions that has the following simple form. There are two players, a "sender" and a "receiver". The sender is informed of the value of an uncertain parameter θ_1 and then chooses an action m (referred to as a *message*, though it may be payoff-relevant). The receiver observes the message (but not the value of θ_1) and takes an action a. Each player's payoff depends upon the value of θ_1, the message m sent by the sender, and the action a taken by the receiver.

Formally, a **signaling game** is a Bayesian extensive game with observable actions $\langle \Gamma, (\Theta_i), (p_i), (u_i) \rangle$ in which Γ is a two-player game form in which first player 1 takes an action then player 2 takes an action, and Θ_2 is a singleton.

The tension in such a game arises from the fact that the receiver controls the action while the sender controls the information. The receiver has an incentive to try to deduce the sender's type from the sender's message, and the sender may have an incentive to mislead the receiver.

A well-known simple example of a signaling game is the following, proposed by Spence (1974).

◇ EXAMPLE 237.2 (*Spence's model of education*) A worker (the sender) knows her talent θ_1 while her employer (the receiver) does not. The value of the worker to the employer is the expectation of θ_1; we assume that the employer pays the worker a wage w that is equal to this expectation. (The economic story that underlies this assumption is that there are many employers who compete for the worker, so that her wage is driven up to the expectation of θ_1.) To model this behavioral assumption we assume that the payoff of the employer is $-(w - \theta_1)^2$ (the expectation of which is maximized when $w = E(\theta_1)$). The worker's message is the amount e of education that she obtains and her payoff is $w - e/\theta$ (reflecting the assumption that the larger is θ the easier it is for a worker to acquire education). Assume that the worker's talent is either θ_1^L or $\theta_1^H > \theta_1^L$, and denote the probabilities of these values by p^L and p^H. Restrict attention to pure strategy equilibria and denote the choices (messages) of the two types by e^L and e^H. This game has two types of perfect Bayesian equilibrium.

Pooling Equilibrium In one type of equilibrium both types choose the same level of education ($e^L = e^H = e^*$) and the wage is $w^* = p^H\theta_1^H + p^L\theta_1^L$. The possible values of e^* are determined as follows. If a worker chooses a value of e different from e^* then in an equilibrium the employer must pay her a wage $w(e)$ for which $w(e) - e/\theta_1^K \leq w^* - e^*/\theta_1^K$ for $K = L, H$. The easiest way to satisfy this inequality is by making the employer believe that every deviation originates from a type θ_1^L worker, so that $w(e) = \theta_1^L$ for $e \neq e^*$. The most profitable deviation for the worker is then to choose $e^L = 0$, so that we need $\theta_1^L \leq w^* - e^*/\theta_1^L$, which is equivalent to $e^* \leq \theta_1^L p^H(\theta_1^H - \theta_1^L)$.

Separating Equilibrium In another type of equilibrium the two types of worker choose different levels of education. In this case $e^L = 0$ (since the wage paid to a type θ_1^L worker is θ_1^L, independent of e^L). For it to be unprofitable for either type to mimic the other we need

$$\theta_1^L \geq \theta_1^H - e^H/\theta_1^L \quad \text{and} \quad \theta_1^H - e^H/\theta_1^H \geq \theta_1^L,$$

which are equivalent to $\theta_1^L(\theta_1^H - \theta_1^L) \leq e^H \leq \theta_1^H(\theta_1^H - \theta_1^L)$. Since $\theta_1^H > \theta_1^L$, a separating equilibrium thus always exists; the messages $e^L = 0$ and $e^H \in [\theta_1^L(\theta_1^H - \theta_1^L), \theta_1^H(\theta_1^H - \theta_1^L)]$ are supported as a part of an equilibrium in which any action other than e^H leads the employer to conclude that the worker's type is θ_1^L.

? EXERCISE 238.1 Verify that the perfect Bayesian equilibria that we have described are also sequential equilibria.

Example 246.1 in the next section shows how a refinement of the notion of sequential equilibrium excludes most of these equilibria.

12.3.2 Modeling Reputation

In Section 6.5 we study two finite horizon games that highlight the fact that in a subgame perfect equilibrium a player maintains the assumption that another player intends to adhere to his equilibrium strategy even after that player has deviated from this strategy many times. For example, in the unique subgame perfect equilibrium of the chain-store game (a finite horizon extensive game with perfect information) every challenger believes that the chain-store will acquiesce to its entry even after a history in which the chain-store has fought every one of a large number of entrants.

One way to capture the idea that after such a history a player may begin to entertain doubts about the intentions of his opponents is to study a model in which at the very beginning of the game there is a small chance that the opponents have motives different from those captured in the original extensive form. (The aberrant players that are thus included, with small probability, in the strategic calculations of their opponents are often referred to as "crazy" or "irrational", although their payoffs, not their strategic reasoning, diverge from the standard.) In such a "perturbed" game the regular types may find it advantageous to imitate the aberrant types at the beginning of the game: the short-term loss from doing so may be more than outweighed by the long-term gain from maintaining their opponents' doubts about their motivations. Thus such a game can capture the idea that people may act as if they are "crazy" because doing so leads their opponents to respond in such a way that even according to their real, "sane", preferences they are better off. The following example illustrates this approach.

◇ EXAMPLE 239.1 (*A perturbation of the chain-store game*) Consider the variant of the chain-store game in which there is small probability, at the beginning of the game, that the chain-store prefers to fight than to accommodate entry. Precisely, consider the Bayesian extensive game with observable actions $\langle \Gamma, (\Theta_i), (p_i), (u_i) \rangle$ in which Γ is the game form of the chain-store game (Section 6.5.1), $\Theta_{CS} = \{R(egular), T(ough)\}$, Θ_k is a singleton for every potential competitor $k = 1, \ldots, K$, $p_{CS}(R) = 1 - \epsilon$, $p_{CS}(T) = \epsilon$, and the payoff functions u_i are defined as follows. For any terminal history h of Γ, let h_k be the sequence of actions in period k. The payoff of each challenger k is independent of the type of the chain-store and is given by

$$u_k(\theta, h) = \begin{cases} b & \text{if } h_k = (In, C) \\ b - 1 & \text{if } h_k = (In, F) \\ 0 & \text{if } h_k = Out, \end{cases}$$

where $0 < b < 1$. The payoff $u_{CS}(\theta, h)$ of the chain-store is the sum of its payoffs in the K periods, where its payoff in period k is given by

$$
\begin{cases}
0 & \text{if } h_k = (In, C) \text{ and } \theta_{CS} = R, \text{ or } h_k = (In, F) \text{ and } \theta_{CS} = T \\
-1 & \text{if } h_k = (In, F) \text{ and } \theta_{CS} = R, \text{ or } h_k = (In, C) \text{ and } \theta_{CS} = T \\
a & \text{if } h_k = Out,
\end{cases}
$$

where $a > 1$. In other words, for both types of chain-store the best outcome in any period is that the challenger stays out; the regular chain-store prefers to accommodate an entrant than to fight (fighting is costly), while the tough chain-store prefers to fight than to accommodate.

We do not characterize all perfect Bayesian equilibria of the game but merely describe an equilibrium that differs radically from the unique subgame perfect equilibrium of the perfect information game. This equilibrium has the following features: so long as no challenger enters the challengers maintain their original belief that the chain-store is tough with probability ϵ; entry that is accommodated leads the challengers to switch to believing that the chain-store is definitely not tough; entry that is fought leads the challengers to maintain or increase the probability that they assign to the chain-store being tough. Consequently it is optimal for a regular chain-store, as well as a tough one, to threaten to fight any entry that occurs, at least until the horizon impends. This threat deters all entry until the horizon gets close, when the regular chain-store's threats become less firm: it cooperates with entrants with positive probability, behavior that is consistent with the entrants beginning to enter with positive probability. Once a challenger enters and the chain-store cooperates with it, the challengers switch to believing that the chain-store is certainly regular and henceforth always enter.

Precisely, the equilibrium is given as follows. The actions prescribed by the strategy $\sigma_{CS}(R)$ of the regular chain-store and by the strategy σ_k of each challenger k after any history depend on $\mu_{CS}(h)(T)$, the probability assigned by the challengers after the history h to the chain-store being tough. The chain-store has to move only after histories that end with entry by a challenger. For any such history h, denote by $t(h)$ the number of challengers who have moved, so that $\sigma_{CS}(R)(h)$ prescribes the response of the chain-store to challenger $t(h)$. The strategy of a regular chain-store is then given by

$$
\sigma_{CS}(R)(h) = \begin{cases}
C & \text{if } t(h) = K \\
F & \text{if } t(h) \le K - 1 \text{ and } \mu_{CS}(h)(T) \ge b^{K - t(h)} \\
m_{CS}^h & \text{if } t(h) \le K - 1 \text{ and } \mu_{CS}(h)(T) < b^{K - t(h)}
\end{cases}
$$

if $P(h) = CS$, where m_{CS}^h is the mixed strategy in which F is used with probability $[(1 - b^{K-t(h)})\mu_{CS}(h)(T)]/[(1 - \mu_{CS}(h)(T))b^{K-t(h)}]$ and C is used with the complementary probability; the strategy of a tough chain-store is given by

$$\sigma_{CS}(T)(h) = F \text{ if } P(h) = CS.$$

The strategy of challenger k is given by

$$\sigma_k(h) = \begin{cases} Out & \text{if } \mu_{CS}(h)(T) > b^{K-k+1} \\ m_k & \text{if } \mu_{CS}(h)(T) = b^{K-k+1} \\ In & \text{if } \mu_{CS}(h)(T) < b^{K-k+1} \end{cases}$$

if $P(h) = k$ (so that $t(h) = k - 1$), where m_k is the mixed strategy in which Out is used with probability $1/a$ and In is used with probability $1 - 1/a$. The challengers' beliefs are as follows: $\mu_{CS}(\varnothing)(T) = \epsilon$, and for any history h with $P(h) = k$ we have $\mu_{CS}(h, In) = \mu_{CS}(h)$ and the probability $\mu_{CS}(h, h_k)(T)$ assigned by challenger $k+1$ to the chain-store being tough is

$$\begin{cases} \mu_{CS}(h)(T) & \text{if } h_k = Out \\ \max\{b^{K-k}, \mu_{CS}(h)(T)\} & \text{if } h_k = (In, F) \text{ and } \mu_{CS}(h)(T) > 0 \\ 0 & \text{if } h_k = (In, C) \text{ or } \mu_{CS}(h)(T) = 0. \end{cases}$$

To understand this equilibrium consider Figure 242.1, which shows, for each value of k, the belief of challenger k at the beginning of period k along the equilibrium path that the chain-store is tough. The number k^* is the smallest value of k for which $\epsilon < b^{K-k^*+1}$. Along the equilibrium path through period k^* the challengers maintain their original belief that the chain-store is tough with probability ϵ; all challengers through $k^* - 1$ stay out. (If, contrary to its strategy, one of them enters then the regular chain-store, as well as the tough one, responds by fighting, after which the beliefs of subsequent challengers that the chain-store is tough are also maintained at ϵ.) Since $\epsilon < b^{K-k^*+1}$, challenger k^* enters. The regular chain-store responds by randomizing between fighting and cooperating (since $\mu_{CS}(h)(T) = \epsilon < b^{K-k^*}$). The probabilities it uses are such that after it fights, the probability (calculated using Bayes' rule) that it is tough is b^{K-k^*}, the point on the graph of b^{K-k+1} for period $k^* + 1$. (This has the implication that the closer $\mu_{CS}(h)(T)$ is to the graph of b^{K-k+1}, the higher is the probability that the regular chain-store fights in the event of entry.) If the chain-store cooperates (as the regular one does with positive probability) the probability that the challengers assign to the chain-store's being tough becomes zero.

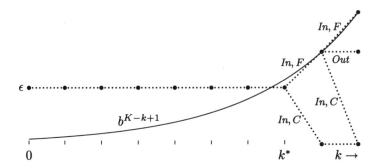

Figure 242.1 The belief of the challengers at the beginning of each period that the chain-store is tough along an equilibrium path of the perfect Bayesian equilibrium for the perturbation of the chain-store game described in the text. Possible beliefs along the path are indicated by small disks. Possible transitions are indicated by dotted lines; the label adjacent to each dotted line indicates the sequence of actions that induces the transition. (Note that only transitions that can occur with positive probability along the equilibrium path are indicated; more are possible off the equilibrium path.)

If the chain-store fights the entrant in period k^* then the probability that challenger $k^* + 1$ assigns to the chain-store's being tough rises to b^{K-k^*}, so that challenger $k^* + 1$ randomizes between entering and not. If challenger $k^* + 1$ does not enter then the belief that the chain-store is tough remains the same and challenger $k^* + 2$ definitely enters. If challenger $k^* + 1$ enters and the chain-store fights then the probability assigned to the chain-store's being tough again rises to the graph of b^{K-k+1}. If challenger $k^* + 1$ enters and the chain-store cooperates then the probability assigned to the chain-store's being tough falls to zero and challenger $k^* + 2$ enters. The same pattern continues until the end of the game: in any period in which the challenger's belief lies below the graph the challenger enters; if the chain-store responds by fighting then the belief of the subsequent challenger rises to the graph. In any period in which the challenger's belief is on the graph the challenger randomizes; if it does not enter then the belief is unchanged, while if it enters and is fought then the belief again rises to the graph. In every case the result of the chain-store's cooperating is that the probability that the challengers assign to its being tough falls to zero.

Note that if the belief of any challenger k is given by a point on the graph then after a history h that ends with the decision by k to enter the probability that the chain-store fights is $\mu_{CS}(h)(R) \cdot \sigma_{CS}(R)(h)(F) + \mu_{CS}(h)(T) \cdot 1 = (1 - b^{K-k+1})[(1 - b^{K-k})b^{K-k+1}]/[(1 - b^{K-k+1})b^{K-k}] +$

$b^{K-k+1} = b$, making challenger k indifferent between entering and staying out. The probability with which the challenger chooses to enter makes the chain-store's expected payoff 0 regardless of its future actions. Similarly, if the belief of any challenger k is given by a point below (above) the graph then the probability that the chain-store fights is less (greater) than b, making it optimal for challenger k to enter (stay out).

The number of periods remaining after the first entry of a challenger ($K - k^*$ in the case just described) is independent of the length of the game. Thus the longer the game, the more periods in which no challenger enters.

[?] EXERCISE 243.1 Complete the proof that the pair $((\sigma_i), (\mu_i))$ described above is a perfect Bayesian equilibrium of the game.

Sometimes it is said that the regular chain-store "builds reputation" in this equilibrium. Note, however, that along the equilibrium path no reputation is built: no entry takes place until the final few periods, so that even though the regular chain-store *would* fight entry were it to occur, it does not get the opportunity to do so. This response of a regular chain-store to a deviation by a challenger at the beginning of the game is necessary in order to maintain the doubt that the challengers hold about the motivation of the chain-store, a doubt required to deter them from entering. The considerations of the regular chain-store after such (out-of-equilibrium) entry near the beginning of the game are like those of a player who wants to build, or at least maintain, a reputation.

12.4 Refinements of Sequential Equilibrium

The concept of sequential equilibrium permits great (though as we have seen not complete) freedom regarding the beliefs that players hold when they observe actions that are not consistent with the equilibrium strategies. An advantage of including beliefs as part of the specification of an equilibrium is that it allows us to discuss further restrictions on these beliefs. Many such restrictions have been proposed; the new solution concepts that arise are referred to in the literature as refinements of sequential equilibrium. We give only a very brief introduction to the subject.

The notion of sequential equilibrium essentially bases beliefs on the equilibrium strategies and imposes only "structural" restrictions on out-of-equilibrium beliefs. The refinements of sequential equilibrium intro-

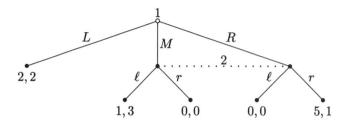

Figure 244.1 This game has a sequential equilibrium with the outcome (L, ℓ), though player 2's belief in such an equilibrium has the undesirable feature that it assigns a positive probability to player 1's having chosen the action M, which is strictly dominated by L.

duce new strategic considerations, as demonstrated by the following example.

◇ EXAMPLE 244.1 The game in Figure 244.1 has a sequential equilibrium with the outcome (R, r). It also has a sequential equilibrium with the outcome (L, ℓ), in which player 2 believes, in the event that his information set is reached, that with high probability player 1 chose M. However, if player 2's information set is reached then a reasonable argument for him may be that since the action M for player 1 is strictly dominated by L it is not rational for player 1 to choose M and hence she must have chosen R. This argument excludes any belief that supports (L, ℓ) as a sequential equilibrium outcome.

The next example further illustrates the strategic considerations introduced in the previous example.

◇ EXAMPLE 244.2 (*Beer or Quiche*) Consider the game in Figure 245.1, a signaling game in which there are two types of player 1, *strong* and *weak*, the probabilities of these types are 0.9 and 0.1 respectively, the set of messages is $\{B, Q\}$ (the consumption of beer or quiche for breakfast), and player 2 has two actions, *F(ight)* or *N(ot)*. Player 1's payoff is the sum of two elements: she obtains two units if player 2 does not fight and one unit if she consumes her preferred breakfast (B if she is strong and Q if she is weak). Player 2's payoff does not depend on player 1's breakfast; it is 1 if he fights the weak type or if he does not fight the strong type.

This game has two types of sequential equilibrium, as follows.

  • Both types of player 1 choose B, and player 2 fights if he observes Q and not if he observes B. If player 2 observes Q then he assigns probability of at least 0.5 that player 1 is weak.

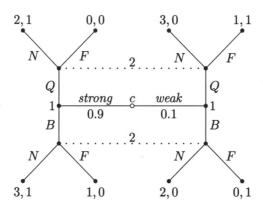

Figure 245.1 The game *Beer or Quiche* (Example 244.2).

• Both types of player 1 choose Q, and player 2 fights if he observes
B and not if he observes Q. If player 2 observes B then he assigns
probability of at least 0.5 that player 1 is weak.

The following argument suggests that an equilibrium of the second
type is not reasonable. If player 2 observes that player 1 chose B then
he should conclude that player 1 is strong, as follows. If player 1 is
weak then she should realize that the choice of B is worse for her than
following the equilibrium (in which she obtains the payoff 3), whatever
the response of player 2. Further, if player 1 is strong and if player 2
concludes from player 1 choosing B that she is strong and consequently
chooses N, then player 1 is indeed better off than she is in the equilibrium
(in which she obtains 2). Thus it is reasonable for a strong type of
player 1 to deviate from the equilibrium, anticipating that player 2 will
reason that indeed she is strong, so that player 2's belief that player 1 is
weak with positive probability when she observes B is not reasonable.

The argument in this example is weaker than that in the previous
example. In the previous example the argument uses only the fact that
the action M is dominated and thus is independent of the equilibrium
that is eliminated. By contrast, in the game here the argument is relative
to the equilibrium that is eliminated. Unless the supposition is that the
players behave according to an equilibrium in which both types choose
Q, there is no basis for the argument that the message B must come from
a strong type. This raises a criticism: if the basis of the argument is that
the situation in which both types of player 1 choose Q is an equilibrium
then perhaps player 2 should conclude after observing a deviation simply

that player 1 is not rational or does not understand the structure of the game, rather than assume that she is rationally trying to send him a strategic signal.

◊ EXAMPLE 246.1 (*Spence's model of education*) We first argue that all the pooling equilibria of the game in Example 237.2 are eliminated by arguments like those in the previous example. Let e satisfy $w^* - e^*/\theta_1^L > \theta_1^H - e/\theta_1^L$ and $w^* - e^*/\theta_1^H < \theta_1^H - e/\theta_1^H$. (Such a value of e clearly exists.) If a worker of type θ_1^H deviates and chooses e (which exceeds e^*) then the firms should conclude that the deviation comes from type θ_1^H since type θ_1^L is worse off if she so deviates even if she persuades the firms that she is of type θ_1^H, while type θ_1^H is better off if she so deviates. Thus the firms should respond to such a deviation by paying a wage of θ_1^H, which makes the deviation profitable for a worker of type θ_1^H.

Now consider separating equilibria. In such an equilibrium $e^L = 0$ and $\theta_1^H - e^H/\theta_1^L \leq \theta_1^L$. If $\theta_1^H - e^H/\theta_1^L < \theta_1^L$ then a worker of type θ_1^H can deviate by slightly reducing the value of e, arguing that she is not of type θ_1^L, who would lose from such a deviation whatever best response the firm used (that is, even if she were paid θ_1^H). Thus in all sequential equilibria that survive this argument, the level e^H of education of type θ_1^H solves the equation $\theta_1^H - e^H/\theta_1^L = \theta_1^L$.

[?] EXERCISE 246.2 (*Pre-trial negotiation*) Player 1 is involved in an accident with player 2. Player 1 knows whether she is negligent or not, but player 2 does not know; if the case comes to court the judge learns the truth. Player 1 sends a "take-it-or-leave-it" pre-trial offer of compensation that must be either 3 or 5, which player 2 either accepts or rejects. If he accepts the offer the parties do not go to court. If he rejects it the parties go to court and player 1 has to pay 5 to player 2 if he is negligent and 0 otherwise; in either case player 1 has to pay the court expenses of 6. The payoffs are summarized in Figure 247.1. Formulate this situation as a signaling game and find its sequential equilibria. Suggest a criterion for ruling out unreasonable equilibria: (Consult Banks and Sobel (1987).)

12.5 Trembling Hand Perfect Equilibrium

The notions of subgame perfect equilibrium and sequential equilibrium treat the requirement of sequential rationality as part of the players' strategic reasoning; they invoke the assumption that the players are rational not only in selecting their actions on the equilibrium path but also

	Y	N
3	$-3,3$	$-6,0$
5	$-5,5$	$-6,0$

Player 1 is non-negligent.

	Y	N
3	$-3,3$	$-11,5$
5	$-5,5$	$-11,5$

Player 1 is negligent.

Figure 247.1 The payoffs in Exercise 246.2.

in forming beliefs about the other players' plans regarding events that do not occur in equilibrium. The solution concepts that we study in this section follow a different route: they treat the players' rationality with respect to out-of-equilibrium events as the result of each player's taking into account that the other players could make uncorrelated mistakes (their hands may tremble) that lead to these unexpected events. The basic idea is that each player's actions be optimal not only given his equilibrium beliefs but also given a perturbed belief that allows for the possibility of slight mistakes. These mistakes are not modeled as part of the description of the game. Rather, a strategy profile is defined to be stable if it satisfies sequential rationality given some beliefs that are generated by a strategy profile that is a perturbation of the equilibrium strategy profile, embodying "small" mistakes. Note that the perturbed strategy profile is common to all players and the equilibrium strategy profile is required to be sequentially rational only with respect to a single such profile.

The requirement that a player's strategy be optimal not only against the other players' equilibrium strategies but also against a perturbation of these strategies that incorporates the possibility of small mistakes is powerful even in strategic games. We begin by studying such games; subsequently we turn back to extensive games with imperfect information.

12.5.1 Strategic Games

Recall that we say that a player's strategy in a strategic game is *completely mixed* if it assigns positive probability to each of the player's actions.

$$
\begin{array}{c|c|c|c|}
 & A & B & C \\
\hline
A & 0,0 & 0,0 & 0,0 \\
\hline
B & 0,0 & 1,1 & 2,0 \\
\hline
C & 0,0 & 0,2 & 2,2 \\
\hline
\end{array}
$$

Figure 248.1 A strategic game in which there are Nash equilibria $((A, A)$ and $(C, C))$ that are not trembling hand perfect.

▶ DEFINITION 248.1 A **trembling hand perfect equilibrium** of a finite strategic game is a mixed strategy profile σ with the property that there exists a sequence $(\sigma^k)_{k=0}^{\infty}$ of completely mixed strategy profiles that converges to σ such that for each player i the strategy σ_i is a best response to σ_{-i}^k for all values of k.

Let σ be a trembling hand perfect equilibrium. Since each player's expected payoff is continuous in the vector of the other players' mixed strategies it follows that for each player i the strategy σ_i is a best response to σ_{-i}, so that every trembling hand perfect equilibrium is a Nash equilibrium. Note that the definition requires only that each player's strategy be a best response to *some* sequence of perturbed strategy profiles in which the probabilities of mistakes converge to zero; all players' strategies must be best responses to the same sequence of strategy profiles, but they need not be best responses to all such sequences.

The game in Figure 248.1 shows that not all Nash equilibria are trembling hand perfect: (B, B) is the only trembling hand perfect equilibrium of the game.

In Section 4.3 we defined the notion of a weakly dominated action in a strategic game; a player has no reason to use such an action, although, depending on the other players' behavior, he may have no reason *not* to use such an action either. The notion of Nash equilibrium does not rule out the use of such actions (see, for example, the actions A and C in the game in Figure 248.1), but the notion of trembling hand perfect equilibrium does, since a weakly dominated strategy is not a best response to a vector of completely mixed strategies.

In a *two-player* game we have the following stronger result.

■ PROPOSITION 248.2 *A strategy profile in a finite two-player strategic game is a trembling hand perfect equilibrium if and only if it is a mixed*

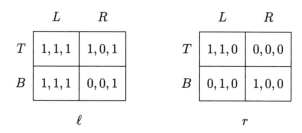

Figure 249.1 A three-player strategic game in which there is a Nash equilibrium $((B, L, \ell))$ that is not trembling hand perfect but in which every player's strategy is undominated.

strategy Nash equilibrium and the strategy of neither player is weakly dominated.

Proof. It remains to show only that a mixed strategy Nash equilibrium in which the action of each player is not weakly dominated is trembling hand perfect. Let σ^* be a mixed strategy Nash equilibrium in which the strategy σ_i^* of neither player i is weakly dominated. By the result in Exercise 64.2, the strategy σ_i^* of each player i is a best response to a completely mixed strategy, say σ_j' of player $j \neq i$. For any $\epsilon > 0$ let $\sigma_j(\epsilon) = (1-\epsilon)\sigma_j^* + \epsilon\sigma_j'$. This strategy is completely mixed and converges to σ_j^*; further, σ_i^* is a best response to it. Thus σ^* is a trembling hand perfect equilibrium. $\qquad\square$

That the same is not true for a game with more than two players is demonstrated by the three-player game in Figure 249.1. In this game the Nash equilibrium (B, L, ℓ) is undominated but is not trembling hand perfect (player 1's payoff to T exceeds her payoff to B whenever players 2 and 3 assign small enough positive probability to R and r respectively).

The following result shows that every strategic game has a trembling hand perfect equilibrium.

■ PROPOSITION 249.1 *Every finite strategic game has a trembling hand perfect equilibrium.*

Proof. Define a perturbation of the game by letting the set of actions of each player i be the set of mixed strategies of player i that assign probability of at least ϵ_i^j to each action j of player i, for some collection (ϵ_i^j) with $\epsilon_i^j > 0$ for each i and j. (That is, constrain each player to use each action available to him with some minimal probability.) Every such perturbed game has a Nash equilibrium by Proposition 20.3. Consider a sequence of such perturbed games in which $\epsilon_i^j \to 0$ for all i and j; by the

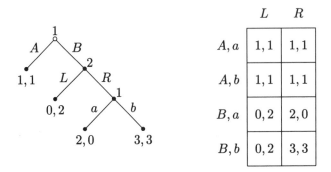

Figure 250.1 An extensive game (left) whose strategic form (right) has a trembling hand perfect equilibrium $(((A, a), L))$ that is not a subgame perfect equilibrium.

compactness of the set of strategy profiles, some sequence of selections from the sets of Nash equilibria of the games in the sequence converges, say to σ^*. It may be verified that σ^* corresponds to a trembling hand perfect equilibrium of the game.   ☐

12.5.2 Extensive Games

We now extend the idea of trembling hand perfection to the model of an extensive game. The game in Figure 250.1 shows that a straightforward generalization of Definition 248.1 has an unsatisfactory feature. This game has a unique subgame perfect equilibrium $((B, b), R)$. However, the strategy pair $((A, a), L)$ is a trembling hand perfect equilibrium of the strategic form of the game, since the strategy (A, a) of player 1 is a best response to any strategy of player 2 for which the probability of L is close enough to 1, and L is a best response to any strategy of player 1 for which the probability of (A, a) is close enough to 1 and the probability of (B, a) is sufficiently high compared with the probability of (B, b). The point is that when evaluating the optimality of her strategy player 1 does not consider the possibility that she herself will make mistakes when carrying out this strategy. If she does allow for mistakes, and considers that in attempting to carry out her strategy she may choose B rather than A at the start of the game (in addition to considering that player 2 may make a mistake and choose R rather than L) then it is no longer optimal for her to choose a at her second information set.

These considerations lead us to study the trembling hand perfect equilibria not of the strategic form but of the **agent strategic form** of the game, in which there is one player for each information set in the extensive game: each player in the extensive game is split into a number of *agents*, one for each of his information sets, all agents of a given player having the same payoffs. (Note that any mixed strategy profile σ in the agent strategic form corresponds to the behavioral strategy profile β in which $\beta_i(I_i)$ is the mixed strategy of player i's agent at the information set I_i.) Thus we make the following definition.

▸ DEFINITION 251.1 A **trembling hand perfect equilibrium** of a finite extensive game is a behavioral strategy profile that corresponds to a trembling hand perfect equilibrium of the agent strategic form of the game.

The behavioral strategy profile $((A, a), L)$ is not a trembling hand perfect equilibrium of the extensive game in Figure 250.1 since for any pair of completely mixed strategies of player 1's first agent and player 2 the unique best response of player 1's second agent is the pure strategy b. More generally, we can show that every trembling hand perfect equilibrium of a finite extensive game with perfect recall corresponds to the behavioral strategy profile of a sequential equilibrium.

■ PROPOSITION 251.2 *For every trembling hand perfect equilibrium β of a finite extensive game with perfect recall there is a belief system μ such that (β, μ) is a sequential equilibrium of the game.*

Proof. Let (β^k) be the sequence of completely mixed behavioral strategy profiles that corresponds to the sequence of mixed strategy profiles in the agent strategic form of the game that is associated with the equilibrium β. At each information set I_i of each player i in the game, define the belief $\mu(I_i)$ to be the limit of the beliefs defined from β^k using Bayes' rule; (β, μ) is then a consistent assessment. Since every agent's information set is reached with positive probability and every agent's strategy is a best response to every β^k it follows from the one deviation property for sequential equilibrium (see Exercise 227.1) that every such strategy is also a best response to β when the beliefs at each information set are defined by μ. Thus (β, μ) is a sequential equilibrium. ☐

The converse of this result does not hold since in a game with simultaneous moves every Nash equilibrium is the strategy profile of a sequential equilibrium, but only those Nash equilibria in which no player's strategy is weakly dominated can be trembling hand perfect. (In the simultaneous-move extensive game whose strategic form is given in Figure 248.1, for example, (A, A) and (C, C) are the strategy profiles of sequential equilibria but are not trembling hand perfect equilibria.) How-

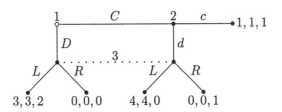

Figure **252.1** The game Example 252.1 (*Selten's horse*).

ever, the converse is "almost" true: for *almost* every game the strategy
profile of *almost* every sequential equilibrium is a trembling hand perfect
equilibrium (see Kreps and Wilson (1982b, Theorems 1 and 3)).

The next example illustrates the concept of trembling hand perfect
equilibrium for the game that we studied in Example 225.2.

◇ EXAMPLE 252.1 (*Selten's horse*) As we saw in Example 225.2, the game
in Figure 252.1 (and Figure 225.1) has two types of Nash equilibria.
Equilibria of the first type, in which player 1 chooses D, player 2 chooses
c with probability at least $\frac{1}{3}$, and player 3 chooses L, do not correspond
to sequential equilibria; they are not trembling hand perfect, since if
player 1 chooses C with positive probability and player 3 chooses L with
probability close to 1 then d is better than c for player 2. Equilibria of the
second type, in which player 1 chooses C, player 2 chooses c, and player 3
chooses R with probability at least $\frac{3}{4}$, correspond to sequential equilibria
and are also trembling hand perfect: take $\sigma_1^\epsilon(D) = \epsilon$, $\sigma_2^\epsilon(d) = 2\epsilon/(1-\epsilon)$,
and $\sigma_3^\epsilon(R) = \sigma_3(R)$ if $\sigma_3(R) < 1$ and $\sigma_3^\epsilon(R) = 1 - \epsilon$ if $\sigma_3(R) = 1$.

The game in Figure 253.1 shows that the set of trembling hand perfect
equilibria of an extensive game is not a subset of the set of trembling
hand perfect equilibria of its strategic form, and that in a trembling
hand perfect equilibrium of an extensive game a player may use a weakly
dominated strategy. The strategy profile $((L, r), R)$ is a trembling hand
perfect equilibrium of the game (take a sequence of strategy profiles in
which player 1's second agent trembles more than player 2 does), but it
is not a trembling hand perfect equilibrium of the strategic form of the
game (since player 1's strategy (L, r) is weakly dominated by (R, r)).

? EXERCISE 252.2 Show that the notion of trembling hand perfect equilibrium of an extensive game (like the notion of sequential equilibrium)
is not invariant to the coalescing of moves (one of the principles studied
in Section 11.2). (Use the game in Figure 253.1.)

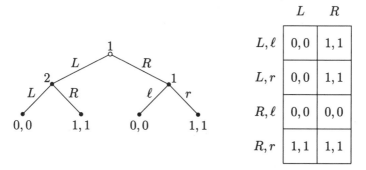

L   R

	L	R
L,ℓ	0,0	1,1
L,r	0,0	1,1
R,ℓ	0,0	0,0
R,r	1,1	1,1

Figure 253.1 An extensive game (left) that has a trembling hand perfect equilibrium that does not correspond to any trembling hand perfect equilibrium of its strategic form (right).

The next exercise gives an extensive game in *all* of whose trembling hand perfect equilibria at least one player uses a weakly dominated strategy (so that no such equilibrium is a trembling hand perfect equilibrium of the strategic form).

[?] EXERCISE 253.1 Two people are engaged in the following game to select either a good outcome or a bad outcome. First each of them names either himself or the other person as the one who will make the choice. If they both name the same person then that person selects the outcome. If each of them chooses himself then chance selects each of them with equal probability to make the choice. If each of them chooses the other then the good outcome is automatically chosen. At no point in the procedure is either person informed of the person initially selected by the other person. Each person's payoff from the good outcome is 2, regardless of who chooses it; his payoff from the bad outcome is 1 if the other person chooses it and 0 if he himself chooses it. Show that the set of trembling hand perfect equilibria of this extensive game is disjoint from the set of behavioral strategy profiles associated with the trembling hand perfect equilibria of its strategic form; interpret the equilibria.

We conclude the chapter by noting that it follows from Proposition 249.1 that every finite extensive game with perfect recall has a trembling hand perfect equilibrium and hence, by Proposition 251.2, a sequential equilibrium.

■ COROLLARY 253.2 *Every finite extensive game with perfect recall has a trembling hand perfect equilibrium and thus also a sequential equilibrium.*

Notes

The main contributors to the extension of the notion of subgame perfect equilibrium to games with imperfect information are Kreps, Selten, and Wilson, who developed the two main solution concepts discussed in this chapter: trembling hand perfect equilibrium (Selten (1975)) and sequential equilibrium (Kreps and Wilson (1982b)).

Section 12.2 relies on ideas and examples that appear in Kreps and Wilson (1982b) and many subsequent papers. For a review of the concept of sequential equilibrium see Kreps (1990b). Example 223.1 and Exercise 229.1 are due to Kreps and Ramey (1987) and Battigalli (1988). Example 225.2 is due to Selten (1975), Exercise 226.1 to Kreps and Wilson (1982b), Example 226.2 to Kohlberg and Mertens (1986), Exercise 227.1 to Hendon, Jacobsen, and Sloth (1996), Example 228.2 to Kreps and Ramey (1987), Example 229.2 to Battigalli (1996), and Example 230.1 to Kohlberg and Reny (1997).

The discussion of perfect Bayesian equilibrium in Section 12.3 is based on Fudenberg and Tirole (1991b), which contains Proposition 234.1 and Example 234.3. Example 236.1 is based on Madrigal, Tan, and Werlang (1987). Sections 12.3.1 and 12.4 are based on Cho and Kreps (1987). The model of reputation in Section 12.3.2 is based on Kreps and Wilson (1982a) (see also Milgrom and Roberts (1982)); Fudenberg and Maskin (1986, Section 5) show that the type of irrationality that is incorporated in the model may dictate the equilibrium outcome. Exercise 246.2 is due to Banks and Sobel (1987).

Most of the material in Section 12.5, which discusses the notion of trembling hand perfect equilibrium, is taken from Selten (1975). Proposition 248.2 was discovered independently by Cave, Kohlberg, and van Damme. Proposition 251.2 is due to Kreps and Wilson (1982b). The game in Exercise 253.1 is taken from Mertens (1995).

Battigalli (1996) studies sequential equilibrium and perfect Bayesian equilibrium and gives an alternative characterization of consistency for a class of games. Kohlberg and Reny (1997) formulate an equivalent definition of sequential equilibrium using systems of "relative" probabilities. As we have mentioned, there are many refinements of the notion of sequential equilibrium; especially noteworthy is the work of Kohlberg and Mertens (1986). Myerson (1978) studies a variant of the notion of trembling hand perfect equilibrium called "proper equilibrium".

Kohlberg (1990) and van Damme (1992) are surveys of the literature on refinements of Nash equilibrium.

IV   Coalitional Games

The primitives of the models we study in Parts I, II, and III (often referred to as "noncooperative" games) are the players' sets of possible actions and their preferences over the possible outcomes, where an outcome is a profile of actions; each action is taken by a single player autonomously. In this part we study the model of a *coalitional game*. One primitive of this model is the collection of sets of joint actions that each group of players (coalition) can take independently of the remaining players. An outcome of a coalitional game is a specification of the coalition that forms and the joint action it takes. (More general models, in which many coalitions may form simultaneously, are discussed in the literature.) The other primitive of the model of a coalitional game is the profile of the players' preferences over the set of all possible outcomes. Thus although actions are taken by coalitions, the theory is based (as are the theories in the other parts of the book) on the individuals' preferences.

A solution concept for coalitional games assigns to each game a set of outcomes. As before, each solution concept we study captures the consequences of a natural line of reasoning for the participants in a game; it defines a set of arrangements that are stable in some sense. In general the stability requirement is that the outcome be immune to deviations of a certain sort by groups of players; by contrast, most (though not all) solutions for noncooperative games require immunity to deviations by individual players. Many variants of the solution concepts we study are analyzed in the literature; we consider a sample designed to illustrate the main ideas.

A coalitional model is distinguished from a noncooperative model primarily by its focus on what groups of players can achieve rather than on what individual players can do and by the fact that it does not consider the details of how groups of players function internally. If we wish to model the possibility of coalition formation in a noncooperative game then we must specify how coalitions form and how their members choose joint actions. These details are absent from a coalitional game, so that the outcome of such a game does not depend on them.

To illustrate the differences between the two modeling approaches, consider the following situation. Each of a group of individuals owns a bundle of inputs and has access to a technology for producing a valuable single output. Each individual's inputs are unproductive in his own technology but productive in some other individual's technology. A noncooperative model of this situation specifies precisely the set of actions that is available to each individual: perhaps each individual can announce a price vector at which he is willing to trade inputs, or perhaps he can propose a distribution of inputs for the whole of the society. A coalitional model, by contrast, starts from the sets of payoff vectors that each group of individuals can jointly achieve. A coalition may use contracts, threats, or promises to achieve a high level of production; these institutions are not modeled explicitly in a coalitional game.

We do not view either of the two approaches as superior or more basic. Each of them reflects different kinds of strategic considerations and contributes to our understanding of strategic reasoning. The study of the interconnections between noncooperative and cooperative models can also be illuminating.

13 The Core

The core is a solution concept for coalitional games that requires that no set of players be able to break away and take a joint action that makes all of them better off. After defining the concept and giving conditions for its nonemptiness, we explore its connection with the concept of a competitive equilibrium in a model of a market.

13.1 Coalitional Games with Transferable Payoff

We begin with a simple version of a coalitional game in which each group of players is associated with a single number, interpreted as the payoff that is available to the group; there are no restrictions on how this payoff may be divided among the members of the group.

▶ DEFINITION 257.1 A **coalitional game with transferable payoff** consists of

- a finite set N (the set of **players**)
- a function v that associates with every nonempty subset S of N (a **coalition**) a real number $v(S)$ (the **worth** of S).

For each coalition S the number $v(S)$ is the total payoff that is available for division among the members of S. That is, the set of joint actions that the coalition S can take consists of all possible divisions of $v(S)$ among the members of S. (Later, in Section 13.5, we define a more general notion of a coalitional game in which each coalition is associated with a *set* of payoff vectors that is not necessarily the set of all possible divisions of some fixed amount.)

In many situations the payoff that a coalition can achieve depends on the actions taken by the other players. However, the interpretation of a

coalitional game that best fits our discussion is that it models a situation
in which the actions of the players who are not part of S do not influence
$v(S)$. In the literature other interpretations are given to a coalitional
game; for example, $v(S)$ is sometimes interpreted to be the most payoff
that the coalition S can guarantee independently of the behavior of the
coalition $N \setminus S$. These other interpretations alter the interpretation of
the solutions concepts defined; we do not discuss them here.

Throughout this chapter and the next we assume that the coalitional
games with transferable payoff that we study have the property that the
worth of the coalition N of all players is at least as large as the sum of the
worths of the members of any partition of N. This assumption ensures
that it is optimal that the coalition N of all players form, as is required
by our interpretations of the solution concepts we study (though the
formal analysis is meaningful without the assumption).

▶ DEFINITION 258.1 A coalitional game $\langle N, v \rangle$ with transferable payoff
is **cohesive** if

$$v(N) \geq \sum_{k=1}^{K} v(S_k) \quad \text{for every partition } \{S_1, \ldots, S_K\} \text{ of } N.$$

(This is a special case of the condition of *superadditivity*, which requires
that $v(S \cup T) \geq v(S) + v(T)$ for all coalitions S and T with $S \cap T = \varnothing$.)

13.2 The Core

The idea behind the core is analogous to that behind a Nash equilibrium
of a noncooperative game: an outcome is stable if no deviation is profitable. In the case of the core, an outcome is stable if no coalition can
deviate and obtain an outcome better for all its members. For a coalitional game with transferable payoff the stability condition is that no
coalition can obtain a payoff that exceeds the sum of its members' current payoffs. Given our assumption that the game is cohesive we confine
ourselves to outcomes in which the coalition N of all players forms.

Let $\langle N, v \rangle$ be a coalitional game with transferable payoff. For any profile $(x_i)_{i \in N}$ of real numbers and any coalition S we let $x(S) = \sum_{i \in S} x_i$.
A vector $(x_i)_{i \in S}$ of real numbers is an S-**feasible payoff** **vector** if
$x(S) = v(S)$. We refer to an N-feasible payoff vector as a **feasible**
payoff profile.

▶ DEFINITION 258.2 The **core of the coalitional game with trans-**
ferable payoff $\langle N, v \rangle$ is the set of feasible payoff profiles $(x_i)_{i \in N}$ for

which there is no coalition S and S-feasible payoff vector $(y_i)_{i \in S}$ for which $y_i > x_i$ for all $i \in S$.

A definition that is obviously equivalent is that the core is the set of feasible payoff profiles $(x_i)_{i \in N}$ for which $v(S) \leq x(S)$ for every coalition S. Thus the core is the set of payoff profiles satisfying a system of weak linear inequalities and hence is closed and convex.

The following examples indicate the wide range of situations that may be modeled as coalitional games and illustrate the notion of the core.

◇ EXAMPLE 259.1 (*A three-player majority game*) Suppose that three players can obtain one unit of payoff, any two of them can obtain $\alpha \in [0, 1]$ independently of the actions of the third, and each player alone can obtain nothing, independently of the actions of the remaining two players. We can model this situation as the coalitional game $\langle N, v \rangle$ in which $N = \{1, 2, 3\}$, $v(N) = 1$, $v(S) = \alpha$ whenever $|S| = 2$, and $v(\{i\}) = 0$ for all $i \in N$. The core of this game is the set of all nonnegative payoff profiles (x_1, x_2, x_3) for which $x(N) = 1$ and $x(S) \geq \alpha$ for every two-player coalition S. Hence the core is nonempty if and only if $\alpha \leq 2/3$.

◇ EXAMPLE 259.2 An expedition of n people has discovered treasure in the mountains; each pair of them can carry out one piece. A coalitional game that models this situation is $\langle N, v \rangle$, where

$$v(S) = \begin{cases} |S|/2 & \text{if } |S| \text{ is even} \\ (|S| - 1)/2 & \text{if } |S| \text{ is odd.} \end{cases}$$

If $|N| \geq 4$ is even then the core consists of the single payoff profile $(\frac{1}{2}, \ldots, \frac{1}{2})$. If $|N| \geq 3$ is odd then the core is empty.

⸮ EXERCISE 259.3 (*A production economy*) A capitalist owns a factory and each of w workers owns only his own labor power. Workers alone can produce nothing; together with the capitalist, any group of m workers can produce output worth $f(m)$, where $f: \mathbb{R}_+ \to \mathbb{R}_+$ is a concave nondecreasing function with $f(0) = 0$. A coalitional game that models this situation is $\langle N, v \rangle$ where $N = \{c\} \cup W$ (player c being the capitalist and W the set of workers) and

$$v(S) = \begin{cases} 0 & \text{if } c \notin S \\ f(|S \cap W|) & \text{if } c \in S. \end{cases}$$

Show that the core of this game is $\{x \in \mathbb{R}^N : 0 \leq x_i \leq f(w) - f(w - 1)$ for $i \in W$ and $\sum_{i \in N} x_i = f(w)\}$, where $w = |W|$, and interpret the members of this set. (See also Exercises 268.1, 289.1, and 295.2.)

◇ EXAMPLE 260.1 (*A market for an indivisible good*) In a market for an indivisible good the set of buyers is B and the set of sellers is L. Each seller holds one unit of the good and has a reservation price of 0; each buyer wishes to purchase one unit of the good and has a reservation price of 1.

We may model this market as a coalitional game with transferable payoff as follows: $N = B \cup L$ and $v(S) = \min\{|S \cap B|, |S \cap L|\}$ for each coalition S. If $|B| > |L|$ then the core consists of the single payoff profile in which every seller receives 1 and every buyer receives 0. To see this, suppose that the payoff profile x is in the core. Let b be a buyer whose payoff is minimal among the payoffs of all the buyers and let ℓ be a seller whose payoff is minimal among the payoffs of all the sellers. Since x is in the core we have $x_b + x_\ell \geq v(\{b, \ell\}) = 1$ and $|L| = v(N) = x(N) \geq |B|x_b + |L|x_\ell \geq (|B| - |L|)x_b + |L|$, which implies that $x_b = 0$ and $x_\ell \geq 1$ and hence (using $v(N) = |L|$ and the fact that ℓ is the worst-off seller) $x_i = 1$ for every seller i.

[?] EXERCISE 260.2 Calculate and interpret the core of this game when $|B| = |L|$.

◇ EXAMPLE 260.3 (*A majority game*) A group of n players, where $n \geq 3$ is odd, has one unit to divide among its members. A coalition consisting of a majority of the players can divide the unit among its members as it wishes. This situation is modeled by the coalitional game $\langle N, v \rangle$ in which $|N| = n$ and

$$v(S) = \begin{cases} 1 & \text{if } |S| \geq n/2 \\ 0 & \text{otherwise.} \end{cases}$$

This game has an empty core by the following argument. Assume that x is in the core. If $|S| = n - 1$ then $v(S) = 1$ so that $\sum_{i \in S} x_i \geq 1$. Since there are n coalitions of size $n - 1$ we thus have $\sum_{\{S:|S|=n-1\}} \sum_{i \in S} x_i \geq n$. On the other hand

$$\sum_{\{S:|S|=n-1\}} \sum_{i \in S} x_i = \sum_{i \in N} \sum_{\{S:|S|=n-1, S \ni i\}} x_i = \sum_{i \in N} (n-1)x_i = n - 1,$$

a contradiction.

[?] EXERCISE 260.4 (*Convex games*) A coalitional game with transferable payoff $\langle N, v \rangle$ is **convex** if

$$v(S) + v(T) \leq v(S \cup T) + v(S \cap T)$$

for all coalitions S and T. Let $\langle \{1, \ldots, n\}, v \rangle$ be such a game and define the payoff profile x by $x_i = v(S_i \cup \{i\}) - v(S_i)$ for each $i \in N$, where $S_i = \{1, \ldots, i-1\}$ (with $S_1 = \varnothing$). Show that x is in the core of $\langle \{1, \ldots, n\}, v \rangle$.

[?] EXERCISE 261.1 (*Simple games*) A coalitional game with transferable payoff $\langle N, v \rangle$ is **simple** if $v(S)$ is either 0 or 1 for every coalition S, and $v(N) = 1$; a coalition S for which $v(S) = 1$ is called a **winning coalition**. A player who belongs to all winning coalitions is a **veto player**.

  a. Show that if there is no veto player then the core is empty.

  b. Show that if the set of veto players is nonempty then the core is the set of all nonnegative feasible payoff profiles that give zero to all other players.

[?] EXERCISE 261.2 (*Zerosum games*) A coalitional game with transferable payoff $\langle N, v \rangle$ is **zerosum** if $v(S) + v(N \setminus S) = v(N)$ for every coalition S; it is **additive** if $v(S) + v(T) = v(S \cup T)$ for all disjoint coalitions S and T. Show that a zerosum game that is not additive has an empty core.

We remarked earlier that when modeling as a coalitional game a situation in which the actions of any coalition affect its complement there may be several ways to define $v(S)$, each entailing a different interpretation. The next exercise asks you to define $v(S)$ to be the highest payoff that S can guarantee independently of the behavior of $N \setminus S$.

[?] EXERCISE 261.3 (*Pollute the lake*) Each of n factories draws water from a lake and discharges waste into the same lake. Each factory requires pure water. It costs any factory kc to purify its water supply, where k is the number of factories that do not treat their waste before discharging it into the lake; it costs any factory b to treat its waste. Assume that $c \leq b \leq nc$.

  a. Model this situation as a coalitional game under the assumption that the worth $v(S)$ of any coalition S is the highest payoff that S can guarantee (that is, $v(S)$ is the highest payoff of S under the assumption that none of the other factories treats its waste).

  b. Find the conditions under which the game has a nonempty core and the conditions under which the core is a singleton.

  c. Discuss the interpretation of the core of this game, taking into account that the definition of $v(S)$ makes assumptions about the behavior of the players outside S.

13.3   Nonemptiness of the Core

We now derive a condition under which the core of a coalitional game
is nonempty. Since the core is defined by a system of linear inequalities
such a condition could be derived from the conditions for the existence of
a solution to a general system of inequalities. However, since the system
of inequalities that defines the core has a special structure we are able
to derive a more specific condition.

Denote by $\mathcal{C}$ the set of all coalitions, for any coalition S denote by $\mathbb{R}^S$
the $|S|$-dimensional Euclidian space in which the dimensions are indexed
by the members of S, and denote by $1_S \in \mathbb{R}^N$ the characteristic vector
of S given by

$$(1_S)_i = \begin{cases} 1 & \text{if } i \in S \\ 0 & \text{otherwise.} \end{cases}$$

A collection $(\lambda_S)_{S \in \mathcal{C}}$ of numbers in $[0, 1]$ is a **balanced collection of
weights** if for every player i the sum of λ_S over all the coalitions that
contain i is 1: $\sum_{S \in \mathcal{C}} \lambda_S 1_S = 1_N$. As an example, let $|N| = 3$. Then
the collection (λ_S) in which $\lambda_S = \frac{1}{2}$ if $|S| = 2$ and $\lambda_S = 0$ otherwise is
a balanced collection of weights; so too is the collection (λ_S) in which
$\lambda_S = 1$ if $|S| = 1$ and $\lambda_S = 0$ otherwise. A game $\langle N, v \rangle$ is **balanced** if
$\sum_{S \in \mathcal{C}} \lambda_S v(S) \leq v(N)$ for every balanced collection of weights.

One interpretation of the notion of a balanced game is the following.
Each player has one unit of time, which he must distribute among all the
coalitions of which he is a member. In order for a coalition S to be active
for the fraction of time λ_S, all its members must be active in S for this
fraction of time, in which case the coalition yields the payoff $\lambda_S v(S)$.
In this interpretation the condition that the collection of weights be
balanced is a feasibility condition on the players' allocation of time, and
a game is balanced if there is no feasible allocation of time that yields
the players more than $v(N)$.

The following result is referred to as the Bondareva–Shapley theorem.

■ PROPOSITION 262.1 *A coalitional game with transferable payoff has a
nonempty core if and only if it is balanced.*

Proof. Let $\langle N, v \rangle$ be a coalitional game with transferable payoff. First let
x be a payoff profile in the core of $\langle N, v \rangle$ and let $(\lambda_S)_{S \in \mathcal{C}}$ be a balanced
collection of weights. Then

$$\sum_{S \in \mathcal{C}} \lambda_S v(S) \leq \sum_{S \in \mathcal{C}} \lambda_S x(S) = \sum_{i \in N} x_i \sum_{S \ni i} \lambda_S = \sum_{i \in N} x_i = v(N),$$

so that $\langle N, v \rangle$ is balanced.

Now assume that $\langle N, v \rangle$ is balanced. Then there is no balanced collection $(\lambda_S)_{S \in \mathcal{C}}$ of weights for which $\sum_{S \in \mathcal{C}} \lambda_S v(S) > v(N)$. Therefore the convex set $\{(1_N, v(N) + \epsilon) \in \mathbb{R}^{|N|+1} : \epsilon > 0\}$ is disjoint from the convex cone

$$\{y \in \mathbb{R}^{|N|+1} : y = \sum_{S \in \mathcal{C}} \lambda_S(1_S, v(S)) \text{ where } \lambda_S \geq 0 \text{ for all } S \in \mathcal{C}\},$$

since if not then $1_N = \sum_{S \in \mathcal{C}} \lambda_S 1_S$, so that $(\lambda_S)_{S \in \mathcal{C}}$ is a balanced collection of weights and $\sum_{S \in \mathcal{C}} \lambda_S v(S) > v(N)$. Thus by the separating hyperplane theorem (see, for example, Rockafeller (1970, Theorem 11.3)) there is a nonzero vector $(\alpha_N, \alpha) \in \mathbb{R}^{|N|} \times \mathbb{R}$ such that

$$(\alpha_N, \alpha) \cdot y \geq 0 > (\alpha_N, \alpha) \cdot (1_N, v(N) + \epsilon) \qquad (263.1)$$

for all y in the cone and all $\epsilon > 0$. Since $(1_N, v(N))$ is in the cone, we have $\alpha < 0$.

Now let $x = \alpha_N/(-\alpha)$. Since $(1_S, v(S))$ is in the cone for all $S \in \mathcal{C}$, we have $x(S) = x \cdot 1_S \geq v(S)$ by the left-hand inequality in (263.1), and $v(N) \geq 1_N x = x(N)$ from the right-hand inequality. Adding a vector of nonnegative numbers to x to get $v(N) = x(N)$, we obtain a payoff profile that is in the core of $\langle N, v \rangle$. □

▣ EXERCISE 263.2 Let $N = \{1, 2, 3, 4\}$. Show that the game $\langle N, v \rangle$ in which

$$v(S) = \begin{cases} 1 & \text{if } S = N \\ \frac{3}{4} & \text{if } S = \{1, 2\}, \{1, 3\}, \{1, 4\}, \text{ or } \{2, 3, 4\} \\ 0 & \text{otherwise} \end{cases}$$

has an empty core, by using the fact that there exists a balanced collection $(\lambda_S)_{S \in \mathcal{C}}$ of weights in which $\lambda_S = 0$ for all coalitions S that are not equal to $\{1, 2\}$, $\{1, 3\}$, $\{1, 4\}$, or $\{2, 3, 4\}$.

13.4 Markets with Transferable Payoff

13.4.1 Definition

In this section we apply the concept of the core to a classical model of an economy. Each of the agents in the economy is endowed with a bundle of goods that can be used as inputs in a production process that the agent can operate. All production processes produce the same output, which can be transferred between the agents. Formally, a **market with transferable payoff** consists of

- a finite set N (the set of *agents*)

- a positive integer ℓ (the number of input goods)
- for each agent $i \in N$ a vector $\omega_i \in \mathbb{R}^\ell_+$ (the *endowment* of agent i)
- for each agent $i \in N$ a continuous, nondecreasing, and concave function $f_i \colon \mathbb{R}^\ell_+ \to \mathbb{R}_+$ (the *production function* of agent i).

An *input vector* is a member of $\mathbb{R}^\ell_+$; a profile $(z_i)_{i \in N}$ of input vectors for which $\sum_{i \in N} z_i = \sum_{i \in N} \omega_i$ is an *allocation*.

In such a market the agents may gain by cooperating: if their endowments are complementary then in order to maximize total output they may need to exchange inputs. However, the agents' interests conflict as far as the distribution of the benefits of cooperation is concerned. Thus a game-theoretic analysis is called for.

We can model a market with transferable payoff $\langle N, \ell, (\omega_i), (f_i) \rangle$ as a coalitional game with transferable payoff $\langle N, v \rangle$ in which N is the set of agents and for each coalition S we have

$$v(S) = \max_{(z_i)_{i \in S}} \left\{ \sum_{i \in S} f_i(z_i) \colon z_i \in \mathbb{R}^\ell_+ \text{ and } \sum_{i \in S} z_i = \sum_{i \in S} \omega_i \right\}. \qquad (264.1)$$

That is, $v(S)$ is the maximal total output that the members of S can produce by themselves. We define the **core of a market** to be the core of the associated coalitional game.

Note that our assumptions that all agents produce the same good and the production of any coalition S is independent of the behavior of $N \setminus S$ are essential.

13.4.2   Nonemptiness of the Core

We now use the Bondareva–Shapley theorem (262.1) to show that every market with transferable payoff has a nonempty core.

■ PROPOSITION 264.2 *Every market with transferable payoff has a non-empty core.*

Proof. Let $\langle N, \ell, (\omega_i), (f_i) \rangle$ be a market with transferable payoff and let $\langle N, v \rangle$ be the coalitional game defined in (264.1). By the Bondareva–Shapley theorem it suffices to show that $\langle N, v \rangle$ is balanced. Let $(\lambda_S)_{S \in \mathcal{C}}$ be a balanced collection of weights. We must show that $\sum_{S \in \mathcal{C}} \lambda_S v(S) \le v(N)$. For each coalition S let $(z_i^S)_{i \in S}$ be a solution of the problem (264.1) defining $v(S)$. For each $i \in N$ let $z_i^* = \sum_{S \in \mathcal{C}, S \ni i} \lambda_S z_i^S$. We have

$$\sum_{i \in N} z_i^* = \sum_{i \in N} \sum_{S \in \mathcal{C}, S \ni i} \lambda_S z_i^S = \sum_{S \in \mathcal{C}} \sum_{i \in S} \lambda_S z_i^S = \sum_{S \in \mathcal{C}} \lambda_S \sum_{i \in S} z_i^S =$$

$$\sum_{S \in \mathcal{C}} \lambda_S \sum_{i \in S} \omega_i = \sum_{i \in N} \omega_i \sum_{S \in \mathcal{C}, S \ni i} \lambda_S = \sum_{i \in N} \omega_i,$$

where the last equality follows from the fact that $(\lambda_S)_{S \in \mathcal{C}}$ is a balanced collection of weights. It follows from the definition of $v(N)$ that $v(N) \geq \sum_{i \in N} f_i(z_i^*)$; the concavity of each function f_i and the fact that the collection of weights is balanced implies that

$$\sum_{i \in N} f_i(z_i^*) \geq \sum_{i \in N} \sum_{S \in \mathcal{C}, S \ni i} \lambda_S f_i(z_i^S) = \sum_{S \in \mathcal{C}} \lambda_S \sum_{i \in S} f_i(z_i^S) = \sum_{S \in \mathcal{C}} \lambda_S v(S),$$

completing the proof. □

◇ EXAMPLE 265.1 Consider the market with transferable payoff in which $N = K \cup M$, there are two input goods ($\ell = 2$), $\omega_i = (1, 0)$ if $i \in K$, $\omega_i = (0, 1)$ if $i \in M$, and $f_i(a, b) = \min\{a, b\}$ for every $i \in N$. Then $v(S) = \min\{|K \cap S|, |M \cap S|\}$. By Proposition 264.2 the core is nonempty. If $|K| < |M|$ then it consists of a single point, in which each agent in K receives the payoff of 1 and each agent in M receive the payoff of 0; the proof is identical to that for the market with an indivisible good in Example 260.1.

? EXERCISE 265.2 Consider the market with transferable payoff like that of the previous example in which there are five agents, $\omega_1 = \omega_2 = (2, 0)$, and $\omega_3 = \omega_4 = \omega_5 = (0, 1)$.

a. Find the coalitional form of this market and calculate the core.

b. Suppose that agents 3, 4, and 5 form a syndicate: they enter coalitions only as a block, so that we have a three-player game. Does the core predict that the formation of the syndicate benefits its members? Interpret your answer.

13.4.3 The Core and the Competitive Equilibria

Classical economic theory defines the solution of "competitive equilibrium" for a market. We now show that the core of a market contains its competitive equilibria.

We begin with the simple case in which all agents have the same production function f and there is only one input. Let $\omega^* = \sum_{i \in N} \omega_i / |N|$, the average endowment. Given the concavity of f, the allocation in which each agent receives the amount ω^* of the input maximizes the total output. Let p^* be the slope of a tangent to the production function at ω^* and let g be the affine function with slope p^* for which $g(\omega^*) = f(\omega^*)$ (see Figure 266.1). Then $(g(\omega_i))_{i \in N}$ is in the core since

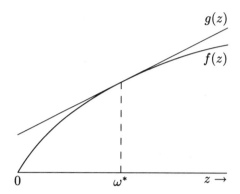

Figure 266.1 The production function f of each agent and the function g in a market in which there is a single input.

$v(S) = |S|f((\sum_{i \in S} \omega_i)/|S|) \leq |S|g((\sum_{i \in S} \omega_i)/|S|) = \sum_{i \in S} g(\omega_i)$ and $v(N) = |N|f((\sum_{i \in N} \omega_i)/|N|) = |N|f(\omega^*) = |N|g(\omega^*) = \sum_{i \in N} g(\omega_i)$.

The payoff profile $(g(\omega_i))_{i \in N}$ can be achieved by each agent trading input for output at the price p^* (each unit of input costs p^* units of output): if trade at this price is possible then agent i maximizes his payoff by choosing the amount z of input to solve $\max_z (f(z) - p^*(z - \omega_i))$, the solution of which is ω^*. In terms of the next definition, the pair $(p^*, (z_i^*)_{i \in N})$ where $z_i^* = \omega^*$ for all $i \in N$ is a competitive equilibrium of the market.

We define a **competitive equilibrium** of a market with transferable payoff as a pair $(p^*, (z_i^*)_{i \in N})$ consisting of a vector $p^* \in \mathbb{R}_+^\ell$ (the vector of *input prices*) and an allocation $(z_i^*)_{i \in N}$ such that for each agent i the vector z_i^* solves the problem

$$\max_{z_i \in \mathbb{R}_+^\ell} (f_i(z_i) - p^*(z_i - \omega_i)). \qquad (266.1)$$

If $(p^*, (z_i^*)_{i \in N})$ is a competitive equilibrium then we refer to $f_i(z_i^*) - p^*(z_i^* - \omega_i)$, the value of the maximum in (266.1), as a **competitive payoff of agent** i. The idea is that the agents can trade inputs at fixed prices, which are expressed in terms of units of output. If after buying and selling inputs agent i holds the bundle z_i then his net expenditure, in units of output, is $p^*(z_i - \omega_i)$; he can produce $f_i(z_i)$ units of output, so that his net payoff is $f_i(z_i) - p^*(z_i - \omega_i)$. A price vector p^* generates a competitive equilibrium if, when each agent chooses his trades to maximize his payoff, the resulting profile $(z_i^*)_{i \in N}$ of input vectors is feasible in the sense that it is an allocation.

We now show that any profile of competitive payoffs in a market with transferable payoff is in the core.

■ PROPOSITION 267.1 *Every profile of competitive payoffs in a market with transferable payoff is in the core of the market.*

Proof. Let $\langle N, \ell, (\omega_i), (f_i) \rangle$ be a market with transferable payoff, let $\langle N, v \rangle$ be the associated coalitional game, let $(p^*, (z_i^*)_{i \in N})$ be a competitive equilibrium of the market, and assume contrary to the result that the profile of associated competitive payoffs is not in the core. Then there is a coalition S and a vector $(z_i)_{i \in S}$ such that $\sum_{i \in S} z_i = \sum_{i \in S} \omega_i$ and $\sum_{i \in S} f_i(z_i) > \sum_{i \in S}(f_i(z_i^*) - p^* z_i^* + p^* \omega_i)$. It follows that $\sum_{i \in S}(f_i(z_i) - p^* z_i) > \sum_{i \in S}(f_i(z_i^*) - p^* z_i^*)$ and hence for at least one agent $i \in S$ we have $f_i(z_i) - p^* z_i > f_i(z_i^*) - p^* z_i^*$, contradicting the fact that z_i^* is a solution of (266.1). Finally, $v(N) = \sum_{i \in N} f_i(z_i^*)$ since for any $(z_i)_{i \in N}$ such that $\sum_{i \in N} z_i = \sum_{i \in N} \omega_i$ we have $\sum_{i \in N} f_i(z_i) \leq \sum_{i \in N}(f_i(z_i^*) - p^* z_i^* + p^* \omega_i) = \sum_{i \in N} f_i(z_i^*)$. □

Proposition 267.1 provides an alternative route to show that the core of a market with transferable payoff is nonempty, since every market with transferable payoff has a competitive equilibrium, as the following exercise shows.

? EXERCISE 267.2 Let $\langle N, \ell, (\omega_i), (f_i) \rangle$ be a market with transferable payoff in which every component of $\sum_{i \in N} \omega_i$ is positive, let $X_i = \{(z_i, y_i) \in \mathbb{R}_+^\ell \times \mathbb{R} : y_i \leq f_i(z_i)\}$ for each $i \in N$, and let $\{z_i^*\}_{i \in N}$ be a solution of

$$\max_{\{z_i\}_{i \in N}} \left\{ \sum_{i \in N} f_i(z_i) : \text{subject to } \sum_{i \in N} z_i \leq \sum_{i \in N} \omega_i \right\}.$$

Show that the coefficients of the hyperplane that separates $\sum_{i \in N} X_i$ from $\{(z, y) \in \mathbb{R}^\ell \times \mathbb{R} : z \leq \sum_{i \in N} z_i^* \text{ and } y \geq \sum_{i \in N} f_i(z_i^*)\}$ define competitive prices.

The notion of competitive equilibrium is intended to capture a world in which the bargaining power of each agent is small. In a market that contains only a few agents some may have strong bargaining positions, and the core may contain outcomes very different from the competitive equilibrium. However, in a large market, where each agent's action has only a small effect on the outcome, we might expect the core to contain only outcomes that are similar to the competitive equilibrium. The following exercise illustrates this idea in a special case; in Section 13.6.2 we study the idea in a more general context.

[?] EXERCISE 268.1 (*A production economy*) Let $\langle N, \ell, (\omega_i), (f_i) \rangle$ be a market with transferable payoff in which $N = \{1, \ldots, k+1\}$, $\ell = 2$, $\omega_1 = (1, 0)$, $\omega_i = (0, 1)$ for $i \neq 1$, and $f_i = f$ for all $i \in N$, with $f(0, m) = 0$ for all m, $f(1, 0) = 0$, and $\lim_{m \to \infty} f(1, m) < \infty$. Suppose that the input goods are indivisible. The associated coalitional game is the same as that in Exercise 259.3. Show that for all $\epsilon > 0$ there is an integer $k^*(\epsilon)$ such that for all $k > k^*(\epsilon)$ no member of the core gives player 1 a payoff less than $f(1, k) - \epsilon$. Give an economic interpretation of this result.

13.5 Coalitional Games without Transferable Payoff

In a coalitional game with transferable payoff each coalition S is characterized by a single number $v(S)$, with the interpretation that $v(S)$ is a payoff that may be distributed in any way among the members of S. We now study a more general concept, in which each coalition cannot necessarily achieve all distributions of some fixed payoff; rather, each coalition S is characterized by an arbitrary set $V(S)$ of consequences.

▶ DEFINITION 268.2 A **coalitional game** (without transferable payoff) consists of

- a finite set N (the set of **players**)

- a set X (the set of **consequences**)

- a function V that assigns to every nonempty subset S of N (a **coalition**) a set $V(S) \subseteq X$

- for each player $i \in N$ a preference relation $\succsim_i$ on X.

Any coalitional game with transferable payoff $\langle N, v \rangle$ (Definition 257.1) can be associated with a general coalitional game $\langle N, X, V, (\succsim_i)_{i \in N} \rangle$ as follows: $X = \mathbb{R}^N$, $V(S) = \{x \in \mathbb{R}^N : \sum_{i \in S} x_i = v(S) \text{ and } x_j = 0 \text{ if } j \in N \setminus S\}$ for each coalition S, and $x \succsim_i y$ if and only if $x_i \geq y_i$. Under this association the set of coalitional games with transferable payoff is a subset of the set of all coalitional games.

The definition of the core of a general coalitional game is a natural extension of our definition for the core of a game with transferable payoff (Definition 258.2).

▶ DEFINITION 268.3 The **core of the coalitional game** $\langle N, V, X, (\succsim_i)_{i \in N} \rangle$ is the set of all $x \in V(N)$ for which there is no coalition S and $y \in V(S)$ for which $y \succ_i x$ for all $i \in S$.

Under conditions like that of balancedness for a coalitional game with transferable payoff (see Section 13.3) the core of a general coalitional game is nonempty (see Scarf (1967), Billera (1970), and Shapley (1973)). We do not discuss these conditions here.

13.6 Exchange Economies

13.6.1 Definitions

A generalization of the notion of a market with transferable payoff is the following. An **exchange economy** consists of

- a finite set N (the set of *agents*)
- a positive integer ℓ (the number of goods)
- for each agent $i \in N$ a vector $\omega_i \in \mathbb{R}_+^\ell$ (the *endowment* of agent i) such that every component of $\sum_{i \in N} \omega_i$ is positive
- for each agent $i \in N$ a nondecreasing, continuous, and quasi-concave preference relation $\succsim_i$ over the set $\mathbb{R}_+^\ell$ of bundles of goods.

The interpretation is that ω_i is the bundle of goods that agent i owns initially. The requirement that every component of $\sum_{i \in N} \omega_i$ be positive means that there is a positive quantity of every good available in the economy. Goods may be transferred between the agents, but there is no payoff that is freely transferable.

An **allocation** is a distribution of the total endowment in the economy among the agents: that is, a profile $(x_i)_{i \in N}$ with $x_i \in \mathbb{R}_+^\ell$ for all $i \in N$ and $\sum_{i \in N} x_i = \sum_{i \in N} \omega_i$. A **competitive equilibrium** of an exchange economy is a pair $(p^*, (x_i^*)_{i \in N})$ consisting of a vector $p^* \in \mathbb{R}_+^\ell$ with $p^* \neq 0$ (the **price** vector) and an allocation $(x_i^*)_{i \in N}$ such that for each agent i we have $p^* x_i^* \leq p^* \omega_i$ and

$$x_i^* \succsim_i x_i \text{ for any } x_i \text{ for which } p^* x_i \leq p^* \omega_i. \qquad (269.1)$$

If $(p^*, (x_i^*)_{i \in N})$ is a competitive equilibrium then $(x_i^*)_{i \in N}$ is a **competitive allocation**.

As in the case of a competitive equilibrium of a market with transferable payoff, the idea is that the agents can trade goods at fixed prices. Here there is no homogeneous output in terms of which the prices are expressed; rather, we can think of p_j^* as the "money" price of good j. Given any price vector p, each agent i chooses a bundle that is most desirable (according to his preferences) among all those that are affordable (i.e. satisfy $p x_i \leq p \omega_i$). Typically an agent chooses a bundle that contains more of some goods and less of others than he initially owns: he "demands" some goods and "supplies" others. The requirement in the

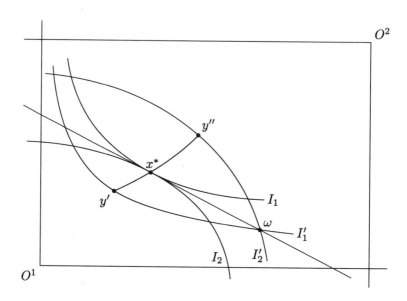

Figure 270.1 An Edgeworth box, illustrating an exchange economy in which there are two agents and two goods. A competitive equilibrium price ratio is given by the slope of the line through ω and x^*; x^* corresponds to a competitive allocation. The core is the set of all allocations that correspond to points on the line joining y' and y''.

definition of competitive equilibrium that the profile of chosen bundles be an allocation means that for every good the sum of the individuals' demands is equal to the sum of their supplies.

A standard result in economic theory is that an exchange economy in which every agent's preference relation is increasing has a competitive equilibrium (see, for example, Arrow and Hahn (1971, Theorem 5 on p. 119)[1]). Note that an economy may possess many such equilibria.

An exchange economy that contains two agents ($|N| = 2$) and two goods ($\ell = 2$) can be conveniently represented in a diagram like that in Figure 270.1, which is known as an *Edgeworth box*. Bundles of goods consumed by agent 1 are measured from the origin O^1 in the bottom left, while bundles consumed by agent 2 are measured from the origin O^2 in the top right. The width of the box formed by the two pairs of axes is the total endowment of good 1 in the economy and the height

---

[1] Arrow and Hahn's result is for the more general notion of an economy with production. To apply it here, let the production set of each firm f be $Y_f = \{0\}$. Note that if every agent's preference relation is increasing then every agent is resource related (in the sense of Arrow and Hahn) to every other agent.

of the box is the total endowment of good 2. Thus each point x in the box corresponds to an allocation in which agent i receives the bundle x measured from O^i; the point labeled ω corresponds to the pair of endowments. The curved lines labeled I_i and I_i' are *indifference curves* of agent i: if x and y are points on one of these curves then $x \sim_i y$. The straight line passing through ω and x^* is (relative to O^i) the set of all bundles x_i for which $px_i = p\omega_i$. The point x^* corresponds to a competitive allocation since the most preferred bundle of agent i in the set $\{x_i : px_i \leq p\omega_i\}$ is x^* when measured from origin O^i. The ratio of the competitive prices is the negative of the slope of the straight line through ω and x^*.

An exchange economy is closely related to a market (as defined in Section 13.4). In a market, payoff can be directly transferred between agents, while in an exchange economy only goods can be directly transferred. Thus we model an exchange economy as a coalitional game without transferable payoff. Precisely, we associate the exchange economy $\langle N, \ell, (\omega_i), (\succsim_i) \rangle$ with the coalitional game $\langle N, X, V, (\succsim_i) \rangle$ where

- $X = \{(x_i)_{i \in N} : x_i \in \mathbb{R}_+^\ell \text{ for all } i \in N\}$;
- $V(S) = \{(x_i)_{i \in N} \in X : \sum_{i \in S} x_i = \sum_{i \in S} \omega_i \text{ and } x_j = \omega_j \text{ for all } j \in N \setminus S\}$ for each coalition S;
- each preference relation $\succsim_i$ is defined by $(x_j)_{j \in N} \succsim_i (y_j)_{j \in N}$ if and only if $x_i \succsim_i y_i$.

The third condition expresses the assumption that each agent cares only about his own consumption. We define the **core of an exchange economy** to be the core of the associated coalitional game.

13.6.2   The Core and the Competitive Equilibria

For the coalitional game $\langle N, X, V, (\succsim_i) \rangle$ associated with the exchange economy $\langle N, \ell, (\omega_i), (\succsim_i) \rangle$ the set $V(N)$ is the set of all allocations and for each $j \in N$ we have $V(\{j\}) = \{(\omega_i)_{i \in N}\}$. Thus the core of a two-agent economy is the set of all allocations $(x_i)_{i \in N}$ such that $x_j \succsim_j \omega_j$ for each agent j and there is no allocation $(x_i')_{i \in N}$ such that $x_j' \succ_j x_j$ for both agents j. For example, in the Edgeworth box in Figure 270.1 the core corresponds to the locus of points in the area bounded by I_1' and I_2' for which an indifference curve of agent 1 and an indifference curve of agent 2 share a common tangent (i.e. it is the curved line passing through y', x^*, and y''). In particular, the core contains the competitive allocation. We now show that this is a general property.

■ PROPOSITION 272.1 *Every competitive allocation in an exchange economy is in the core.*

Proof. Let $E = \langle N, \ell, (\omega_i), (\succsim_i) \rangle$ be an exchange economy, let $(p^*, (x_i^*)_{i \in N})$ be a competitive equilibrium of E, and assume that $(x_i^*)_{i \in N}$ is not in the core of E. Then there is a coalition S and $(y_i)_{i \in S}$ with $\sum_{i \in S} y_i = \sum_{i \in S} \omega_i$ such that $y_i \succ_i x_i^*$ for all $i \in S$; using (269.1) we have $p^* y_i > p^* \omega_i$ for all $i \in S$. Hence $p^* \sum_{i \in S} y_i > p^* \sum_{i \in S} \omega_i$, contradicting $\sum_{i \in S} y_i = \sum_{i \in S} \omega_i$. □

Note that it follows from this result that an economy that has a competitive equilibrium has a nonempty core.

By examining an Edgeworth box we can see that in a two-good two-agent economy the core may be large. However, we now show that as the number of agents increases, the core shrinks to the set of competitive allocations. That is, in a large enough economy the predictions of the competitive equilibrium—a concept that is based on agents who trade at fixed prices—are very close to those of the core—a concept that is based on the ability of a group of agents to improve its lot by forming an autonomous subeconomy, without reference to prices. Put differently, in a large enough economy the only outcomes that are immune to deviations by groups of agents are competitive equilibrium allocations.

To state the result precisely, let E be an exchange economy in which there are n agents. For any positive integer k let kE be the economy derived from E in which there are kn agents—k copies of each agent in E. We refer to an agent j in kE who is a copy of agent i in E as being of **type** $i = \iota(j)$. The comparison between the core of E and that of kE is facilitated by the following result.

■ LEMMA 272.2 (Equal treatment in the core) *Let E be an exchange economy in which the preference relation of every agent is increasing and strictly quasi-concave, and let k be a positive integer. In any allocation in the core of kE all agents of the same type obtain the same bundle.*

Proof. Let $E = \langle N, \ell, (\omega_i), (\succsim_i) \rangle$ and let x be an allocation in the core of kE in which there are two agents of type t^* whose bundles are different. We now show that there is a distribution of the endowment of the coalition consisting of the worst-off agent of each type that makes every member of the coalition better off than he is in x. Precisely, for each type t select one agent, i_t, in kE who is least well off (according to $\succsim_t$) in x among all agents of type t, and let S be the coalition (of size $|N|$) of these agents. For each type t let z_t be the average bundle of the agents of type t in the allocation x: $z_t = \sum_{\{j:\iota(j)=t\}} x_j/k$. Then we have

- $\sum_{t \in N} z_t = \sum_{t \in N} \omega_t$;
- $z_t \succsim_t x_{i_t}$ (otherwise, $z_t \prec_t x_j$ whenever $\iota(j) = t$, so that by the quasi-concavity of $\succsim_t$ we have $z_t \prec_t z_t$, a contradiction);
- $z_{t^*} \succ_{t^*} x_{i_{t^*}}$ (by the strict quasi-concavity of the preference relations).

That is, (i) it is feasible for the coalition S to assign to each agent $j \in S$ the bundle $z_{\iota(j)}$ (since $\sum_{j \in S} z_{\iota(j)} = \sum_{t \in N} z_t = \sum_{j \in S} \omega_j$), (ii) for every agent $j \in S$ the bundle $z_{\iota(j)}$ is at least as desirable as x_j, and (iii) for the agent $j \in S$ of type t^* the bundle $z_{\iota(j)}$ is preferable to x_j.

Since each agent's preference relation is increasing we can modify the allocation $(z_t)_{t \in N}$ by reducing t^*'s bundle by a small amount and distributing this amount equally among the other members of S so that we have a profile $(z'_t)_{t \in N}$ with $\sum_{t \in N} z'_t = \sum_{t \in N} \omega_t$ and $z'_{\iota(j)} \succ_{\iota(j)} x_j$ for all $j \in S$. This contradicts the fact that x is in the core of kE. □

Given this result, for any positive integer k we can identify the core of kE with a profile of $|N|$ bundles, one for each type. Under this identification it is clear that the core of kE is a subset of the core of E. We now show that the core of kE shrinks to the set of competitive allocations of E as k increases.

- PROPOSITION 273.1 *Let E be an exchange economy in which every agent's preference relation is increasing and strictly quasi-concave and every agent's endowment of every good is positive. Let x be an allocation in E. If for every positive integer k the allocation in kE in which every agent of each type t receives the bundle x_t is in the core of kE then x is a competitive allocation of E.*

Proof. Let $E = \langle N, \ell, (\omega_i), (\succsim_i) \rangle$. Let

$$Q = \left\{ \sum_{t \in N} \alpha_t z_t : \sum_{t \in N} \alpha_t = 1, \alpha_t \geq 0, \text{ and } z_t + \omega_t \succ_t x_t \text{ for all } t \right\}.$$

Under our assumptions on preferences Q is convex. We claim that $0 \notin Q$. Suppose to the contrary that $0 = \sum_{t \in N} \alpha_t z_t$ for some (α_t) and (z_t) with $\sum_{t \in N} \alpha_t = 1$, $\alpha_t \geq 0$, and $z_t + \omega_t \succ_t x_t$ for all t. Suppose that every α_t is a rational number. (If not, we need to do some approximation.) Choose an integer K large enough that $K\alpha_t$ is an integer for all t, let S be a coalition in KE that consists of $K\alpha_t$ agents of each type t, and let $x'_i = z_{\iota(i)} + \omega_i$ for each $i \in S$. We have $\sum_{i \in S} x'_i = \sum_{t \in N} K\alpha_t z_t +$

$\sum_{i \in S} \omega_i = \sum_{i \in S} \omega_i$ and $x_i' \succ_i x_i$ for all $i \in S$, contradicting the fact that x is in the core of KE.

Now by the separating hyperplane theorem (see, for example, Rockafeller (1970, Theorem 11.3)) there is a nonzero vector $p \in \mathbb{R}^\ell$ such that $pz \geq 0$ if $z \in Q$. Since all the agents' preferences are increasing, each unit vector is in Q (take $z_t = x_t - \omega_t + 1_{\{m\}}$ and $\alpha_t = 1/|N|$ for each t, where $1_{\{m\}}$ is the mth unit vector in $\mathbb{R}^\ell$). Thus $p \geq 0$.

We now argue that if $y_i \succ_i x_i$ for some $i \in N$ then $py_i > p\omega_i$, so that by (269.1) x is a competitive allocation of E. Suppose that $y_i \succ_i x_i$. Then $y_i - \omega_i \in Q$, so that by the choice of p we have $py_i \geq p\omega_i$. Furthermore, $\theta y_i \succ_i x_i$ for some $\theta < 1$, so that $\theta y_i - \omega_i \in Q$ and hence $\theta p y_i \geq p\omega_i$; also $p\omega_i > 0$ since every component of ω_i is positive. Thus $py_i > p\omega_i$. $\square$

In any competitive equilibrium of kE all agents of the same type consume the same bundle, so that any such equilibrium is naturally associated with a competitive equilibrium of E. Thus the result shows a sense in which the larger k is, the closer are the core and the set of competitive allocations of kE.

[?] EXERCISE 274.1 Consider an exchange economy E in which there are two goods and two agents; agent 1's endowment is $(1, 0)$ and her preferences are represented by the utility function $x_1 + x_2$, while agent 2's endowment is $(0, 1)$ and his preferences are represented by the utility function $\min\{x_1, x_2\}$. For each positive integer k find the core and set of competitive allocations of kE.

Notes

The notion of a coalitional game is due to von Neumann and Morgenstern (1944). In the early 1950s Gillies introduced the notion of the core as a tool to study stable sets (his work is published in Gillies (1959)); Shapley and Shubik developed it as a solution concept. Proposition 262.1 is due to Bondareva (1963) and Shapley (1967). The idea of modeling markets as coalitional games is due to von Neumann and Morgenstern (1944, pp. 583–584); it was developed by Shapley and Shubik (see, for example, Shapley (1959) and Shubik (1959a)). Proposition 264.2 is due to Shapley and Shubik (1969a). The idea of generalizing a coalitional game to situations in which payoff is not transferable is due to Shapley and Shubik (1953) and Luce and Raiffa (1957, pp. 234–235); the formulation that we describe is due to Aumann and Peleg (1960).

Scarf (1967), Billera (1970), and Shapley (1973) discuss the nonemptiness of the core of a coalitional game without transferable payoff. The relation between the core and the set of competitive equilibria of an economy was first noticed by Edgeworth (1881, pp. 35–39). The relation between Edgeworth's work and modern notions in game theory was recognized by Shubik (1959a). Proposition 273.1 is due to Debreu and Scarf (1963); for a diagrammatic proof for a two-agent two-good economy see Varian (1992, pp. 387–392).

Example 259.2 is due to Shapley (inspired by the 1948 movie *The Treasure of the Sierra Madre*). The game in Exercise 259.3 is analyzed by Shapley and Shubik (1967). The market in Example 260.1 is studied by Shapley (1959). Exercise 260.4 is taken from Shapley (1971/72), Exercise 261.3 from Shapley and Shubik (1969b), Exercise 265.2 from Postlewaite and Rosenthal (1974), and Exercise 268.1 from Owen (1982, Theorem IX.3.2).

Aumann (1989) contains an introduction to the theory of coalitional games. Other references include Owen (1982), Shubik (1982), Moulin (1986, 1988), Friedman (1990), and Myerson (1991).

Aumann (1964) provides an alternative formulation of Edgeworth's idea that the core converges to the set of competitive equilibria in a large economy: he studies a model in which there is a continuum of agents and shows that the core coincides with the set of competitive equilibria. Axiomatizations of the core are surveyed by Peleg (1992).

14 Stable Sets, the Bargaining Set, and the Shapley Value

In contrast to the core, the solution concepts we study in this chapter restrict the way that an objecting coalition may deviate, by requiring that each possible deviation either itself be a stable outcome or be balanced by a counterdeviation. These restrictions yield several solutions: stable sets, the bargaining set, the kernel, the nucleolus, and the Shapley value.

14.1 Two Approaches

The definition of the core does not restrict a coalition's credible deviations, beyond imposing a feasibility constraint. In particular it assumes that any deviation is the end of the story and ignores the fact that a deviation may trigger a reaction that leads to a different final outcome. The solution concepts we study in this chapter consider various restrictions on deviations that are motivated by these considerations.

In the first approach we study (in Section 14.2), an objection by a coalition to an outcome consists of an alternative outcome that is itself constrained to be stable. The idea is that a deviation by a coalition will lead via some sequence of events to a stable outcome and that a coalition should choose to deviate on the basis of the ultimate effect of its action, not the proximate effect. This stability condition is self-referential: a stable outcome has the property that no coalition can achieve some other stable outcome that improves the lot of all its members.

In the second approach (studied in Sections 14.3 and 14.4) the chain of events that a deviation unleashes is cut short after two stages: the stability condition is that for every objection to an outcome there is a balancing counterobjection. Different notions of objection and counterobjection give rise to a number of different solution concepts.

The arguments captured by the solution concepts in this chapter are attractive. Nevertheless, it is our impression that there are few persuasive applications of the concepts. Consequently we simply describe the concepts, discuss their interpretations, and give simple examples. Throughout we restrict attention to coalitional games with transferable payoff.

14.2   The Stable Sets of von Neumann and Morgenstern

The idea behind the first solution concept we study is that a coalition S that is unsatisfied with the current division of $v(N)$ can *credibly* object by suggesting a *stable* division x of $v(N)$ that is better for all the members of S and is backed up by a threat to implement $(x_i)_{i \in S}$ on its own (by dividing the worth $v(S)$ among its members). The logic behind the requirement that an objection itself be stable is that otherwise the objection may unleash a process involving further objections by other coalitions, at the end of which some of members of the deviating coalition may be worse off.

This idea leads to a definition in which a set of stable outcomes satisfies two conditions: (i) for every outcome that is not stable some coalition has a credible objection and (ii) no coalition has a credible objection to any stable outcome. Note that this definition is self-referential and admits the possibility that there be many stable sets.

We now turn to the formal definition. Let $\langle N, v \rangle$ be a coalitional game with transferable payoff. As in the previous chapter we assume that $\langle N, v \rangle$ is cohesive (see Definition 258.1). An **imputation** of $\langle N, v \rangle$ is a feasible payoff profile x for which $x_i \geq v(\{i\})$ for all $i \in N$; let X be the set of all imputations of $\langle N, v \rangle$. We first define objections (which are not necessarily credible).

- An imputation x is an **objection of the coalition S to the imputation** y if $x_i > y_i$ for all $i \in S$ and $x(S) \leq v(S)$, in which case we write $x \succ_S y$.

(In the literature it is sometimes said that "x dominates y via S" if x is an objection of S to y.) Since $\langle N, v \rangle$ is cohesive we have $x \succ_S y$ if and only if there is an S-feasible payoff vector $(x_i)_{i \in S}$ for which $x_i > y_i$ for all $i \in S$. The core of the game $\langle N, v \rangle$ is the set of all imputations to which there is no objection: $\{y \in X:$ there is no coalition S and imputation x for which $x \succ_S y\}$. The solution concept we now study is defined as follows.

▶ DEFINITION 279.1 A subset Y of the set X of imputations of a coalitional game with transferable payoff $\langle N, v \rangle$ is a **stable set** if it satisfies the following two conditions.

Internal stability If $y \in Y$ then for no $z \in Y$ does there exist a coalition S for which $z \succ_S y$.

External stability If $z \in X \setminus Y$ then there exists $y \in Y$ such that $y \succ_S z$ for some coalition S.

This definition can be written alternatively as follows. For any set Y of imputations let $\mathcal{D}(Y)$ be the set of imputations z for which there is a coalition S and an imputation $y \in Y$ such that $y \succ_S z$. Then internal and external stability are equivalent to the conditions $Y \subseteq X \setminus \mathcal{D}(Y)$ and $Y \supseteq X \setminus \mathcal{D}(Y)$, so that a set Y of imputations is a stable set if and only if $Y = X \setminus \mathcal{D}(Y)$.

While the core is a single set of imputations, a game may have more than one stable set (see the examples below) or none at all (as shown by the complex example in Lucas (1969)); each such set may contain many imputations. Von Neumann and Morgenstern (1944) interpret each stable set as corresponding to a *standard of behavior*, the idea being that all the imputations in any given stable set correspond to some mode of behavior while imputations in different stable sets correspond to different modes of behavior.

Some simple properties of stable sets are given in the following result.

■ PROPOSITION 279.2 a. *The core is a subset of every stable set.* b. *No stable set is a proper subset of any other.* c. *If the core is a stable set then it is the only stable set.*

Proof. a. Every member of the core is an imputation and no member is dominated by an imputation, so the result follows from external stability. b. This follows from external stability. c. This follows from (a) and (b).□

◇ EXAMPLE 279.3 (*The three-player majority game*) Consider the game $\langle \{1, 2, 3\}, v \rangle$ in which $v(S) = 1$ if $|S| \geq 2$ and $v(S) = 0$ otherwise. One stable set of this game is

$$Y = \{(1/2, 1/2, 0), (1/2, 0, 1/2), (0, 1/2, 1/2)\}.$$

This corresponds to the "standard of behavior" in which some pair of players shares equally the single unit of payoff that is available. The internal stability of Y follows from the fact that for all x and y in Y only one player prefers x to y. To check external stability, let z be an imputation outside Y. Then there are two players i and j for whom

$z_i < \frac{1}{2}$ and $z_j < \frac{1}{2}$, so that there is an imputation in Y that is an objection of $\{i,j\}$ to z.

For any $c \in [0, \frac{1}{2})$ and any $i \in \{1,2,3\}$ the set

$$Y_{i,c} = \{x \in X : x_i = c\}$$

is also a stable set of the game. This corresponds to a "standard of behavior" in which one of the players is singled out and given a fixed payoff. The internal stability of $Y_{i,c}$ follows from the fact that for any x and y in the set there is only one player who prefers x to y. To show the external stability of $Y_{i,c}$ let $i = 3$ and let z be an imputation outside $Y_{3,c}$. If $z_3 > c$ then $z_1 + z_2 < 1 - c$ and there exists $x \in Y_{3,c}$ such that $x_1 > z_1$ and $x_2 > z_2$, so that $x \succ_{\{1,2\}} z$. If $z_3 < c$ and, say, $z_1 \le z_2$ then $(1-c, 0, c) \succ_{\{1,3\}} z$.

[?] EXERCISE 280.1 (*Simple games*) Let $\langle N, v \rangle$ be a simple game (see Exercise 261.1). Let T be a minimal winning coalition (a winning coalition that has no strict subset that is winning). Show that the set of imputations that assign 0 to all players not in T is a stable set.

[?] EXERCISE 280.2 (*A market for an indivisible good*) For the market described in Example 260.1 with $|B| \ge |L|$ show that the set

$$Y = \{x \in X : x_i = x_j \text{ if } i, j \in L \text{ or } i, j \in B\},$$

is a stable set; interpret it.

[?] EXERCISE 280.3 (*Three-player games*) For a three-player game the set of imputations can be represented geometrically as an equilateral triangle with height $v(N)$ in which each point represents the imputation whose components are the distances to each edge. (Thus the corners correspond to the three imputations that assign $v(N)$ to a single player.) Use such a diagram to find the general form of a stable set of the three-player game in which $v(\{1,2\}) = \beta < 1$, $v(\{1,3\}) = v(\{1,2,3\}) = 1$, and $v(S) = 0$ otherwise. We can interpret this game as a market in which player 1 is a seller and players 2 and 3 are buyers with reservation values β and 1 respectively. Interpret the stable sets of the game in terms of this market.

[?] EXERCISE 280.4 Player i is a **dummy** in $\langle N, v \rangle$ if $v(S \cup \{i\}) - v(S) = v(\{i\})$ for every coalition S of which i is not a member. Show that if player i is a dummy in $\langle N, v \rangle$ then his payoff in any imputation in any stable set is $v(\{i\})$.

[?] EXERCISE 280.5 Let X be an arbitrary set (of outcomes) and let D be a binary relation on X, with the interpretation that if $x \, D \, y$ then x is an objection of some coalition S to y. Generalize the definition of stable sets as follows. The set $Y \subseteq X$ of outcomes is stable if it satisfies the following two conditions.

Internal stability If $y \in Y$ then there exists no $z \in Y$ such that $z \, D \, y$.

External stability If $z \in X \setminus Y$ then there exists $y \in Y$ such that $y \, D \, z$.

Consider an exchange economy (see Section 13.6) in which there are two goods and two agents. Let X be the set of all allocations x for which $x_i \succsim_i \omega_i$ for each agent i. Define the relation D by $x \, D \, y$ if *both* agents prefer x to y. Show that the only (generalized) stable set is the core of the economy.

14.3   The Bargaining Set, Kernel, and Nucleolus

We now turn to the second approach that we described at the start of the chapter. That is, we regard an objection by a coalition to be convincing if no other coalition has a "balancing" counterobjection; we do not require the objection or counterobjection to be themselves stable in any sense. We study three solution concepts that differ in the nature of the objections and counterobjections.

14.3.1   The Bargaining Set

Let x be an imputation in a coalitional game with transferable payoff $\langle N, v \rangle$. Define objections and counterobjections as follows.

- A pair (y, S), where S is a coalition and y is an S-feasible payoff vector, is an **objection of i against j to x** if S includes i but not j and $y_k > x_k$ for all $k \in S$.

- A pair (z, T), where T is a coalition and z is a T-feasible payoff vector, is a **counterobjection to the objection (y, S) of i against j** if T includes j but not i, $z_k \geq x_k$ for all $k \in T \setminus S$, and $z_k \geq y_k$ for all $k \in T \cap S$.

Such an objection is an argument by one player against another. An objection of i against j to x specifies a coalition S that includes i but not j and a division y of $v(S)$ that is preferred by all members of S to x. A counterobjection to (y, S) by j specifies an alternative coalition T that contains j but not i and a division of $v(T)$ that is at least as good as y for all the members of T who are also in S and is at least as good as x for the other members of T. The solution concept that we study is defined as follows.

▶ DEFINITION 282.1 The **bargaining set** of a coalitional game with transferable payoff is the set of all imputations x with the property that for every objection (y, S) of any player i against any other player j to x there is a counterobjection to (y, S) by j.

The bargaining set models the stable arrangements in a society in which any argument that any player i makes against an imputation x takes the following form: "I get too little in the imputation x and j gets too much; I can form a coalition that excludes j in which everybody is better off than in x". Such an argument is ineffective as far as the bargaining set is concerned if player j can respond as follows: "Your demand is not justified; I can form a coalition that excludes you in which everybody is at least as well off as they are in x and the players who participate in your coalition obtain at least what you offer them."

The bargaining set, like the other solution concepts in this section, assumes that the argument underlying an objection for which there is no counterobjection undermines the stability of an outcome. This fact is taken as given, and is not derived from more primitive assumptions about the players' behavior. The appropriateness of the solution in a particular situation thus depends on the extent to which the participants in that situation regard the existence of an objection for which there is no counterobjection as a reason to change the outcome.

Note that an imputation is in the core if and only if no player has an objection against any other player; hence the core is a subset of the bargaining set. We show later (in Corollary 288.3) that the bargaining set of every game is nonempty.

◇ EXAMPLE 282.2 (*The three-player majority game*) Consider the three-player majority game. The core of this game is empty (see Example 259.1) and the game has many stable sets (see Example 279.3). The bargaining set of the game is the singleton $\{(\frac{1}{3}, \frac{1}{3}, \frac{1}{3})\}$, by the following argument. Let x be an imputation and suppose that (y, S) is an objection of i against j to x. Then we must have $S = \{i, h\}$, where h is the third player and $y_h < 1 - x_i$ (since $y_i > x_i$ and $y(S) = v(S) = 1$). For j to have a counterobjection to (y, S) we need $y_h + x_j \leq 1$. Thus for x to be in the bargaining set we require that for all players i, j, and h we have $y_h \leq 1 - x_j$ whenever $y_h < 1 - x_i$, which implies that $1 - x_i \leq 1 - x_j$ or $x_j \leq x_i$ for all i and j, so that $x = (\frac{1}{3}, \frac{1}{3}, \frac{1}{3})$. Obviously this imputation is in the bargaining set.

◇ EXAMPLE 282.3 (*My aunt and I*) Let $\langle\{1, 2, 3, 4\}, v\rangle$ be a simple game (see Exercise 261.1) in which $v(S) = 1$ if and only if S contains one of the coalitions $\{2,3,4\}$ or $\{1,i\}$ for $i \in \{2,3,4\}$. (Player 2 is "I" and player 1 is his aunt.) In this game, player 1 appears to be in a stronger position than the other players since she needs the cooperation of only one player to form a winning coalition. If x is an imputation for which $x_2 < x_3$ then player 2 has an objection against 3 (via $\{1,2\}$) to x for which there is no counterobjection. Thus if x is in the bargaining set then $x_2 = x_3 = x_4 = \alpha$, say. Any objection of player 1 against player 2 to x takes the form $(y, \{1,j\})$ where $j = 3$ or 4 and $y_j < 3\alpha$; there is no counterobjection if and only if $\alpha + 3\alpha + \alpha > 1$, or $\alpha > \frac{1}{5}$. An objection of player 2 against 1 to x must use the coalition $\{2,3,4\}$ and give one of the players 3 or 4 less than $(1-\alpha)/2$; player 1 does not have a counterobjection if and only if $1 - 3\alpha + (1-\alpha)/2 > 1$, or $\alpha < \frac{1}{7}$. Hence the bargaining set is $\{(1 - 3\alpha, \alpha, \alpha, \alpha): \frac{1}{7} \leq \alpha \leq \frac{1}{5}\}$. (Note that by contrast the core is empty.)

We saw (Example 265.1) that the competition inherent in the core can drive to zero the payoff of players holding goods that are in excess supply. The following exercise gives an example that shows how this intense competition is muted in the bargaining set.

[?] EXERCISE 283.1 (*A market*) Consider the coalitional game derived from the market with transferable payoff in Exercise 265.2. Show that the bargaining set of this game is $\{(\alpha, \alpha, \beta, \beta, \beta): 0 \leq \alpha \leq \frac{3}{2}$ and $2\alpha + 3\beta = 3\}$. Contrast this set with the core and give an interpretation.

14.3.2 The Kernel

We now describe another solution that, like the bargaining set, is defined by the condition that to every objection there is a counterobjection; it differs from the bargaining set in the nature of objections and counterobjections that are considered effective.

Let x be an imputation in a coalitional game with transferable payoff $\langle N, v \rangle$; for any coalition S call $e(S, x) = v(S) - x(S)$ the *excess* of S. If the excess of the coalition S is positive then it measures the amount that S has to forgo in order for the imputation x to be implemented; it is the sacrifice that S makes to maintain the social order. If the excess of S is negative then its absolute value measures the amount over and above the worth of S that S obtains when the imputation x is implemented; it is S's surplus in the social order.

A player i objects to an imputation x by forming a coalition S that excludes some player j for whom $x_j > v(\{j\})$ and pointing out that he is dissatisfied with the sacrifice or gain of this coalition. Player j counterobjects by pointing to the existence of a coalition that contains j but not i and sacrifices more (if $e(S, x) > 0$) or gains less (if $e(S, x) < 0$). More precisely, define objections and counterobjections as follows.

- A coalition S is an **objection of i against j to** x if S includes i but not j and $x_j > v(\{j\})$.

- A coalition T is **counterobjection to the objection** S of i against j if T includes j but not i and $e(T, x) \geq e(S, x)$.

▶ DEFINITION 284.1 The **kernel** of a coalitional game with transferable payoff is the set of all imputations x with the property that for every objection S of any player i against any other player j to x there is a counterobjection of j to S.

For any two players i and j and any imputation x define $s_{ij}(x)$ to be the maximum excess of any coalition that contains i but not j:

$$s_{ij}(x) = \max_{S \in \mathcal{C}} \{e(S, x) : i \in S \text{ and } j \in N \setminus S\}.$$

Then we can alternatively define the kernel to be the set of imputations $x \in X$ such that for every pair (i, j) of players either $s_{ji}(x) \geq s_{ij}(x)$ or $x_j = v(\{j\})$.

The kernel models the stable arrangements in a society in which a player makes arguments of the following type against an imputation x: "Here is a coalition to which I belong that excludes player j and sacrifices too much (or gains too little)". Such an argument is ineffective as far as the kernel is concerned if player j can respond by saying "your demand is not justified; I can name a coalition to which I belong that excludes you and sacrifices even more (or gains even less) than the coalition that you name".

Note that the definitions of the core and the bargaining set do not require us to compare the payoffs of different players, while that of the kernel does. Thus the definitions of the former concepts can easily be extended to a general coalitional game $\langle N, X, V, (\succsim_i) \rangle$ (see Definition 268.2). For example, as we saw in Section 13.5, the core is the set of all $x \in V(N)$ for which there is no coalition S and $y \in V(S)$ for which $y \succ_i x$ for all $i \in S$. By contrast, the definition of the kernel cannot be so extended; it assumes that there is meaning to the statement that the excess of one coalition is larger than that of another. Thus the kernel is an appropriate solution concept only in situations in which the payoffs of different players can be meaningfully compared.

We show later that the kernel is nonempty (see Corollary 288.3). Its relation with the bargaining set is as follows.

■ LEMMA 285.1 *The kernel of a coalitional game with transferable payoff is a subset of the bargaining set.*

Proof. Let $\langle N, v \rangle$ be a coalitional game with transferable payoff, let x be an imputation in the kernel, and let (y, S) be an objection in the sense of the bargaining set of player i against j to x: $i \in S$, $j \in N \setminus S$, $y(S) = v(S)$, and $y_k > x_k$ for all $k \in S$. If $x_j = v(\{j\})$ then $(z, \{j\})$ with $z_j = v(\{j\})$ is a counterobjection to (y, S). If $x_j > v(\{j\})$ then since x is in the kernel we have $s_{ji}(x) \geq s_{ij}(x) \geq v(S) - x(S) = y(S) - x(S)$. Let T be a coalition that contains j but not i for which $s_{ji}(x) = v(T) - x(T)$. Then $v(T) - x(T) \geq y(S) - x(S)$, so that $v(T) \geq y(S \cap T) + y(S \setminus T) + x(T \setminus S) - x(S \setminus T) > y(S \cap T) + x(T \setminus S)$, since $y(S \setminus T) > x(S \setminus T)$. Thus there exists a T-feasible payoff vector z with $z_k \geq x_k$ for all $k \in T \setminus S$ and $z_k \geq y_k$ for all $k \in T \cap S$, so that (z, T) is a counterobjection to (y, S). □

◇ EXAMPLE 285.2 (*The three-player majority game*) It follows from our calculation of the bargaining set (Example 282.2), the previous lemma (285.1), and the nonemptiness of the kernel that the kernel of the three-player majority game is $\{(\frac{1}{3}, \frac{1}{3}, \frac{1}{3})\}$. To see this directly, assume that $x_1 \geq x_2 \geq x_3$, with at least one strict inequality. Then $s_{31}(x) = 1 - x_2 - x_3 > 1 - x_2 - x_1 = s_{13}(x)$ and $x_1 > 0 = v(\{1\})$, so that x is not in the kernel.

◇ EXAMPLE 285.3 (*My aunt and I*) The kernel of the game in Example 282.3 is $\{(\frac{2}{5}, \frac{1}{5}, \frac{1}{5}, \frac{1}{5})\}$, by the following argument. Let x be in the kernel. By Lemma 285.1 and the calculation of the bargaining set of the game we have $x = (1 - 3\alpha, \alpha, \alpha, \alpha)$ for some $\frac{1}{7} \leq \alpha \leq \frac{1}{5}$, so that $s_{12}(x) = 2\alpha$ and $s_{21}(x) = 1 - 3\alpha$. Since $1 - 3\alpha > 0$ we need $s_{12}(x) = 2\alpha \geq s_{21}(x) = 1 - 3\alpha$, or $\alpha \geq \frac{1}{5}$; hence $\alpha = \frac{1}{5}$.

14.3.3 The Nucleolus

A solution that is closely related to the kernel is the nucleolus. Let x be an imputation in a coalitional game with transferable payoff. Define objections and counterobjections as follows.

- A pair (S, y) consisting of a coalition S and an imputation y is an **objection** to x if $e(S, x) > e(S, y)$ (i.e. $y(S) > x(S)$).

- A coalition T is a **counterobjection to the objection** (S, y) if $e(T, y) > e(T, x)$ (i.e. $x(T) > y(T)$) and $e(T, y) \geq e(S, x)$.

▶ DEFINITION 286.1 The **nucleolus** of a coalitional game with transferable payoff is the set of all imputations x with the property that for every objection (S, y) to x there is a counterobjection to (S, y).

As for the kernel the idea is that the excess of S is a measure of S's dissatisfaction with x: it is the price that S pays to tolerate x rather than secede from N. In the definition of the kernel an objection is made by a single player, while here an objection is made by a coalition. An objection (S, y) may be interpreted as a statement by S of the form "our excess is too large in x; we suggest the alternative imputation y in which it is smaller". The nucleolus models situations in which such objections cause outcomes to be unstable only if no coalition T can respond by saying "your demand is not justified since our excess under y is larger than it was under x and furthermore exceeds under y what yours was under x". Put differently, an imputation fails to be stable according to the nucleolus if the excess of some coalition S can be reduced without increasing the excess of some coalition to a level at least as large as that of the original excess of S.

This definition of the nucleolus, which is not standard, facilitates a comparison with the kernel and the bargaining set and is easier to interpret than the standard definition, to which we now show it is equivalent.

For any imputation x let $S_1, \ldots, S_{2^{|N|}-1}$ be an ordering of the coalitions for which $e(S_\ell, x) \geq e(S_{\ell+1}, x)$ for $\ell = 1, \ldots, 2^{|N|} - 2$ and let $E(x)$ be the vector of excesses defined by $E_\ell(x) = e(S_\ell, x)$ for all $\ell = 1, \ldots, 2^{|N|} - 1$. Let $B_1(x), \ldots, B_K(x)$ be the partition of the set of all coalitions in which S and S' are in the same cell if and only if $e(S, x) = e(S', x)$. For any $S \in B_k(x)$ let $e(S, x) = e_k(x)$, so that $e_1(x) > e_2(x) > \cdots > e_K(x)$.

We say that $E(x)$ is *lexicographically less than* $E(y)$ if $E_\ell(x) < E_\ell(y)$ for the smallest ℓ for which $E_\ell(x) \neq E_\ell(y)$, or equivalently if there exists k^* such that for all $k < k^*$ we have $|B_k(x)| = |B_k(y)|$ and $e_k(x) = e_k(y)$, and either (i) $e_{k^*}(x) < e_{k^*}(y)$ or (ii) $e_{k^*}(x) = e_{k^*}(y)$ and $|B_{k^*}(x)| < |B_{k^*}(y)|$.

■ LEMMA 286.2 *The nucleolus of a coalitional game with transferable payoff is the set of imputations x for which the vector $E(x)$ is lexicographically minimal.*

Proof. Let $\langle N, v \rangle$ be a coalitional game with transferable payoff and let x be an imputation for which $E(x)$ is lexicographically minimal. To show that x is in the nucleolus, suppose that (S, y) is an objection to x, so that $e(S, y) < e(S, x)$. Let k^* be the maximal value of k such that $e_k(x) = e_k(y)$ and $B_k(x) = B_k(y)$ (not just $|B_k(x)| = |B_k(y)|$) for all $k < k^*$. Since $E(y)$ is not lexicographically less than $E(x)$ we have either (i) $e_{k^*}(y) > e_{k^*}(x)$ or (ii) $e_{k^*}(x) = e_{k^*}(y)$ and $|B_{k^*}(x)| \leq |B_{k^*}(y)|$. In either case there is a coalition $T \in B_{k^*}(y)$ with $e_{k^*}(y) = e(T, y) > e(T, x)$. We now argue that $e(T, y) \geq e(S, x)$, so that T is a counterobjection to (S, y). Since $e(S, y) < e(S, x)$ we have $S \notin \cup_{k=1}^{k^*-1} B_k(x)$ and hence $e_{k^*}(x) \geq e(S, x)$; since $e_{k^*}(y) \geq e_{k^*}(x)$ we have $e(T, y) \geq e(S, x)$.

Now assume that x is in the nucleolus and that $E(y)$ is lexicographically less than $E(x)$. Let k^* be the smallest value of k for which $B_k(x) = B_k(y)$ for all $k < k^*$ and either (i) $e_{k^*}(y) < e_{k^*}(x)$ or (ii) $e_{k^*}(y) = e_{k^*}(x)$ and $B_{k^*}(y) \neq B_{k^*}(x)$ (and hence $|B_{k^*}(y)| \neq |B_{k^*}(x)|$). In either case there exists a coalition $S \in B_{k^*}(x)$ for which $e(S, y) < e(S, x)$. Let $\lambda \in (0, 1)$ and let $z(\lambda) = \lambda x + (1 - \lambda)y$; we have $e(R, z(\lambda)) = \lambda e(R, x) + (1 - \lambda)e(R, y)$ for any coalition R. We claim that the pair $(S, z(\lambda))$ is an objection to x for which there is no counterobjection. It is an objection since $e(S, z(\lambda)) < e(S, x)$. For T to be a counterobjection we need both $e(T, z(\lambda)) > e(T, x)$ and $e(T, z(\lambda)) \geq e(S, x)$. However, if $e(T, z(\lambda)) > e(T, x)$ then $e(T, y) > e(T, x)$, which implies that $T \notin \cup_{k=1}^{k^*} B_k(x)$ and hence $e(S, x) > e(T, x)$. Also, since $T \notin \cup_{k=1}^{k^*-1} B_k(y)$ we have $e(S, x) = e_{k^*}(x) \geq e_{k^*}(y) \geq e(T, y)$. Thus $e(S, x) > e(T, z(\lambda))$. We conclude that there is no counterobjection to $(S, z(\lambda))$. $\quad\square$

The nucleolus is related to the kernel as follows.

■ LEMMA 287.1 *The nucleolus of a coalitional game with transferable payoff is a subset of the kernel.*

Proof. Let $\langle N, v \rangle$ be a coalitional game with transferable payoff and let x be an imputation that is not in the kernel of $\langle N, v \rangle$. We show that x is not in the nucleolus of $\langle N, v \rangle$. Since x is not in the kernel there are players i and j for which $s_{ij}(x) > s_{ji}(x)$ and $x_j > v(\{j\})$. Since $x_j > v(\{j\})$ there exists $\epsilon > 0$ such that $y = x + \epsilon 1_{\{i\}} - \epsilon 1_{\{j\}}$ is an imputation (where $1_{\{k\}}$ is the kth unit vector); choose ϵ small enough that $s_{ij}(y) > s_{ji}(y)$. Note that $e(S, x) < e(S, y)$ if and only if S contains i but not j and $e(S, x) > e(S, y)$ if and only if S contains j but not i. Let k^* be the minimal value of k for which there is a coalition $S \in B_{k^*}(x)$ with $e(S, x) \neq e(S, y)$. Since $s_{ij}(x) > s_{ji}(x)$ the set $B_{k^*}(x)$ contains at least one coalition that contains i but not j and no coalition that contains j but not i. Further, for all $k < k^*$ we have $B_k(y) = B_k(x)$ and $e_k(y) = e_k(x)$. Now, if $B_{k^*}(x)$ contains coalitions that contain both i and j or neither of them then $e_{k^*}(y) = e_{k^*}(x)$ and $B_{k^*}(y)$ is a strict subset of $B_{k^*}(x)$. If not, then since $s_{ij}(y) > s_{ji}(y)$ we have $e_{k^*}(y) < e_{k^*}(x)$. In both cases $E(y)$ is lexicographically less than $E(x)$ and hence x is not in the nucleolus of $\langle N, v \rangle$. □

We now show that the nucleolus of any game is nonempty.

■ PROPOSITION 288.1 *The nucleolus of any coalitional game with transferable payoff is nonempty.*

Proof.   First we argue that for each value of k the function E_k is continuous. This follows from the fact that for any k we have

$$E_k(x) = \min_{T \in \mathcal{C}^{k-1}} \max_{S \in \mathcal{C} \setminus T} e(S, x), \tag{288.2}$$

where $\mathcal{C}^0 = \{\varnothing\}$ and $\mathcal{C}^k$ for $k \geq 1$ is the set of all collections of k coalitions. Since E_1 is continuous the set $X_1 = \arg\min_{x \in X} E_1(x)$ is nonempty and compact. Now, for each integer $k \geq 1$ define $X_{k+1} = \arg\min_{x \in X_k} E_{k+1}(x)$. By induction every such set is nonempty and compact; since $X_{2^{|N|}-1}$ is the nucleolus the proof is complete. □

This result immediately implies that the bargaining set and kernel of any game are nonempty.

■ COROLLARY 288.3 *The bargaining set and kernel of any coalitional game with transferable payoff are nonempty.*

Proof.   This follows from the nonemptiness of the nucleolus (Proposition 288.1) and the facts that the nucleolus is a subset of the kernel (Lemma 287.1) and the kernel is a subset of the bargaining set (Lemma 285.1). □

As we have seen above the bargaining set of a game may contain many imputations; the same is true of the kernel. However, the nucleolus is always a singleton, as the following result shows.

■ PROPOSITION 288.4 *The nucleolus of any coalitional game with transferable payoff is a singleton.*

Proof.   Let $\langle N, v \rangle$ be a coalitional game with transferable payoff. Suppose that the imputations x and y are both in the nucleolus, so that $E(x) = E(y)$. We show that for any coalition S we have $e(S, x) = e(S, y)$ and hence, in particular, for any player i we have $e(\{i\}, x) = e(\{i\}, y)$, so that $x = y$. Assume there is at least one coalition S^* with $e(S^*, x) \neq e(S^*, y)$ and consider the imputation $z = \frac{1}{2}(x + y)$. Since $E_k(x) = E_k(y)$ for all k we have $e_k(x) = e_k(y)$ and $|B_k(x)| = |B_k(y)|$ for all k. But since $e(S^*, x) \neq e(S^*, y)$ there exists a minimal value k^* of k for which $B_{k^*}(x) \neq B_{k^*}(y)$. Now, if $B_{k^*}(x) \cap B_{k^*}(y) \neq \varnothing$ then $B_{k^*}(z) = B_{k^*}(x) \cap B_{k^*}(y) \subset B_{k^*}(x)$; if $B_{k^*}(x) \cap B_{k^*}(y) = \varnothing$ then $e_{k^*}(z) < e_{k^*}(x) = e_{k^*}(y)$. In both cases $E(z)$ is lexicographically less than $E(x)$, contradicting the fact that x is in the nucleolus.   □

[?] EXERCISE 289.1 (*A production economy*) Show that the single imputation in the nucleolus of the game in Exercise 259.3, which models a production economy with one capitalist and w workers, gives each worker $\frac{1}{2}[f(w) - f(w - 1)]$. (Note that since the nucleolus is a singleton you need only to verify that the imputation is in the nucleolus.)

[?] EXERCISE 289.2 (*Weighted majority games*) A **weighted majority game** is a simple game $\langle N, v \rangle$ in which

$$v(S) = \begin{cases} 1 & \text{if } w(S) \geq q \\ 0 & \text{otherwise,} \end{cases}$$

for some $q \in \mathbb{R}$ and $w \in \mathbb{R}_+^N$, where $w(S) = \sum_{i \in S} w_i$ for any coalition S. An interpretation is that w_i is the number of votes that player i has and q is the number of votes needed to win (the quota). A weighted majority game is **homogeneous** if $w(S) = q$ for any minimal winning coalition S and is **zerosum** if for each coalition S either $v(S) = 1$ or $v(N \setminus S) = 1$, but not both. Consider a zerosum homogeneous weighted majority game $\langle N, v \rangle$ in which $w_i = 0$ for every player i who does not belong to any minimal winning coalition. Show that the nucleolus of $\langle N, v \rangle$ consists of the imputation x defined by $x_i = w_i/w(N)$ for all $i \in N$.

14.4 The Shapley Value

The last solution concept that we study in this chapter is the Shapley value. Following our approach in the previous section we begin by characterizing this solution in terms of objections and counterobjections. Then we turn to the standard (axiomatic) characterization.

14.4.1 A Definition in Terms of Objections and Counterobjections

The solution concepts for coalitional games that we have studied so far are defined with reference to single games in isolation. By contrast, the Shapley value of a given game is defined with reference to other games. It is an example of a **value**—a function that assigns a *unique* feasible payoff profile to every coalitional game with transferable payoff, a payoff profile being feasible if the sum of its components is $v(N)$. (The requirement that the payoff profile assigned by the value be feasible is sometimes called *efficiency*.)

Our first presentation of the Shapley value, like our presentations of the solutions studied in the previous section, is in terms of certain types of objections and counterobjections. To define these objections and counterobjections, let $\langle N, v \rangle$ be a coalitional game with transferable payoff and for each coalition S define the *subgame* $\langle S, v^S \rangle$ of $\langle N, v \rangle$ to be the coalitional game with transferable payoff in which $v^S(T) = v(T)$ for any $T \subseteq S$. Let ψ be a value. An *objection* of player i against player j to the division x of $v(N)$ may take one of the following two forms.

- "Give me more since otherwise I will leave the game, causing you to obtain only $\psi_j(N \setminus \{i\}, v^{N \setminus \{i\}})$ rather than the larger payoff x_j, so that you will lose the positive amount $x_j - \psi_j(N \setminus \{i\}, v^{N \setminus \{i\}})$."

- "Give me more since otherwise I will persuade the other players to exclude you from the game, causing me to obtain $\psi_i(N \setminus \{j\}, v^{N \setminus \{j\}})$ rather than the smaller payoff x_i, so that I will gain the positive amount $\psi_i(N \setminus \{j\}, v^{N \setminus \{j\}}) - x_i$."

A *counterobjection* by player j to an objection of the first type is an assertion

- "It is true that if you leave then I will lose, but if *I* leave then *you* will lose at least as much: $x_i - \psi_i(N \setminus \{j\}, v^{N \setminus \{j\}}) \geq x_j - \psi_j(N \setminus \{i\}, v^{N \setminus \{i\}})$."

A *counterobjection* by player j to an objection of the second type is an assertion

- "It is true that if you exclude me then you will gain, but if *I* exclude *you* then I will gain at least as much: $\psi_j(N \setminus \{i\}, v^{N \setminus \{i\}}) - x_j \geq \psi_i(N \setminus \{j\}, v^{N \setminus \{j\}}) - x_i$."

The Shapley value is required to satisfy the property that for every objection of any player i against any other player j there is a counterobjection of player j.

These objections and counterobjections differ from those used to define the bargaining set, kernel, and nucleolus in that they refer to the outcomes of smaller games. It is assumed that these outcomes are derived from the same logic as the payoff of the game itself: that is, the outcomes of the smaller games, like the outcome of the game itself, are given by the value. In this respect the definition of a value shares features with that of stable sets.

The requirement that a value assign to every game a payoff profile with the property that every objection is balanced by a counterobjection is equivalent to the following condition.

▶ DEFINITION 291.1 A value ψ satisfies the **balanced contributions property** if for every coalitional game with transferable payoff $\langle N, v \rangle$ we have

$$\psi_i(N, v) - \psi_i(N \setminus \{j\}, v^{N \setminus \{j\}}) = \psi_j(N, v) - \psi_j(N \setminus \{i\}, v^{N \setminus \{i\}})$$

whenever $i \in N$ and $j \in N$.

We now show that the unique value that satisfies this property is the Shapley value, defined as follows. First define the *marginal contribution of player i* to any coalition S with $i \notin S$ in the game $\langle N, v \rangle$ to be

$$\Delta_i(S) = v(S \cup \{i\}) - v(S).$$

▶ DEFINITION 291.2 The **Shapley value** φ is defined by the condition

$$\varphi_i(N, v) = \frac{1}{|N|!} \sum_{R \in \mathcal{R}} \Delta_i(S_i(R)) \text{ for each } i \in N,$$

where $\mathcal{R}$ is the set of all $|N|!$ orderings of N and $S_i(R)$ is the set of players preceding i in the ordering R.

We can interpret the Shapley value as follows. Suppose that all the players are arranged in some order, all orders being equally likely. Then $\varphi_i(N, v)$ is the expected marginal contribution over all orders of player i to the set of players who precede him. Note that the sum of the marginal contributions of all players in any ordering is $v(N)$, so that the Shapley value is indeed a value.

■ PROPOSITION 291.3 *The unique value that satisfies the balanced contributions property is the Shapley value.*

Proof. First we show that there is at most one value that satisfies the property. Let ψ and ψ' be any two values that satisfy the condition. We prove by induction on the number of players that ψ and ψ' are identical. Suppose that they are identical for all games with less than n players and let $\langle N, v \rangle$ be a game with n players. Since $\psi_i(N \setminus \{j\}, v^{N \setminus \{j\}}) = \psi'_i(N \setminus \{j\}, v^{N \setminus \{j\}})$ for any i, $j \in N$, we deduce from the balanced contributions property that $\psi_i(N, v) - \psi'_i(N, v) = \psi_j(N, v) - \psi'_j(N, v)$

for all i, $j \in N$. Now fixing i and summing over $j \in N$, using the fact that $\sum_{j \in N} \psi_j(N, v) = \sum_{j \in N} \psi'_j(N, v) = v(N)$, we conclude that $\psi_i(N, v) = \psi'_i(N, v)$ for all $i \in N$.

We now verify that the Shapley value φ satisfies the balanced contributions property. Fix a game $\langle N, v \rangle$. We show that $\varphi_i(N, v) - \varphi_j(N, v) = \varphi_i(N \setminus \{j\}, v^{N \setminus \{j\}}) - \varphi_j(N \setminus \{i\}, v^{N \setminus \{i\}})$ for all i, $j \in N$. The left-hand side of this equation is

$$\sum_{S \subseteq N \setminus \{i,j\}} \alpha_S[\Delta_i(S) - \Delta_j(S)] + \beta_S[\Delta_i(S \cup \{j\}) - \Delta_j(S \cup \{i\})],$$

where $\alpha_S = |S|!(|N| - |S| - 1)!/|N|!$ and $\beta_S = (|S| + 1)!(|N| - |S| - 2)!/|N|!$, while the right-hand side is

$$\sum_{S \subseteq N \setminus \{i,j\}} \gamma_S[\Delta_i(S) - \Delta_j(S)],$$

where $\gamma_S = |S|!(|N| - |S| - 2)!/(|N| - 1)!$. The result follows from the facts that $\Delta_i(S) - \Delta_j(S) = \Delta_i(S \cup \{j\}) - \Delta_j(S \cup \{i\})$ and $\alpha_S + \beta_S = \gamma_S$. □

Note that the balanced contributions property links a game only with its subgames. Thus in the derivation of the Shapley value of a game $\langle N, v \rangle$ we could restrict attention to the subgames of $\langle N, v \rangle$, rather than work with the set of all possible games.

14.4.2   An Axiomatic Characterization

We now turn to an axiomatic characterization of the Shapley value. The derivation, unlike that in the previous section, restricts attention to the set of games with a given set of players. Throughout we fix this set to be N and denote a game simply by its worth function v.

To state the axioms we need the following definitions. Player i is a *dummy* in v if $\Delta_i(S) = v(\{i\})$ for every coalition S that excludes i. Players i and j are *interchangeable* in v if $\Delta_i(S) = \Delta_j(S)$ for every coalition S that contains neither i nor j (or, equivalently, $v((S \setminus \{i\}) \cup \{j\}) = v(S)$ for every coalition S that includes i but not j). The axioms are the following.

SYM (*Symmetry*) If i and j are interchangeable in v then $\psi_i(v) = \psi_j(v)$.

DUM (*Dummy player*) If i is a dummy in v then $\psi_i(v) = v(\{i\})$.

ADD (*Additivity*) For any two games v and w we have $\psi_i(v + w) = \psi_i(v) + \psi_i(w)$ for all $i \in N$, where $v + w$ is the game defined by $(v + w)(S) = v(S) + w(S)$ for every coalition S.

Note that the first two axioms impose conditions on single games, while the last axiom links the outcomes of different games. This last axiom is mathematically convenient but hard to motivate: the structure of $v + w$ may induce behavior that is unrelated to that induced by v or w separately. Luce and Raiffa (1957, p. 248) write that the axiom "strikes us as a flaw in the concept of value"; for a less negative view see Myerson (1991, p. 437–438).

■ PROPOSITION 293.1 *The Shapley value is the only value that satisfies* SYM, DUM, *and* ADD.

Proof. We first verify that the Shapley value satisfies the axioms.

SYM: Assume that i and j are interchangeable. For every ordering $R \in \mathcal{R}$ let $R' \in \mathcal{R}$ differ from R only in that the positions of i and j are interchanged. If i precedes j in R then we have $\Delta_i(S_i(R)) = \Delta_j(S_j(R'))$. If j precedes i then $\Delta_i(S_i(R)) - \Delta_j(S_j(R')) = v(S \cup \{i\}) - v(S \cup \{j\})$, where $S = S_i(R) \setminus \{j\}$. Since i and j are interchangeable we have $v(S \cup \{i\}) = v(S \cup \{j\})$, so that $\Delta_i(S_i(R)) = \Delta_j(S_j(R'))$ in this case too. It follows that φ satisfies SYM.

DUM: It is immediate that φ satisfies this condition.

ADD: This follows from the fact that if $u = v + w$ then

$$u(S \cup \{i\}) - u(S) = v(S \cup \{i\}) - v(S) + w(S \cup \{i\}) - w(S).$$

We now show that the Shapley value is the only value that satisfies the axioms. Let ψ be a value that satisfies the axioms. For any coalition T define the game v_T by

$$v_T(S) = \begin{cases} 1 & \text{if } S \supseteq T \\ 0 & \text{otherwise.} \end{cases}$$

Regard a game v as a collection of $2^{|N|} - 1$ numbers $(v(S))_{S \in \mathcal{C}}$. We begin by showing that for any game v there is a unique collection $(\alpha_T)_{T \in \mathcal{C}}$ of real numbers such that $v = \sum_{T \in \mathcal{C}} \alpha_T v_T$. That is, we show that $(v_T)_{T \in \mathcal{C}}$ is an algebraic basis for the space of games. Since the collection $(v_T)_{T \in \mathcal{C}}$ of games contains $2^{|N|} - 1$ members it suffices to show that these games are linearly independent. Suppose that $\sum_{S \in \mathcal{C}} \beta_S v_S = 0$; we need to show that $\beta_S = 0$ for all S. Suppose to the contrary that there exists some coalition T with $\beta_T \neq 0$. Then we can choose such a coalition T for which $\beta_S = 0$ for all $S \subset T$, in which case $\sum_{S \in \mathcal{C}} \beta_S v_S(T) = \beta_T \neq 0$, a contradiction.

Now, by SYM and DUM the value of any game αv_T for $\alpha \geq 0$ is given uniquely by $\psi_i(\alpha v_T) = \alpha/|T|$ if $i \in T$ and $\psi_i(\alpha v_T) = 0$ otherwise. We complete the proof by noting that if $v = \sum_{T \in \mathcal{C}} \alpha_T v_T$ then we have

$v = \sum_{\{T \in \mathcal{C}: \alpha_T > 0\}} \alpha_T v_T - \sum_{\{T \in \mathcal{C}: \alpha_T < 0\}} (-\alpha_T v_T)$ so that by ADD the value of v is determined uniquely. □

◇ EXAMPLE 294.1 (*Weighted majority games*) Consider the weighted majority game v (see Exercise 289.2) with weights $w = (1, 1, 1, 2)$ and quota $q = 3$. In all orderings in which player 4 is first or last his marginal contribution is 0; in all other orderings his marginal contribution is 1. Thus $\varphi(v) = (\frac{1}{6}, \frac{1}{6}, \frac{1}{6}, \frac{1}{2})$. Note that we have $v = v_{\{1,4\}} + v_{\{2,4\}} + v_{\{3,4\}} + v_{\{1,2,3\}} - v_{\{1,2,4\}} - v_{\{1,3,4\}} - v_{\{2,3,4\}}$, from which we can alternatively deduce $\varphi_4(v) = 3 \cdot \frac{1}{2} + 0 - 3 \cdot \frac{1}{3} = \frac{1}{2}$.

? EXERCISE 294.2 Show the following results, which establish that if any one of the three axioms SYM, DUM, and ADD is dropped then there is a value different from the Shapley value that satisfies the remaining two.

  a. For any game v and any $i \in N$ let $\psi_i(v)$ be the average marginal contribution of player i over all the $(|N| - 1)!$ orderings of N in which player 1 is first. Then ψ satisfies DUM and ADD but not SYM.

  b. For any game v let $\psi_i(v) = v(N)/|N|$. Then ψ satisfies SYM and ADD but not DUM.

  c. For any game v let $D(v)$ be the set of dummies in v and let

$$\psi_i(v) = \begin{cases} \frac{1}{|N \setminus D(v)|} \left(v(N) - \sum_{j \in D(v)} v(\{j\}) \right) & \text{if } i \in N \setminus D(v) \\ v(\{i\}) & \text{if } i \in D(v). \end{cases}$$

  Then ψ satisfies SYM and DUM but not ADD.

◇ EXAMPLE 294.3 Consider the game $\langle \{1, 2, 3\}, v \rangle$ in which $v(1, 2, 3) = v(1, 2) = v(1, 3) = 1$ and $v(S) = 0$ otherwise. (This game can be interpreted as a model of a market in which there is a seller (player 1) who holds one unit of a good that she does not value and two potential buyers (players 2 and 3) who each value the good as worth one unit of payoff.) There are six possible orderings of the players. In the four in which player 1 is second or third her marginal contribution is 1 and the marginal contributions of the other two players are 0; in the ordering $(1, 2, 3)$ player 2's marginal contribution is 1, and in $(1, 3, 2)$ player 3's marginal contribution is 1. Thus the Shapley value of the game is $(\frac{2}{3}, \frac{1}{6}, \frac{1}{6})$. By contrast, the core of the game consists of the single payoff profile $(1, 0, 0)$.

◇ EXAMPLE 294.4 (*A market*) Consider the market for an indivisible good in Example 260.1, in which there are b buyers and ℓ sellers, with $\ell < b$.

Consider replications of the market in which there are kb buyers and $k\ell$ sellers for some positive integer k. If k is very large then in most random orderings of the players the fraction of buyers in the set of players who precede any given player i is close to $b/\ell > 1$. In any such ordering the marginal contribution of player i is 1 if she is a seller, so that the Shapley value payoff of a seller is close to 1 (and that of a buyer is close to 0). Precisely, it can be shown that the limit as $k \to \infty$ of the Shapley value payoff of a seller is 1. This is the simplest example of a more general result due to Aumann (1975) that the Shapley value converges to the profile of competitive payoffs as the size of the market increases.

⟨?⟩ EXERCISE 295.1 Find the core and the Shapley value of the game $\langle \{1,2,3,4\}, v \rangle$ in which $v(\{1,2,3,4\}) = 3$, $v(S) = 0$ if S includes at most one of the players in $\{1,2,3\}$, and $v(S) = 2$ otherwise. Explain the source of the difference between the two solutions.

⟨?⟩ EXERCISE 295.2 (*A production economy*) Find the Shapley value of the game in Exercise 259.3 and contrast it with the core and the nucleolus (see Exercise 289.1).

◇ EXAMPLE 295.3 (*A majority game*) Consider a parliament in which there is one party with $m-1$ seats and m parties each with one seat, and a majority is decisive (a generalization of *My aunt and I*). This situation can be modeled as a weighted majority game (see Exercise 289.2) in which $N = \{1,\ldots,m+1\}$, $w_1 = m-1$, $w_i = 1$ for $i \neq 1$, and $q = m$. The marginal contribution of the large party is 1 in all but the $2m!$ orderings in which it is first or last. Hence the Shapley value of the game assigns to the large party the payoff $[(m+1)! - 2m!]/(m+1)! = (m-1)/(m+1)$.

⟨?⟩ EXERCISE 295.4 Consider a parliament in which there are n parties; two of them have $\frac{1}{3}$ of the seats each and the other $n-2$ share the remaining seats equally. Model this situation as a weighted majority game (see Exercise 289.2).

    a. Show that the limit as $n \to \infty$ of the Shapley value payoff of each of the large parties is $\frac{1}{4}$.

    b. Is it desirable according to the Shapley value for the $n-2$ small parties to form a single united party?

⟨?⟩ EXERCISE 295.5 Show that in a convex game (see Exercise 260.4) the Shapley value is a member of the core.

The result in the following exercise suggests an interpretation of the Shapley value that complements those discussed above.

? EXERCISE 296.1 Consider the following variant of the bargaining game of alternating offers studied in Chapter 7. Let $\langle N, v \rangle$ be a coalitional game with transferable payoff in which $v(S) \geq 0$ and $v(S \cup \{i\}) \geq v(S) + v(\{i\})$ for every coalition S and player $i \in N \setminus S$. In each period there is a set $S \subseteq N$ of active players, initially N, one of whom, say player i, is chosen randomly to propose an S-feasible payoff vector $x^{S,i}$. Then the remaining active players, in some fixed order, each either accepts or rejects $x^{S,i}$. If every active player accepts $x^{S,i}$ then the game ends and each player $j \in S$ receives the payoff $x_j^{S,i}$. If at least one active player rejects $x^{S,i}$ then we move to the next period, in which with probability $\rho \in (0, 1)$ the set of active players remains S and with probability $1 - \rho$ it becomes $S \setminus \{i\}$ (i.e. player i is ejected from the game) and player i receives the payoff $v(\{i\})$. Players do not discount the future.

Suppose that there is a collection $(x^{S,i})_{S \in \mathcal{C}, i \in S}$ of S-feasible payoff vectors such that $x_j^{S,i} = \rho \bar{x}_j^S + (1 - \rho) \bar{x}_j^{S \setminus \{i\}}$ for all S, all $i \in S$, and all $j \in S \setminus \{i\}$, where $\bar{x}^S = \sum_{i \in S} x^{S,i} / |S|$ for all S. Show that the game has a subgame perfect equilibrium in which each player $i \in S$ proposes $x^{S,i}$ whenever the set of active players is S. Show further that there is such a collection for which $\bar{x}^S = \varphi(S, v)$ for each $S \in \mathcal{C}$, thus showing that the game has a subgame perfect equilibrium in which the expected payoff of each player i is his Shapley value payoff $\varphi_i(N, v)$. Note that if ρ is close to 1 in this case then every proposal $x^{S,i}$ is close to the Shapley value of the game $\langle S, v \rangle$. (Hart and Mas-Colell (1996) show that every subgame perfect equilibrium in which each player's strategy is independent of history has this property; Krishna and Serrano (1995) study non-stationary equilibria.)

14.4.3 Cost-Sharing

Let N be a set of players and for each coalition S let $C(S)$ be the cost of providing some service to the members of S. How should $C(N)$ be shared among the players? One possible answer is given by the Shapley value $\varphi(C)$ of the game $\langle N, C \rangle$, where $\varphi_i(C)$ is the payment requested from player i. This method of cost-sharing is supported by the axioms presented above, which in the current context can be given the following interpretations. The feasibility requirement $\sum_{i \in N} \varphi_i(C) = C(N)$ says that the total payments requested from the players should equal $C(N)$, the total cost of providing the service. The axioms DUM and SYM have interpretations as principles of "fairness" when applied to the game. DUM says that a player for whom the marginal cost of providing the service is the same, no matter which group is currently receiving the service, should pay that cost. SYM says that two players for whom the marginal cost is the same, no matter which group is currently receiving the service, should pay the same. ADD is somewhat more attractive here than it is in the context of strategic interaction. It says that the payment of any player for two different services should be the sum of the payments for the two services separately.

Notes

Stable sets were first studied by von Neumann and Morgenstern (1944). The idea of the bargaining set is due to Aumann and Maschler (1964); the formulation that we give is that of Davis and Maschler (1963). The kernel and nucleolus are due respectively to Davis and Maschler (1965) and Schmeidler (1969). Proofs of the nonemptiness of the bargaining set (using direct arguments) were first given by Davis, Maschler, and Peleg (see Davis and Maschler (1963, 1967) and Peleg (1963b, 1967)). Our definition of the nucleolus in terms of objections and counterobjections appears to be new. The results in Section 14.3.3 (other than Lemma 286.2) are due to Schmeidler (1969). The Shapley value is due to Shapley (1953), who proved Proposition 293.1. The balanced contributions property (Definition 291.1) is due to Myerson (1977, 1980); see also Hart and Mas-Colell (1989).

The application of the Shapley value to the problem of cost-sharing was suggested by Shubik (1962); the theory has been developed by many authors, including Roth and Verrecchia (1979) and Billera, Heath, and Raanan (1978).

The game *My aunt and I* in Examples 282.3 and 285.3 is studied by Davis and Maschler (1965, Section 6). The result in Exercise 283.1 is due to Maschler (1976). Exercise 289.1 is taken from Moulin (1988, pp. 126–127; see also Exercise 5.3). Weighted majority games were first studied by von Neumann and Morgenstern (1944); the result in Exercise 289.2 is due to Peleg (1968). The game in Exercise 295.1 is due to Zamir, quoted in Aumann (1986, p. 986). Exercise 295.2 is taken from Moulin (1988, p. 111). The result in Exercise 295.4 is due to Milnor and Shapley (1978), that in Exercise 295.5 to Shapley (1971/72), and that in Exercise 296.1 to Hart and Mas-Colell (1996).

Much of the material in this chapter draws on Aumann's (1989) lecture notes, though some of our interpretations of the solution concepts are different from his.

The definitions of stable sets and the bargaining set can be extended straightforwardly to coalitional games without transferable payoff (see, for example, Aumann and Peleg (1960) and Peleg (1963a)). For extensions of the Shapley value to such games see Harsanyi (1963), Shapley (1969), Aumann (1985a), Hart (1985), and Maschler and Owen (1989, 1992).

Harsanyi (1974) studies an extensive game for which a class of subgame perfect equilibria correspond to stable sets. Harsanyi (1981), Gul (1989), and Hart and Mas-Colell (1996) study extensive games that have equilibria corresponding to the Shapley value.

The solution concepts that we study in this chapter can be interpreted as formalizing notions of "fairness"; for an analysis along these lines see Moulin (1988).

Lucas (1992) and Maschler (1992) are surveys that cover the models in Sections 14.2 and 14.3.

15 The Nash Solution

In this chapter we study two-person bargaining problems from the perspective of coalitional game theory. We give a definition of the Nash solution[1] in terms of objections and counterobjections and characterize the solution axiomatically. In addition we explore the connection between the Nash solution and the subgame perfect equilibrium outcome of a bargaining game of alternating offers.

15.1 Bargaining Problems

In Chapter 7 we discuss two-person bargaining using the tools of the theory of extensive games. Here we do so using the approach of coalitional game theory. We define a *bargaining problem* to be a tuple $\langle X, D, \succsim_1, \succsim_2 \rangle$ in which X is a set of possible consequences that the two players can jointly achieve, $D \in X$ is the event that occurs if the players fail to agree, and $\succsim_1$ and $\succsim_2$ are the players' preference relations over $\mathcal{L}(X)$, the set of lotteries over X. We refer to X as the set of possible *agreements* and to D as the *disagreement* outcome. Note that such a tuple can be identified with a coalitional game without transferable payoff $\langle \{1,2\}, \mathcal{L}(X), V, (\succsim_i) \rangle$ in which $V(\{1,2\}) = X$ and $V(\{i\}) = \{D\}$ for $i = 1$, 2 (see Definition 268.2).

The members of X should be thought of as deterministic. Note that we require the players' preference relations to be defined over the set of *lotteries* over X, rather than simply over X itself. That is, each preference relation includes information not only about the player's preferences over the set of possible joint actions but also about his attitude

---

[1]The only connection between the Nash solution and the notion of Nash equilibrium studied in Parts I, II, and III is John Nash.

towards risk. We denote by $p \cdot x \oplus (1 - p) \cdot y$ the lottery that gives x with probability p and y with probability $1 - p$ and by $p \cdot x$ the lottery $p \cdot x \oplus (1 - p) \cdot D$.

Our basic definition of a bargaining problem contains some restrictions, as follows.

▶ DEFINITION 300.1 A **bargaining problem** is a tuple $\langle X, D, \succsim_1, \succsim_2 \rangle$ where

- X (the set of **agreements**) is a compact set (for example, in a Euclidian space)

- D (the **disagreement** outcome) is a member of X

- $\succsim_1$ and $\succsim_2$ are continuous preference relations on the set $\mathcal{L}(X)$ of lotteries over X that satisfy the assumptions of von Neumann and Morgenstern

- $x \succsim_i D$ for all $x \in X$ for $i = 1, 2$, and there exists $x \in X$ such that $x \succ_1 D$ and $x \succ_2 D$

- (convexity) for any $x \in X$, $y \in X$, and $p \in [0, 1]$ there exists $z \in X$ such that $z \sim_i p \cdot x \oplus (1 - p) \cdot y$ for $i = 1, 2$

- (non-redundancy) if $x \in X$ then there is no $x' \in X$ with $x' \neq x$ such that $x \sim_i x'$ for $i = 1, 2$

- (unique best agreements) for each player i there is a unique agreement $B_i \in X$ with $B_i \succsim_i x$ for all $x \in X$

- for each player i we have $B_i \sim_j D$ for $j \neq i$.

The first three of these assumptions guarantee that each player's preference relation over $\mathcal{L}(X)$ can be represented by the expectation of some continuous function over X (the player's von Neumann–Morgenstern utility function). The fourth assumption says that disagreement is the worst possible outcome and that the problem is non-degenerate in the sense that there exists an agreement that is more attractive to both players than disagreement. The assumption of convexity requires that the set of agreements be rich enough that every lottery is equivalent for both players to some (deterministic) agreement. The last three assumptions are made for convenience. The assumption of non-redundancy says that we identify any two agreements between which both players are indifferent. The assumption of unique best agreements implies that the best agreement for each player is strongly Pareto efficient (i.e. there is no agreement that is better for one player and at least as good for the other). The last assumption says that each player is indifferent between disagreement and the outcome in which the other player obtains his favorite agreement.

Given our assumptions on the players' preferences we can associate with any bargaining problem $\langle X, D, \succsim_1, \succsim_2 \rangle$ and any von Neumann–Morgenstern utility functions u_1 and u_2 that represent $\succsim_1$ and $\succsim_2$ a pair $\langle U, d \rangle$ in which $U = \{(u_1(x), u_2(x)): x \in X\}$ and $d = (u_1(D), u_2(D))$; we can choose u_1 and u_2 so that $d = (0, 0)$. Our assumptions imply that U is compact and convex and contains a point y for which $y_i > d_i$ for $i = 1, 2$. In the standard treatment of the Nash solution such a pair $\langle U, d \rangle$, rather than a description like $\langle X, D, \succsim_1, \succsim_2 \rangle$ of the physical agreements and the players' preferences, is taken as the primitive; we find the language of agreements and preferences more natural. Note that bargaining problems with different agreement sets and preference relations can lead to the same pair $\langle U, d \rangle$: a bargaining problem contains more information than such a pair.

Our aim now is to construct reasonable systematic descriptions of the way that bargaining problems may be resolved. The notion of a bargaining solution is a formal expression of such a systematic description.

▸ DEFINITION 301.1 A **bargaining solution** is a function that assigns to every bargaining problem $\langle X, D, \succsim_1, \succsim_2 \rangle$ a unique member of X.

A bargaining solution describes the way in which the agreement (or disagreement) depends upon the parameters of the bargaining problem. The bargaining theory that we study focuses on the effect of the players' risk attitudes on the bargaining outcome. Alternative theories focus on other relevant factors (for example the players' time preferences or their ability to bargain), but such theories require that we change the primitives of the model.

15.2 The Nash Solution: Definition and Characterization

15.2.1 Definition

We now define the solution concept that we study in this chapter.

▸ DEFINITION 301.2 The **Nash solution** is a bargaining solution that assigns to the bargaining problem $\langle X, D, \succsim_1, \succsim_2 \rangle$ an agreement $x^* \in X$ for which

if $p \cdot x \succ_i x^*$ for some $p \in [0, 1]$ and $x \in X$ then $p \cdot x^* \succsim_j x$ for $j \neq i$.
$$(301.3)$$

This definition is equivalent to one whose structure is similar to those of the bargaining set, kernel, and nucleolus given in the previous chapter. To see this, define an **objection** of player i to the agreement $x^* \in X$ to be a pair (x, p) with $x \in X$ and $p \in [0, 1]$ for which $p \cdot x \succ_i x^*$. The interpretation is that x is an alternative agreement that player i proposes and $1 - p$ is the probability that the negotiations will break down if player i presses his objection. The agreement x and the probability p are chosen by player i; the probability p may be determined indirectly by the actions (like threats and intimidations) that player i takes when he presses his demand that the agreement be x. Thus player i makes an argument of the form "I demand the outcome x rather than x^*; I back up this demand by threatening to take steps that will cause us to fail to agree with probability $1 - p$, a threat that is credible since if I carry it out and the outcome is x then I will be better off than I am now". Player j can **counterobject** to (x, p) if $p \cdot x^* \succsim_j x$. The interpretation is that under the risky conditions that player i creates by his objection it is desirable for player j to insist on the original agreement x^*. Thus player j's argument is "If you take steps that will cause us to disagree with probability $1 - p$ then it is still desirable for me to insist on x^* rather than agreeing to x". Given these definitions of objection and counterobjection the Nash solution is the set of all agreements x^* with the property that player j can counterobject to every objection of player i to x^*.

15.2.2 Characterization

We now show that the Nash solution is well-defined and has a simple characterization: the Nash solution of the bargaining problem $\langle X, D, \succsim_1, \succsim_2 \rangle$ is the agreement that maximizes the product $u_1(x)u_2(x)$, where u_i is a von Neumann–Morgenstern utility function that represents $\succsim_i$ for $i = 1, 2$.

■ PROPOSITION 302.1

a. *The agreement $x^* \in X$ is a Nash solution of the bargaining problem $\langle X, D, \succsim_1, \succsim_2 \rangle$ if and only if*

$$u_1(x^*)u_2(x^*) \geq u_1(x)u_2(x) \text{ for all } x \in X,$$

where u_i is a von Neumann–Morgenstern utility function that represents $\succsim_i$ and satisfies $u_i(D) = 0$ for $i = 1, 2$.

b. *The Nash solution is well-defined.*

Proof. We first prove (a). Suppose that $u_1(x^*)u_2(x^*) \geq u_1(x)u_2(x)$ for all $x \in X$. Then $u_i(x^*) > 0$ for $i = 1, 2$ (since X contains an agreement y for which $u_i(y) > 0$ for $i = 1, 2$). Now, if $pu_i(x) > u_i(x^*)$ for some $p \in [0,1]$ then $pu_i(x)u_j(x^*) > u_i(x^*)u_j(x^*) \geq u_i(x)u_j(x)$ and hence $pu_j(x^*) > u_j(x)$ (since $u_i(x) > 0$), or $p \cdot x^* \succ_j x$.

Now suppose that x^* satisfies (301.3): if $p \cdot x \succ_i x^*$ for some $p \in [0,1]$ and $x \in X$ then $p \cdot x^* \succsim_j x$. Let $x \in X$ be such that $u_i(x) > 0$ for $i = 1, 2$ and $u_i(x) > u_i(x^*)$ for some i. (For any other value of x we obviously have $u_1(x^*)u_2(x^*) \geq u_1(x)u_2(x)$.) Then if $p > u_i(x^*)/u_i(x)$ for some $p \in [0,1]$ we have $pu_j(x^*) \geq u_j(x)$, so that, since $u_j(x) > 0$, we have $p \geq u_j(x)/u_j(x^*)$. Hence $u_i(x^*)/u_i(x) \geq u_j(x)/u_j(x^*)$ and thus $u_1(x^*)u_2(x^*) \geq u_1(x)u_2(x)$.

To prove (b), let $U = \{(u_1(x), u_2(x)): x \in X\}$. By (a), the agreement x^* is a Nash solution of $\langle X, D, \succsim_1, \succsim_2 \rangle$ if and only if $(v_1, v_2) = (u_1(x^*), u_2(x^*))$ maximizes $v_1 v_2$ over U. Since U is compact this problem has a solution; since the function $v_1 v_2$ is strictly quasi-concave on the interior of $\mathbb{R}^2_+$ and U is convex the solution is unique. Finally, by the assumption of non-redundancy there is a unique agreement $x^* \in X$ that yields the pair of maximizing utilities. $\square$

The simplicity of this characterization is attractive and accounts for the widespread application of the Nash solution. The characterization also allows us to illustrate the Nash solution geometrically, as in Figure 304.1. Although the maximization of a product of utilities is a simple mathematical operation it lacks a straightforward interpretation; we view it simply as a technical device. Originally Nash defined the solution in terms of this characterization; we find Definition 301.2 preferable since it has a natural interpretation.

15.2.3 Comparative Statics of Risk Aversion

A main goal of Nash's theory is to provide a relationship between the players' attitudes towards risk and the outcome of the bargaining. Thus a first test of the plausibility of the theory is whether this relationship accords with our intuition. We compare two bargaining problems that differ only in that one player's preference relation in one of the problems is more risk-averse than it is in the other; we verify that the outcome of the former problem is worse for the player than that of the latter.

Define the preference relation $\succsim_1'$ to be *at least as risk-averse as* $\succsim_1$ if $\succsim_1$ and $\succsim_1'$ agree on X and whenever $x \sim_1 L$ for some $x \in X$ and

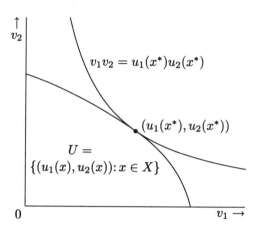

Figure 304.1 A geometric characterization of the Nash solution x^* of the bargaining problem $\langle X, D, \succsim_1, \succsim_2 \rangle$. For $i = 1, 2$ the function u_i is a von Neumann–Morgenstern utility function that represents $\succsim_i$ and satisfies $u_i(D) = 0$.

$L \in \mathcal{L}(X)$ we have $x \succsim'_1 L$. (This definition is equivalent to the standard definition that is given in terms of utility representations.)

- PROPOSITION 304.1 *Let x and x' be the Nash solutions of the bargaining problems $\langle X, D, \succsim_1, \succsim_2 \rangle$ and $\langle X, D, \succsim'_1, \succsim_2 \rangle$ respectively, where $\succsim'_1$ is at least as risk-averse as $\succsim_1$. Then $x \succsim_1 x'$.*

Proof. Assume to the contrary that $x' \succ_1 x$. By the convexity of the bargaining problems there exists an agreement $z \in X$ such that $z \sim_i \frac{1}{2} \cdot x' \oplus \frac{1}{2} \cdot x$ for $i = 1, 2$. Let z^* be a Pareto efficient agreement for which $z^* \succsim_i z$ for $i = 1, 2$. By the characterization of the Nash solution (Proposition 302.1a), the agreements x and x' are Pareto efficient, so that $x \prec_1 z^* \prec_1 x'$ and $x' \prec_2 z^* \prec_2 x$. Now, since x is the Nash solution of $\langle X, D, \succsim_1, \succsim_2 \rangle$ we have $u_1(x)u_2(x) > u_1(x')u_2(x')$, where u_i is a von Neumann–Morgenstern utility function with $u_i(D) = 0$ that represents $\succsim_i$ for $i = 1, 2$. By the quasi-concavity of the function $H(v_1, v_2) = v_1 v_2$ we have $u_1(z)u_2(z) > u_1(x')u_2(x')$ and hence $u_1(z^*)u_2(z^*) > u_1(x')u_2(x')$. Since $x' \succ_1 z^*$ it follows that $1 > u_1(z^*)/u_1(x') > u_2(x')/u_2(z^*)$, so that there exists $p \in [0,1]$ such that $u_1(z^*)/u_1(x') > p > u_2(x')/u_2(z^*)$ and hence $p \cdot z^* \succ_2 x'$ and $z^* \succ_1 p \cdot x'$. Since the preference relation $\succsim'_1$ is at least as risk-averse as $\succsim_1$ we also have $z^* \succ'_1 p \cdot x'$, so that (z^*, p) is an objection of player 2 to x' for which there is no counterobjection, contradicting the fact that x' is the Nash solution of $\langle X, D, \succsim'_1, \succsim_2 \rangle$. $\qquad\square$

15.3 An Axiomatic Definition

15.3.1 Axioms

A beauty of the Nash solution is that it is uniquely characterized by three simple axioms (properties). In the following statements of these axioms F denotes an arbitrary bargaining solution.

PAR (*Pareto efficiency*) There is no agreement $x \in X$ such that $x \succsim_i F(X, D, \succsim_1, \succsim_2)$ for $i = 1, 2$ with strict preference for at least one i.

The standard justification of PAR is that an inefficient outcome is not likely since it leaves room for renegotiation that makes both players better off. The fact that the Nash solution satisfies PAR follows immediately from Proposition 302.1a.

To state the next axiom we need first to define a symmetric bargaining problem. Informally, a bargaining problem is symmetric if there is a relabeling of the set of agreements that interchanges the players' preference relations: player 1's preference relation in the relabeled problem coincides with player 2's preference relation in the original problem, and vice versa. To state this definition differently, consider the language that consists of the names of the preference relations and the name of the disagreement point, but not the names of the agreements. A problem is symmetric if any definition of an agreement by means of a formula in this language defines the same agreement if we interchange the names of the players.

▸ DEFINITION 305.1 A bargaining problem $\langle X, D, \succsim_1, \succsim_2 \rangle$ is **symmetric** if there is a function $\phi \colon X \to X$ with $\phi(D) = D$ and $\phi(x) = y$ if and only if $\phi(y) = x$, such that $L_1 \succsim_i L_2$ if and only if $\phi(L_1) \succsim_j \phi(L_2)$ for $i \neq j$ and for any lotteries L_1 and L_2 in $\mathcal{L}(X)$, where $\phi(L)$ is the lottery in which each prize x in the support of L is replaced by the prize $\phi(x)$.

We refer to the function $\phi \colon X \to X$ in this definition as the *symmetry function*. An example of a symmetric bargaining problem is that in which two risk-neutral players split a pie, obtaining nothing if they disagree (consider the symmetry function given by $\phi(x_1, x_2) = (x_2, x_1)$).

SYM (*Symmetry*) If $\langle X, D, \succsim_1, \succsim_2 \rangle$ is symmetric with symmetry function ϕ then $\phi(F(X, D, \succsim_1, \succsim_2)) = F(X, D, \succsim_1, \succsim_2)$.

The justification of this axiom is that we seek a solution in which all asymmetries between the players are included in the description of the bargaining problem. Thus if players 1 and 2 are indistinguishable in a certain problem then the agreement assigned to that problem should not discriminate between them.

■ LEMMA 306.1 *The Nash solution satisfies* SYM.

Proof. Let x^* be the Nash solution of the symmetric bargaining problem $\langle X, D, \succsim_1, \succsim_2 \rangle$ with symmetry function ϕ. Suppose that $\phi(x^*)$ is not the Nash solution of the bargaining problem $\langle X, D, \succsim_2, \succsim_1 \rangle$. Then some player i has an objection (x, p) to $\phi(x^*)$ for which there is no counterobjection by player j: $p \cdot x \succ_i \phi(x^*)$ and $p \cdot \phi(x^*) \prec_j x$. But then $\phi(p \cdot x) = p \cdot \phi(x) \succ_j \phi(\phi(x^*)) = x^*$ and $\phi(p \cdot \phi(x^*)) = p \cdot x^* \prec_i \phi(x)$, so that $(\phi(x), p)$ is an objection by player j to x^* for which there is no counterobjection by player i, contradicting the fact that x^* is the Nash solution. □

The final axiom is the most problematic.

IIA (*Independence of irrelevant alternatives*) Let $x^* = F(X, D, \succsim_1, \succsim_2)$ and let $\succsim_i'$ be a preference relation that agrees with $\succsim_i$ on X and satisfies

- if $x \succsim_i x^*$ and $p \cdot x \sim_i x^*$ for some $x \in X$ and $p \in [0, 1]$ then $p \cdot x \precsim_i' x^*$
- if $x \precsim_i x^*$ and $x \sim_i p \cdot x^*$ for some $x \in X$ and $p \in [0, 1]$ then $x \sim_i' p \cdot x^*$.

Then $F(X, D, \succsim_i, \succsim_j) = F(X, D, \succsim_i', \succsim_j)$.

A player whose preference relation is $\succsim_i'$ is more apprehensive than one whose preference relation is $\succsim_i$ about the risk of demanding alternatives that are better than x^* but has the same attitudes to alternatives that are worse than x^*. The axiom requires that the outcome when player i has the preference relation $\succsim_i'$ is the same as that when player i has the preference relation $\succsim_i$. The idea is that if x^* survives player i's objections originally then it should survive them also in a problem in which he is less eager to make them (i.e. fewer pairs (x, p) are objections of player i); it should continue also to survive player j's objections since player i's ability to counterobject has not been changed.

Note that despite its name, the axiom involves a comparison of two problems in which the sets of alternatives are the same; it is the players' preferences that are different. (The name derives from the fact that the axiom is analogous to an axiom presented by Nash that does involve a comparison of two problems with different sets of agreements.) Note also that the axiom differs from PAR and SYM in that it involves a comparison of bargaining problems, while PAR and SYM impose conditions on the solutions of single bargaining problems.

■ LEMMA 307.1 *The Nash solution satisfies* IIA.

Proof. Let x^* be the Nash solution of the bargaining problem $\langle X, D, \succsim_i, \succsim_j \rangle$ and let $\succsim_i'$ be a preference relation that satisfies the hypotheses of IIA. Consider the bargaining problem $\langle X, D, \succsim_i', \succsim_j \rangle$. We show that for every objection of either i or j to x^* in $\langle X, D, \succsim_1', \succsim_2 \rangle$ there is a counterobjection, so that x^* is the Nash solution of $\langle X, D, \succsim_1', \succsim_2 \rangle$.

First suppose that player i has an objection to x^*: $p \cdot x \succ_i' x^*$ for some $x \in X$ and $p \in [0, 1]$. Then $x \succ_i' x^*$ and hence $x \succ_i x^*$ (since $\succsim_i$ and $\succsim_i'$ agree on X). Thus from the first part of IIA we have $p \cdot x \succ_i x^*$ (if $p \cdot x \precsim_i x^*$ then there exists $q \geq p$ such that $q \cdot x \sim_i x^*$ and thus $q \cdot x \precsim_i' x^*$, so that $p \cdot x \precsim_i' x^*$). Since x^* is the Nash solution of $\langle X, D, \succsim_i, \succsim_j \rangle$ we thus have $p \cdot x^* \succsim_j x$.

Now suppose that player j has an objection to x^*: $p \cdot x \succsim_j x^*$ for some $x \in X$ and $p \in [0, 1]$. Since x^* is Pareto efficient we have $x^* \succsim_i x$ and since x^* is the Nash solution of $\langle X, D, \succsim_i, \succsim_j \rangle$ we have $p \cdot x^* \succsim_i x$. Thus from the second part of IIA we have $p \cdot x^* \succsim_i' x$.   □

15.3.2 Characterization

The following result completes the characterization of the Nash solution in terms of the axioms PAR, SYM, and IIA discussed above.

■ PROPOSITION 307.2 *The Nash solution is the only bargaining solution that satisfies* PAR, SYM, *and* IIA.

Proof. We have shown that the Nash solution satisfies the three axioms; we now show uniqueness.

Step 1. Let x^* be the Nash solution of the bargaining problem $\langle X, D, \succsim_1, \succsim_2 \rangle$. If $x \sim_i p \cdot x^*$ then $[1/(2-p)] \cdot x \precsim_j x^*$.

Proof. For each player i choose the von Neumann–Morgenstern utility function u_i that represents $\succsim_i$ and satisfies $u_i(x^*) = 1$ and $u_i(D) = 0$. We first argue that for every agreement $y \in X$ we have $u_1(y) + u_2(y) \leq 2$. To see this, suppose to the contrary that for some $y \in X$ we have $u_1(y) + u_2(y) = 2 + \epsilon$ with $\epsilon > 0$. By the convexity of the bargaining problem, for every $p \in [0, 1]$ there is an agreement $z(p) \in X$ with $u_i(z(p)) = pu_i(y) + (1-p)u_i(x^*) = pu_i(y) + 1 - p$ for $i = 1, 2$, so that $u_1(z(p))u_2(z(p)) = 1 + \epsilon p + p^2[u_1(y)u_2(y) - 1 - \epsilon]$. Thus for p close enough to 0 we have $u_1(z(p))u_2(z(p)) > 1 = u_1(x^*)u_2(x^*)$, contradicting the fact that x^* is the Nash solution of the problem. Now, if $x \sim_i p \cdot x^*$ we have $u_i(x) = p$ and hence $u_j(x) \le 2 - p$, so that $[1/(2-p)] \cdot x \precsim_j x^*$.

Step 2. Any bargaining solution that satisfies PAR, SYM, and IIA is the Nash solution.

Proof. Let x^* be the Nash solution of the bargaining problem $\langle X, D, \succsim_1, \succsim_2 \rangle$ and let F be a bargaining solution that satisfies PAR, SYM, and IIA. Let $\succsim_1'$ and $\succsim_2'$ be preference relations that coincide with $\succsim_1$ and $\succsim_2$ on X and satisfy the following conditions. For any Pareto efficient agreement $x \in X$ we have

- if $x \succ_1 x^*$ and $x \sim_2 p \cdot x^*$ for some $p \in [0, 1]$ then $x \sim_2' p \cdot x^*$ and $x^* \sim_1' [1/(2 - p)] \cdot x$.

- if $x \prec_1 x^*$ and $x \sim_1 p \cdot x^*$ for some $p \in [0, 1]$ then $x \sim_1' p \cdot x^*$ and $x^* \sim_2' [1/(2 - p)] \cdot x$

(These conditions completely describe a pair of preference relations satisfying the assumptions of von Neumann and Morgenstern since for every $x \in X$ and each player i there is some Pareto efficient agreement x' for which $x \sim_i x'$.) Let u_i be the von Neumann–Morgenstern utility function that represents $\succsim_i'$ and satisfies $u_i(D) = 0$ and $u_i(x^*) = 1$ for $i = 1$, 2. Then $u_1(x) + u_2(x) = 2$ for all Pareto efficient agreements $x \in X$.

It is easy to verify that the problem $\langle X, D, \succsim_1', \succsim_2' \rangle$ is convex. (One way to do so is to verify that the set of pairs of utilities is the triangle $\{(v_1, v_2): v_1 + v_2 \le 2 \text{ and } v_i \ge 0 \text{ for } i = 1, 2\}$. To show this, use the fact that since B_i is Pareto efficient and $B_i \sim_j D$ we have $u_j(B_i) = 0$ and $u_i(B_i) = 2$.) To see that the problem is symmetric, define $\phi: X \to X$ by $\phi(D) = D$ and $[p \cdot B_1 \sim_1' x \text{ and } q \cdot B_2 \sim_2' x]$ if and only if $[p \cdot B_2 \sim_2' \phi(x)$ and $q \cdot B_1 \sim_1' \phi(x)]$. This function ϕ assigns an agreement with the utilities (v_1, v_2) to an agreement with utilities (v_2, v_1). Thus, an efficient agreement that is a fixed point of ϕ yields the pair of utilities $(1, 1)$ and hence by non-redundancy is x^*. Thus by SYM and PAR we have $F(X, D, \succsim_1', \succsim_2') = x^*$.

Now, the pair of problems $\langle X, D, \succsim_1', \succsim_2' \rangle$ and $\langle X, D, \succsim_1, \succsim_2' \rangle$ and the pair of problems $\langle X, D, \succsim_1, \succsim_2 \rangle$ and $\langle X, D, \succsim_1, \succsim_2' \rangle$ satisfy the hypothesis of IIA since by Step 1 we have $[1/(2-p)] \cdot x \precsim_j x^*$ if $x \sim_i p \cdot x^*$. Therefore $F(X, D, \succsim_1, \succsim_2) = F(X, D, \succsim_1', \succsim_2') = x^*$. □

As noted earlier, Nash defined a bargaining problem to be a pair $\langle U, d \rangle$, where $U \subseteq \mathbb{R}^2$ is a compact convex set (the set of pairs of payoffs to agreements) and $d \in U$ (the pair of payoffs in the event of disagreement). A *bargaining solution* in this context is a function that assigns a point in U to every bargaining problem $\langle U, d \rangle$. Nash showed that there is a unique bargaining solution that satisfies axioms similar to those considered above and that this solution assigns to the bargaining problem $\langle U, d \rangle$ the pair (v_1, v_2) of payoffs in U for which the product $(v_1 - d_1)(v_2 - d_2)$ is highest. The following exercise asks you to prove this result.

? EXERCISE 309.1 Show, following the line of the proof of the previous result, that in the standard Nash bargaining model (as presented in the previous paragraph) there is a unique bargaining solution that satisfies analogs of PAR and SYM and the following two axioms, in which f denotes a bargaining solution.

(*Covariance with positive affine transformations*) Let $\langle U, d \rangle$ be a bargaining problem, let $\alpha_i > 0$ and β_i be real numbers, let

$$U' = \{(v_1', v_2') : v_i' = \alpha_i v_i + \beta_i \text{ for } i = 1, 2 \text{ for some } (v_1, v_2) \in U\},$$

and let $d_i' = \alpha_i d_i + \beta_i$ for $i = 1, 2$. Then $f_i(U', d') = \alpha_i f_i(U, d) + \beta_i$ for $i = 1, 2$.

(*Independence of irrelevant alternatives*) If $U \subseteq U'$ and $f(U', d) \in U$ then $f(U', d) = f(U, d)$.

15.3.3 Is Any Axiom Superfluous?

We have shown that the axioms PAR, SYM, and IIA uniquely define the Nash solution; we now show that none of these axioms is superfluous. We do so by exhibiting, for each axiom, a bargaining solution that is different from Nash's and satisfies the remaining two axioms.

PAR: Consider the solution defined by $F(X, D, \succsim_1, \succsim_2) = D$. This satisfies SYM and IIA and differs from the Nash solution.

? EXERCISE 309.2 Show that there is a solution F different from the Nash solution that satisfies SYM, IIA, and $F(X, D, \succsim_1, \succsim_2) \succ_i D$ for $i = 1, 2$ (*strict individual rationality*). Roth (1977) shows that in the standard Nash bargaining model (as presented in the previous exercise) the axioms SYM, IIA, and strict individual rationality are sufficient to characterize the Nash solution. Account for the difference.

SYM: For each $\alpha \in (0, 1)$ consider the solution (an *asymmetric Nash solution*) that assigns to $\langle X, D, \succsim_1, \succsim_2 \rangle$ the agreement x^* for which $(u_1(x^*))^\alpha (u_2(x^*))^{1-\alpha} \geq (u_1(x))^\alpha (u_2(x))^{1-\alpha}$ for all $x \in X$, where u_1 and u_2 represent $\succsim_1$ and $\succsim_2$ and satisfy $u_i(D) = 0$ for $i = 1, 2$.

[?] EXERCISE 310.1 Show that any asymmetric Nash solution is well-defined
(the agreement that it selects does not depend on the utility functions
chosen to represent the preferences), satisfies PAR and IIA, and, for
$\alpha \neq \frac{1}{2}$, differs from the Nash solution.

IIA: Let $\langle X, D, \succsim_1, \succsim_2 \rangle$ be a bargaining problem and let u_i be a utility function that represents $\succsim_i$ and satisfies $u_i(D) = 0$ for $i = 1, 2$.
The *Kalai–Smorodinsky solution* assigns to $\langle X, D, \succsim_1, \succsim_2 \rangle$ the Pareto
efficient agreement x for which $u_1(x)/u_2(x) = u_1(B_1)/u_2(B_2)$.

[?] EXERCISE 310.2 Show that the Kalai–Smorodinsky solution is well-
defined, satisfies SYM and PAR, and differs from the Nash solution.

15.4 The Nash Solution and the Bargaining Game of Alternating Offers

We now show that there is a close relationship between the Nash solution
and the subgame perfect equilibrium outcome of the bargaining game
of alternating offers studied in Chapter 7, despite the different methods
that are used to derive them.

Fix a bargaining problem $\langle X, D, \succsim_1, \succsim_2 \rangle$ and consider the version of
the bargaining game of alternating offers described in Section 7.4.4, in
which the set of agreements is X, the preference relations of the players
are $\succsim_1$ and $\succsim_2$, and the outcome that results if negotiations break down
at the end of a period, an event with probability $\alpha \in (0, 1)$, is D. Under
assumptions analogous to A1–A4 (Section 7.3.1) this game has a unique
subgame perfect equilibrium outcome: player 1 proposes $x^*(\alpha)$, which
player 2 accepts, where $(x^*(\alpha), y^*(\alpha))$ is the pair of Pareto efficient agree-
ments that satisfies $(1 - \alpha) \cdot x^*(\alpha) \sim_1 y^*(\alpha)$ and $(1 - \alpha) \cdot y^*(\alpha) \sim_2 x^*(\alpha)$
(see Exercise 130.2).

■ PROPOSITION 310.3 *Let $\langle X, D, \succsim_1, \succsim_2 \rangle$ be a bargaining problem. The
agreements $x^*(\alpha)$ and $y^*(\alpha)$ proposed by the players in every subgame
perfect equilibrium of the variant of the bargaining game of alternating
offers associated with $\langle X, D, \succsim_1, \succsim_2 \rangle$ in which there is a probability α
of breakdown after any rejection both converge to the Nash solution of
$\langle X, D, \succsim_1, \succsim_2 \rangle$ as $\alpha \to 0$.*

Proof. Let u_i represent the preference relation $\succsim_i$ and satisfy $u_i(D) = 0$
for $i = 1, 2$. From the conditions defining $x^*(\alpha)$ and $y^*(\alpha)$ we have
$u_1(x^*(\alpha))u_2(x^*(\alpha)) = u_1(y^*(\alpha))u_2(y^*(\alpha))$. Since $x^*(\alpha) \succ_1 y^*(\alpha)$ for
all $\alpha \in [0, 1)$ we have $x^*(\alpha) \succsim_1 z^* \succsim_1 y^*(\alpha)$, where z^* is the Nash

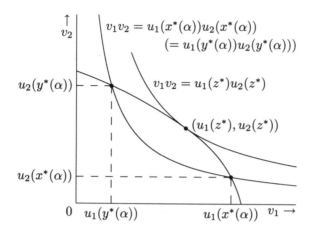

Figure 311.1 An illustration of the proof of Proposition 310.3.

solution of $\langle X, D, \succsim_1, \succsim_2 \rangle$ (see Figure 311.1). For any sequence $(\alpha_k)_{k=1}^{\infty}$ converging to 0 we have $u_i(x^*(\alpha_k)) - u_i(y^*(\alpha_k)) \to 0$ for $i = 1$, 2 by the definition of $x^*(\alpha_k)$ and $y^*(\alpha_k)$, so that $u_i(x^*(\alpha_k))$ and $u_i(y^*(\alpha_k))$ converge to $u_i(z^*)$ for $i = 1$, 2 and thus $x^*(\alpha_k)$ and $y^*(\alpha_k)$ converge to z^* (using non-redundancy). $\square$

15.5   An Exact Implementation of the Nash Solution

We now return to the implementation approach described in Chapter 10. A byproduct of the result in the previous section is that the bargaining game of alternating offers with risk of breakdown approximately SPE-implements the Nash solution. We now describe an extensive game with perfect information that *exactly* implements it. From the point of view of a planner this game has the advantage that it is simpler, in the sense that it involves a small number of stages. However, it has the disadvantage of being more remote from familiar bargaining procedures.

Fix a set X and an event D and assume the planner wants to implement the Nash solution for all pairs $(\succsim_1, \succsim_2)$ for which $\langle X, D, \succsim_1, \succsim_2 \rangle$ is a bargaining problem. Consider the extensive game form (with perfect information and chance moves) consisting of the following stages.

- Player 1 chooses $y \in X$.

- Player 2 chooses $x \in X$ and $p \in [0, 1]$.

- With probability $1 - p$ the game ends, with the outcome D, and with probability p it continues.

- Player 1 chooses either x or the lottery $p \cdot y$; this choice is the outcome.

■ PROPOSITION 312.1 *The game form described above SPE-implements the Nash solution.*

?| EXERCISE 312.2 Let x^* be the Nash solution of $\langle X, D, \succsim_1, \succsim_2 \rangle$. Show that x^* is the unique subgame perfect equilibrium outcome of the game form when the players' preferences are $(\succsim_1, \succsim_2)$.

Notes

The seminal paper on the topic of this chapter is Nash (1950b).

Our presentation follows Rubinstein, Safra, and Thomson (1992). Zeuthen (1930, Ch. IV) contains an early model in which negotiators bear in mind the risk of a breakdown when making demands. The connection between the Nash solution and the subgame perfect equilibrium outcome of a bargaining game of alternating offers was first pointed out by Binmore (1987) and was further investigated by Binmore, Rubinstein, and Wolinsky (1986). The exact implementation of the Nash solution in Section 15.5 is due to Howard (1992).

The comparative static result of Section 15.2.3 concerning the effect of the players' degree of risk aversion on the solution was first explored by Kihlstrom, Roth, and Schmeidler (1981). Harsanyi and Selten (1972) study the asymmetric Nash solutions described in Section 15.3.3 (axiomatizations appear in Kalai (1977) and Roth (1979, p. 16)) and Kalai and Smorodinsky (1975) axiomatize the Kalai–Smorodinsky solution. Exercise 309.2 is based on Roth (1977).

Several other papers (e.g. Roemer (1988)) study models in which the set of physical agreements, rather than the resulting set of utility pairs (as in Nash's model), is a primitive. Roth (1979) and Kalai (1985) are surveys of the field of axiomatic bargaining theory.

List of Results

This is a list of the main results in the book, stated informally. It is designed to give an overview of the properties of the solutions that we study. Not all conditions are included in the statements; refer to the complete statements in the text for details.

Strategic Games

Nash Equilibrium and Mixed Strategy Equilibrium

- (*Nash equilibrium existence*) Every game in which the action set of each player is compact and convex and the preference relation of each player is continuous and quasi-concave has a Nash equilibrium

  Proposition 20.3

- A symmetric game has a symmetric Nash equilibrium   Exercise 20.4

- In a strictly competitive game that has a Nash equilibrium, a pair of actions is a Nash equilibrium if and only if each action is a maxminimizer

  Proposition 22.2

- (*Mixed strategy equilibrium existence*) Every finite game has a mixed strategy Nash equilibrium   Proposition 33.1

- A mixed strategy profile is a mixed strategy Nash equilibrium of a finite game if and only if every player is indifferent between all actions in the support of his equilibrium strategy   Lemma 33.2

- A strategy profile in a finite two-player strategic game is a trembling hand perfect equilibrium if and only if it is mixed strategy Nash equilibrium and the strategy of neither player is weakly dominated

  Proposition 248.2

■ (*Trembling hand perfect equilibrium existence*) Every finite strategic game has a trembling hand perfect equilibrium   Proposition 249.1

Correlated Equilibrium

■ Every mixed strategy Nash equilibrium corresponds to a correlated equilibrium   Proposition 45.3

■ Every convex combination of correlated equilibrium payoff profiles is a correlated equilibrium payoff profile   Proposition 46.2

■ Every correlated equilibrium outcome is the outcome of a correlated equilibrium in which the set of states is the set of action profiles
                                                                  Proposition 47.1

Rationalizability

■ Every action used with positive probability in a correlated equilibrium is rationalizable   Lemma 56.2

■ An action is a never-best response if and only if it is strictly dominated
                                                                  Lemma 60.1

■ An action that is not weakly dominated is a best response to a completely mixed belief   Exercise 64.2

■ Actions that survive iterated elimination of strictly dominated actions are rationalizable   Proposition 61.2

Knowledge

■ (*Individuals cannot agree to disagree*) If two individuals have the same prior and their posterior beliefs are common knowledge then these beliefs are the same   Proposition 75.1

■ If each player is rational, knows the other players' actions, and has a belief consistent with his knowledge, then the action profile is a Nash equilibrium   Proposition 77.1

■ If there are two players and each player knows that the other player is rational, knows the other player's belief, and has a belief consistent with his knowledge, then the pair of beliefs is a mixed strategy Nash equilibrium   Proposition 78.1

■ If it is common knowledge that each player is rational and that each players' belief is consistent with his knowledge then each player's action is rationalizable   Proposition 80.1

■ If all players are rational in all states, every player's belief in every state is derived from a common prior, and each player's action is the same in all states in any given member of his information partition, then the information partitions and actions correspond to a correlated equilibrium   Exercise 81.1

Extensive Games with Perfect Information

Basic Theory

■ A strategy profile is a subgame perfect equilibrium of a finite horizon game if and only if it has the one deviation property
Lemma 98.2, Exercise 102.1, Exercise 103.3

■ (*Subgame perfect equilibrium existence: Kuhn's theorem*) Every finite game has a subgame perfect equilibrium
Proposition 99.2, Exercise 102.1

■ All players are indifferent among all subgame perfect equilibria of a finite game that satisfies the no indifference condition, and all equilibria are interchangeable   Exercise 100.2

Bargaining Games

■ A bargaining game of alternating offers that satisfies A1–A4 has a unique subgame perfect equilibrium outcome   Proposition 122.1

■ In a subgame perfect equilibrium of a bargaining game of alternating offers, a player is worse off the more impatient he is   Proposition 126.1

Infinitely Repeated Games

■ (*Nash folk theorem for limit of means*) Every feasible enforceable payoff profile of the constituent game is a Nash equilibrium payoff profile of the limit of means infinitely repeated game   Proposition 144.3

■ (*Nash folk theorem for discounting*) Every feasible strictly enforceable payoff profile of the constituent game is close to a Nash equilibrium payoff profile of the discounting infinitely repeated game for a discount factor close enough to 1   Proposition 145.2

■ (*Perfect folk theorem for limit of means*) Every feasible strictly enforceable payoff profile of the constituent game is a subgame perfect equilibrium payoff profile of the limit of means infinitely repeated game

Proposition 146.2

■ (*Perfect folk theorem for overtaking*) For every strictly enforceable outcome of the constituent game there is a subgame perfect equilibrium of the overtaking infinitely repeated game consisting of a repetition of the outcome   Proposition 149.1

■ (*Perfect folk theorem for discounting*) For every feasible strictly enforceable outcome of a full-dimensional constituent game there is a discount factor close enough to 1 for which there is a subgame perfect equilibrium of the discounting infinitely repeated game consisting of a repetition of the outcome   Proposition 151.1

■ A strategy profile is a subgame perfect equilibrium of a discounted infinitely repeated game if and only if it has the one deviation property

Lemma 153.1

■ For any subgame perfect equilibrium outcome of a discounted infinitely repeated game there is a strategy profile that generates the same outcome in which the sequence of action profiles that follows a deviation depends only on the identity of the deviant (not on the history or on the nature of the deviation)   Proposition 154.1

■ In every equilibrium of a machine game of a discounted infinitely repeated game there is a one-to-one correspondence between the actions chosen by the two machines in the repeated game   Lemma 170.1

■ Every equilibrium of a machine game of a discounted infinitely repeated game consists of an introductory phase, in which all the states are distinct, followed by a cycling phase, in each cycle of which each state appears at most once   Proposition 171.1

Finitely Repeated Games

■ If the payoff profile in every Nash equilibrium of the constituent game is the profile of minmax payoffs then every Nash equilibrium of the finitely repeated game generates a sequence of Nash equilibria of the constituent game   Proposition 155.1

■ (*Nash folk theorem for finitely repeated games*) If the constituent game has a Nash equilibrium in which every player's payoff exceeds his minmax payoff then for any strictly enforceable outcome there is a Nash equilibrium of the finitely repeated game in which each player's payoff is close to his payoff from the outcome   Proposition 156.1

- If the constituent game has a unique Nash equilibrium payoff profile then every subgame perfect equilibrium of the finitely repeated game generates a sequence of Nash equilibria of the constituent game
  Proposition 157.2

- (*Perfect folk theorem for finitely repeated games*) If the constituent game is full dimensional and for every player there are two Nash equilibria that yield different payoffs then for any strictly enforceable outcome a sufficiently long finitely repeated game has a subgame perfect equilibrium in which each player's payoff is close to his payoff from the outcome
  Proposition 160.1

Implementation Theory

- (*Gibbard–Satterthwaite theorem*) In an environment in which there are at least three consequences and any preference ordering is possible, any choice rule that is DSE-implementable and satisfies the condition that for any consequence there is a preference profile for which the choice rule induces that consequence is dictatorial   Proposition 181.2

- (*Revelation principle for DSE-implementation*) If a choice rule is DSE-implementable then it is truthfully DSE-implementable.   Lemma 181.4

- (*Revelation principle for Nash-implementation*) If a choice rule is Nash-implementable then it is truthfully Nash-implementable.   Lemma 185.2

- If a choice rule is Nash-implementable then it is monotonic
  Proposition 186.2

- In an environment in which there are at least three players, a choice rule that is monotonic and has no veto power is Nash-implementable
  Proposition 187.2

- In an environment in which there are at least three players, who can be required to pay monetary fines, every choice function is virtually SPE-implementable   Proposition 193.1

Extensive Games with Imperfect Information

- For any mixed strategy of a player in a finite extensive game with perfect recall there is an outcome-equivalent behavioral strategy
  Proposition 214.1

■ Every sequential equilibrium of the extensive game associated with a finite Bayesian game with observable actions induces a perfect Bayesian equilibrium of the Bayesian game   Proposition 234.1

■ Every trembling hand perfect equilibrium of a finite extensive game with perfect recall is associated with a sequential equilibrium
                                                         Proposition 251.2

■ (*Trembling hand perfect equilibrium and sequential equilibrium existence*) Every finite extensive game with perfect recall has a trembling hand perfect equilibrium and hence a sequential equilibrium   Corollary 253.2

Coalitional Games

Core

■ A coalitional game with transferable payoff has a nonempty core if and only if it is balanced   Proposition 262.1

■ Every market with transferable payoff has a nonempty core
                                                         Proposition 264.2

■ Every profile of competitive payoffs in a market with transferable payoff is in the core of the market   Proposition 267.1

■ Every competitive allocation in an exchange economy is in the core
                                                         Proposition 272.1

■ If every agent's preference relation is increasing and strictly quasi-concave and every agent's endowment of every good is positive, the core converges to the set of competitive allocations   Proposition 273.1

Stable Sets

■ The core is a subset of every stable set; no stable set is a proper subset of any other; if the core is a stable set then it is the only stable set
                                                         Proposition 279.2

Bargaining Set, Kernel, Nucleolus

■ In a coalitional game with transferable payoff the nucleolus is a member of the kernel, which is a subset of the bargaining set
                                              Lemmas 285.1 and 287.1

■ The nucleolus of any coalitional game with transferable payoff is a singleton   Proposition 288.4

Shapley Value

■ The unique value that satisfies the balanced contributions property is
the Shapley value   Proposition 291.3

■ The Shapley value is the only value that satisfies axioms of symmetry,
dummy, and additivity   Proposition 293.1

Nash Solution

■ The definition of the Nash solution of a bargaining problem in terms of
objections and counterobjections is equivalent to the definition of it as
the agreement that maximizes the product of the players' von Neumann–
Morgenstern utilities   Proposition 302.1

■ In the Nash solution a player is worse off the more risk-averse he is
                                                               Proposition 304.1

■ The Nash solution is the only bargaining solution that satisfies axioms of
Pareto efficiency, symmetry, and independence of irrelevant alternatives
                                                               Proposition 307.2

■ The agreements proposed by the players in every subgame perfect equilibrium outcome of the variant of a bargaining game of alternating offers
in which there is a risk of breakdown converge to the Nash solution
                                                               Proposition 310.3

References

The numbers in brackets after each reference indicate the pages on which the reference is cited.

Abreu, D. (1988), "On the Theory of Infinitely Repeated Games with Discounting", *Econometrica* **56**, 383–396. [160]

Abreu, D., P. K. Dutta, and L. Smith (1994), "The Folk Theorem for Repeated Games: A NEU Condition", *Econometrica* **62**, 939–948. [160]

Abreu, D., and H. Matsushima (1992), "Virtual Implementation in Iteratively Undominated Strategies: Complete Information", *Econometrica* **60**, 993–1008. [195]

Abreu, D., and A. Rubinstein (1988), "The Structure of Nash Equilibrium in Repeated Games with Finite Automata", *Econometrica* **56**, 1259–1281. [175]

Arrow, K. J., E. W. Barankin, and D. Blackwell (1953), "Admissible Points of Convex Sets", pp. 87–91 in *Contributions to the Theory of Games*, Volume II (Annals of Mathematics Studies, 28) (H. W. Kuhn and A. W. Tucker, eds.), Princeton: Princeton University Press. [52, 65]

Arrow, K. J., and F. H. Hahn (1971), *General Competitive Analysis*. San Francisco: Holden-Day. [270]

Aumann, R. J. (1959), "Acceptable Points in General Cooperative *n*-Person Games", pp. 287–324 in *Contributions to the Theory of Games*, Volume IV (Annals of Mathematics Studies, 40) (A. W. Tucker and R. D. Luce, eds.), Princeton: Princeton University Press. [160]

Aumann, R. J. (1964), "Markets with a Continuum of Traders", *Econometrica* **32**, 39–50. [275]

Aumann, R. J. (1974), "Subjectivity and Correlation in Randomized Strategies", *Journal of Mathematical Economics* **1**, 67–96. [51]

Aumann, R. J. (1975), "Values of Markets with a Continuum of Traders", *Econometrica* **43**, 611–646. [294]

Aumann, R. J. (1976), "Agreeing to Disagree", *Annals of Statistics* **4**, 1236–1239. [84]

Aumann, R. J. (1985a), "An Axiomatization of the Non-Transferable Utility Value", *Econometrica* **53**, 599–612. [298]

Aumann, R. J. (1985b), "What Is Game Theory Trying to Accomplish?", pp. 28–76 in *Frontiers of Economics* (K. J. Arrow and S. Honkapohja, eds.), Oxford: Basil Blackwell. [8]

Aumann, R. J. (1986), "Rejoinder", *Econometrica* **54**, 985–989. [297]

Aumann, R. J. (1987a), "Correlated Equilibrium as an Expression of Bayesian Rationality", *Econometrica* **55**, 1–18. [51, 85]

Aumann, R. J. (1987b), "Game Theory", pp. 460–482 in *The New Palgrave*, Volume 2 (J. Eatwell, M. Milgate, and P. Newman, eds.), London: Macmillan. [8]

Aumann, R. J. (1989), *Lectures on Game Theory*. Boulder: Westview Press. [275, 297]

Aumann, R. J., and A. Brandenburger (1995), "Epistemic Conditions for Nash Equilibrium", *Econometrica* **63**, 1161–1180. [80, 84]

Aumann, R. J., and M. Maschler (1964), "The Bargaining Set for Cooperative Games", pp. 443–476 in *Advances in Game Theory* (Annals of Mathematics Studies, 52) (M. Dresher, L. S. Shapley, and A. W. Tucker, eds.), Princeton: Princeton University Press. [297]

Aumann, R. J., and M. Maschler (1972), "Some Thoughts on the Minimax Principle", *Management Science* **18**, P-54–P-63. [218]

Aumann, R. J., and B. Peleg (1960), "Von Neumann–Morgenstern Solutions to Cooperative Games without Side Payments", *Bulletin of the American Mathematical Society* **66**, 173–179. [275, 298]

Aumann, R. J., and L. S. Shapley (1994), "Long-Term Competition—A Game-Theoretic Analysis", pp. 1–15 in *Essays in Game Theory* (N. Megiddo, ed.), New York: Springer-Verlag. [160]

Bacharach, M. (1985), "Some Extensions of a Claim of Aumann in an Axiomatic Model of Knowledge", *Journal of Economic Theory* **37**, 167–190. [85]

Banks, J. S., and J. Sobel (1987), "Equilibrium Selection in Signaling Games", *Econometrica* **55**, 647–661. [246, 254]

Barberá, S. (1983), "Strategy-Proofness and Pivotal Voters: A Direct Proof of the Gibbard–Satterthwaite Theorem", *International Economic Review* **24**, 413–417. [195]

Battigalli, P. (1988), "Implementable Strategies, Prior Information and the Problem of Credibility in Extensive Games", *Rivista Internazionale di Scienze Economiche e Commerciali* **35**, 705–733. [254]

Battigalli, P. (1996), "Strategic Independence and Perfect Bayesian Equilibria", *Journal of Economic Theory*, **70**, 201–234. [254]

Battigalli, P., M. Gilli, and M. C. Molinari (1992), "Learning and Convergence to Equilibrium in Repeated Strategic Interactions: An Introductory Survey", *Ricerche Economiche* **46**, 335–377. [52]

Benoît, J.-P., and V. Krishna (1985), "Finitely Repeated Games", *Econometrica* **53**, 905–922. [160, 161]

Benoît, J.-P., and V. Krishna (1987), "Nash Equilibria of Finitely Repeated Games", *International Journal of Game Theory* **16**, 197–204. [160]

Ben-Porath, E., and E. Dekel (1992), "Signaling Future Actions and the Potential for Sacrifice", *Journal of Economic Theory* **57**, 36–51. [115]

Ben-Porath, E., and B. Peleg (1987), "On the Folk Theorem and Finite Automata", Research Memorandum 77, Center for Research in Mathematical Economics and Game Theory, Hebrew University, Jerusalem. [161]

Bernheim, B. D. (1984), "Rationalizable Strategic Behavior", *Econometrica* **52**, 1007–1028. [64]

Billera, L. J. (1970), "Some Theorems on the Core of an *n*-Person Game without Side-Payments", *SIAM Journal on Applied Mathematics* **18**, 567–579. [269, 275]

Billera, L. J., D. C. Heath, and J. Raanan (1978), "Internal Telephone Billing Rates—A Novel Application of Non-Atomic Game Theory", *Operations Research* **26**, 956–965. [297]

Binmore, K. G. (1985), "Bargaining and Coalitions", pp. 269–304 in *Game-Theoretic Models of Bargaining* (A. E. Roth, ed.), Cambridge: Cambridge University Press. [131]

Binmore, K. G. (1987), "Nash Bargaining Theory II", pp. 61–76 in *The Economics of Bargaining* (K. G. Binmore and P. Dasgupta, eds.), Oxford: Blackwell. [312]

Binmore, K. G. (1987/88), "Modeling Rational Players, Parts I and II", *Economics and Philosophy* **3**, 179–214 and **4**, 9–55. [5, 8]

Binmore, K. G. (1992), *Fun and Games*. Lexington, Massachusetts: D. C. Heath. [8]

Binmore, K. G., and A. Brandenburger (1990), "Common Knowledge and Game Theory", pp. 105–150 in *Essays on the Foundations of Game Theory* (K. G. Binmore), Oxford: Blackwell. [85]

Binmore, K. G., A. Rubinstein, and A. Wolinsky (1986), "The Nash Bargaining Solution in Economic Modelling", *Rand Journal of Economics* **17**, 176–188. [131, 312]

Bondareva, O. N. (Бондарева, О. Н.) (1963), "Некоторые Применения Методов Линейного Программирования к Теории Кооперативных Игр", *Проблемы Кибернетики (Problemy Kibernetiki)* **10**, 119–139. [274]

Borel, E. (1921), "La Théorie du Jeu et les Equations Intégrales à Noyau Symétrique", *Comptes Rendus Hebdomadaires des Séances de l'Académie des Sciences (Paris)* **173**, 1304–1308. (Translated as "The Theory of Play and Integral Equations with Skew Symmetric Kernels", *Econometrica* **21** (1953), pp. 97–100.) [29, 51]

Borel, E. (1924), *Eléments de la Théorie des Probabilités* (Third Edition). Paris: Librairie Scientifique, J. Hermann. (Pp. 204–221 translated as "On Games That Involve Chance and the Skill of the Players", *Econometrica* **21** (1953), 101–115.) [51]

Borel, E. (1927), "Sur les Systèmes de Formes Linéaires à Déterminant Symétrique Gauche et la Théorie Générale du Jeu", *Comptes Rendus Hebdomadaires des Séances de l'Académie des Sciences (Paris)* **184**, 52–54. (Translated as "On Systems of Linear Forms of Skew Symmetric Determinant and the General Theory of Play", *Econometrica* **21** (1953), 116–117.) [51]

Brams, S. J., D. M. Kilgour, and M. D. Davis (1993), "Unraveling in Games of Sharing and Exchange", pp. 195–212 in *Frontiers of Game Theory* (K. G. Binmore, A. Kirman, and P. Tani, eds.), Cambridge, Mass.: MIT Press. [30]

Brams, S. J., and A. D. Taylor (1994), "Divide the Dollar: Three Solutions and Extensions", *Theory and Decision* **37**, 211–231. [65]

Brandenburger, A. (1992), "Knowledge and Equilibrium in Games", *Journal of Economic Perspectives* **6**, 83–101. [84]

Brown, G. W. (1951), "Iterative Solution of Games by Fictitious Play", pp. 374–376 in *Activity Analysis of Production and Allocation* (T. C. Koopmans, ed.), New York: Wiley. [52]

Carnap, R. (1966), *Philosophical Foundations of Physics*. New York: Basic Books. [5]

Cho, I.-K., and D. M. Kreps (1987), "Signaling Games and Stable Equilibria", *Quarterly Journal of Economics* **102**, 179–221. [254]

Chung, K. L. (1974), *A Course in Probability Theory* (Second Edition). New York: Academic Press. [7]

Clarke, E. H. (1971), "Multipart Pricing of Public Goods", *Public Choice* **11**, 17–33. [183]

Cournot, A. A. (1838), *Recherches sur les Principes Mathématiques de la Théorie des Richesses*. Paris: Hachette. (English translation: *Researches into the Mathematical Principles of the Theory of Wealth*, New York: Macmillan, 1897.) [30]

Crawford, V. P. (1990), "Equilibrium without Independence", *Journal of Economic Theory* **50**, 127–154. [52]

Davis, M., and M. Maschler (1963), "Existence of Stable Payoff Configurations for Cooperative Games", *Bulletin of the American Mathematical Society* **69**, 106–108. [297]

Davis, M., and M. Maschler (1965), "The Kernel of a Cooperative Game", *Naval Research Logistics Quarterly* **12**, 223–259. [297]

Davis, M., and M. Maschler (1967), "Existence of Stable Payoff Configurations for Cooperative Games", pp. 39–52 in *Essays in Mathematical Economics* (M. Shubik, ed.), Princeton: Princeton University Press. [297]

Debreu, G., and H. E. Scarf (1963), "A Limit Theorem on the Core of an Economy", *International Economic Review* **4**, 235–246. [275]

Derman, C. (1970), *Finite State Markovian Decision Processes*. New York: Academic Press. [168]

Diamond, P. A. (1965), "The Evaluation of Infinite Utility Streams", *Econometrica* **33**, 170–177. [160]

Dubey, P., and M. Kaneko (1984), "Information Patterns and Nash Equilibria in Extensive Games: I", *Mathematical Social Sciences* **8**, 111–139. [115]

Edgeworth, F. Y. (1881), *Mathematical Psychics*. London: Kegan Paul. [275]

Elmes, S., and P. J. Reny (1994), "On the Strategic Equivalence of Extensive Form Games", *Journal of Economic Theory* **62**, 1–23. [218]

Fishburn, P. C., and A. Rubinstein (1982), "Time Preference", *International Economic Review* **23**, 677–694. [119, 131]

Forges, F. (1992), "Repeated Games of Incomplete Information: Non-Zero-Sum", pp. 155–177 in *Handbook of Game Theory with Economic Applications*, Volume 1 (R. J. Aumann and S. Hart, eds.), Amsterdam: North-Holland. [161]

Friedman, J. W. (1971), "A Non-Cooperative Equilibrium for Supergames", *Review of Economic Studies* **38**, 1–12. [160]

Friedman, J. W. (1985), "Cooperative Equilibria in Finite Horizon Noncooperative Supergames", *Journal of Economic Theory* **35**, 390–398. [161]

Friedman, J. W. (1990), *Game Theory with Applications to Economics* (Second Edition). New York: Oxford University Press. [8, 275]

Fudenberg, D. (1992), "Explaining Cooperation and Commitment in Repeated Games", pp. 89–131 in *Advances in Economic Theory*, Volume I (J.-J. Laffont, ed.), Cambridge: Cambridge University Press. [161]

Fudenberg, D., and D. K. Levine (1989), "Reputation and Equilibrium Selection in Games with a Patient Player", *Econometrica* **57**, 759–778. [161]

Fudenberg, D., and E. S. Maskin (1986), "The Folk Theorem in Repeated Games with Discounting or with Incomplete Information", *Econometrica* **54**, 533–554. [160, 161, 254]

Fudenberg, D., and E. S. Maskin (1991), "On the Dispensability of Public Randomization in Discounted Repeated Games", *Journal of Economic Theory* **53**, 428–438. [161]

Fudenberg, D., and J. Tirole (1991a), *Game Theory*. Cambridge, Mass.: MIT Press. [8]

Fudenberg, D., and J. Tirole (1991b), "Perfect Bayesian Equilibrium and Sequential Equilibrium", *Journal of Economic Theory* **53**, 236–260. [254]

Gabay, D., and H. Moulin (1980), "On the Uniqueness and Stability of Nash-Equilibria in Noncooperative Games", pp. 271–293 in *Applied Stochastic Control in Econometrics and Management Science* (A. Bensoussan, P. Kleindorfer, and C. S. Tapiero, eds.), Amsterdam: North-Holland. [64]

Gale, D. (1953), "A Theory of N-Person Games with Perfect Information", *Proceedings of the National Academy of Sciences of the United States of America* **39**, 496–501. [64]

Geanakoplos, J. (1992), "Common Knowledge", *Journal of Economic Perspectives* **6**, 53–82. [85]

Geanakoplos, J. (1994), "Common Knowledge", pp. 1437–1496 in *Handbook of Game Theory*, Volume 2 (R. J. Aumann and S. Hart, eds.), Amsterdam: North-Holland. [85]

Gibbard, A. (1973), "Manipulation of Voting Schemes: A General Result", *Econometrica* **41**, 587–601. [195]

Gibbons, R. (1992), *Game Theory for Applied Economists.* Princeton: Princeton University Press. [8]

Gillies, D. B. (1959), "Solutions to General Non-Zero-Sum Games", pp. 47–85 in *Contributions to the Theory of Games*, Volume IV (Annals of Mathematics Studies, 40) (A. W. Tucker and R. D. Luce, eds.), Princeton: Princeton University Press. [274]

Glazer, J., and C. A. Ma (1989), "Efficient Allocation of a 'Prize'— King Solomon's Dilemma", *Games and Economic Behavior* **1**, 222–233. [195]

Glazer, J. and M. Perry (1996), "Virtual Implementation in Backwards Induction", *Games and Economic Behavior* **15**, 27–32. [195]

Glicksberg, I. L. (1952), "A Further Generalization of the Kakutani Fixed Point Theorem, with Application to Nash Equilibrium Points", *Proceedings of the American Mathematical Society* **3**, 170–174. [30, 33]

Green, J., and J.-J. Laffont (1977), "Characterization of Satisfactory Mechanisms for the Revelation of Preferences for Public Goods", *Econometrica* **45**, 427–438. [195]

Groves, T. (1973), "Incentives in Teams", *Econometrica* **41**, 617–631. [183]

Groves, T., and M. Loeb (1975), "Incentives and Public Inputs", *Journal of Public Economics* **4**, 211–226. [195]

Guilbaud, G. T. (1961), "Faut-il Jouer au Plus Fin? (Notes sur l'Histoire de la Théorie des Jeux)", pp. 171–182 in *La Décision*, Paris: Éditions du Centre National de la Recherce Scientifique. [51]

Gul, F. (1989), "Bargaining Foundations of Shapley Value", *Econometrica* **57**, 81–95. [298]

Halpern, J. Y. (1986), "Reasoning about Knowledge: An Overview", pp. 1–17 in *Theoretical Aspects of Reasoning about Knowledge* (J. Y. Halpern, ed.), Los Altos, California: Morgan Kaufmann. [85]

Harsanyi, J. C. (1963), "A Simplified Bargaining Model for the n-Person Cooperative Game", *International Economic Review* **4**, 194–220. [298]

Harsanyi, J. C. (1967/68), "Games with Incomplete Information Played by 'Bayesian' Players, Parts I, II, and III", *Management Science* **14**, 159–182, 320–334, and 486–502. [28, 30]

Harsanyi, J. C. (1973), "Games with Randomly Disturbed Payoffs: A New Rationale for Mixed-Strategy Equilibrium Points", *International Journal of Game Theory* **2**, 1–23. [38, 41, 42, 51]

Harsanyi, J. C. (1974), "An Equilibrium-Point Interpretation of Stable Sets and a Proposed Alternative Definition", *Management Science (Theory Series)* **20**, 1472–1495. [298]

Harsanyi, J. C. (1981), "The Shapley Value and the Risk-Dominance Solutions of Two Bargaining Models for Characteristic-Function Games", pp. 43–68 in *Essays in Game Theory and Mathematical Economics* (R. J. Aumann, J. C. Harsanyi, W. Hildenbrand, M. Maschler, M. A. Perles, J. Rosenmüller, R. Selten, M. Shubik, and G. L. Thompson), Mannheim: Bibliographisches Institut. [298]

Harsanyi, J. C., and R. Selten (1972), "A Generalized Nash Solution for Two-Person Bargaining Games with Incomplete Information", *Management Science* **18**, P-80–P-106. [312]

Hart, S. (1985), "An Axiomatization of Harsanyi's Nontransferable Utility Solution", *Econometrica* **53**, 1295–1313. [298]

Hart, S., and A. Mas-Colell (1989), "Potential, Value, and Consistency", *Econometrica* **57**, 589–614. [297]

Hart, S. and A. Mas-Colell (1996), "Bargaining and Value", *Econometrica* **64**, 357–380. [296, 297, 298]

Hendon, E., H. J. Jacobsen, and B. Sloth (1996), "The One-Shot-Deviation Principle for Sequential Rationality" *Games and Economic Behavior* **12**, 274–282. [254]

Hintikka, J. (1962), *Knowledge and Belief.* Ithaca: Cornell University Press. [84]

Hotelling, H. (1929), "Stability in Competition", *Economic Journal* **39**, 41–57. [30]

Howard, J. V. (1992), "A Social Choice Rule and Its Implementation in Perfect Equilibrium", *Journal of Economic Theory* **56**, 142–159. [312]

Isbell, J. R. (1957), "Finitary Games", pp. 79–96 in *Contributions to the Theory of Games*, Volume III (Annals of Mathematics Studies, 39) (M. Dresher, A. W. Tucker, and P. Wolfe, eds.), Princeton: Princeton University Press. [218]

Kakutani, S. (1941), "A Generalization of Brouwer's Fixed Point Theorem", *Duke Mathematical Journal* **8**, 457–459. [19]

Kalai, E. (1977), "Nonsymmetric Nash Solutions and Replications of 2-Person Bargaining", *International Journal of Game Theory* **6**, 129–133. [312]

Kalai, E. (1985), "Solutions to the Bargaining Problem", pp. 77–105 in *Social Goals and Social Organization* (L. Hurwicz, D. Schmeidler, and H. Sonnenschein, eds.), Cambridge: Cambridge University Press. [312]

Kalai, E., and M. Smorodinsky (1975), "Other Solutions to Nash's Bargaining Problem", *Econometrica* **43**, 513–518. [312]

Kihlstrom, R. E., A. E. Roth, and D. Schmeidler (1981), "Risk Aversion and Solutions to Nash's Bargaining Problem", pp. 65–71 in *Game Theory and Mathematical Economics* (O. Moeschlin and D. Pallaschke, eds.), Amsterdam: North-Holland. [312]

Kohlberg, E. (1990), "Refinement of Nash Equilibrium: The Main Ideas", pp. 3–45 in *Game Theory and Applications* (T. Ichiishi, A. Neyman, and Y. Tauman, eds.), San Diego: Academic Press. [254]

Kohlberg, E., and J.-F. Mertens (1986), "On the Strategic Stability of Equilibria", *Econometrica* **54**, 1003–1037. [115, 254]

Kohlberg, E., and P. J. Reny (1997), "Independence on Relative Probability Spaces and Consistent Assessments in Game Trees", *Journal of Economic Theory* **75**, 280–313. [254]

Krelle, W. (1976), *Preistheorie* (Part 2). Tübingen: J. C. B. Mohr (Paul Siebeck). [131]

Kreps, D. M. (1988), *Notes on the Theory of Choice*. Boulder: Westview Press. [8]

Kreps, D. M. (1990a), *A Course in Microeconomic Theory*. Princeton: Princeton University Press. [8, 225]

Kreps, D. M. (1990b), *Game Theory and Economic Modelling*. Oxford: Clarendon Press. [8, 254]

Kreps, D. M., and G. Ramey (1987), "Structural Consistency, Consistency, and Sequential Rationality", *Econometrica* **55**, 1331–1348. [254]

Kreps, D. M., and R. B. Wilson (1982a), "Reputation and Imperfect Information", *Journal of Economic Theory* **27**, 253–279. [115, 254]

Kreps, D. M., and R. B. Wilson (1982b), "Sequential Equilibria", *Econometrica* **50**, 863–894. [252, 254]

Krishna, V. (1989), "The Folk Theorems for Repeated Games", Working
Paper 89-003, Division of Research, Harvard Business School.
[160, 161]

Krishna, V., and R. Serrano (1995), "Perfect Equilibria of a Model of
N-Person Noncooperative Bargaining", *International Journal of
Game Theory* **24**, 259–272. [296]

Kuhn, H. W. (1950), "Extensive Games", *Proceedings of the National
Academy of Sciences of the United States of America* **36**, 570–576.
[114, 115, 217, 218]

Kuhn, H. W. (1953), "Extensive Games and the Problem of Information", pp. 193–216 in *Contributions to the Theory of Games*, Volume II (Annals of Mathematics Studies, 28) (H. W. Kuhn and
A. W. Tucker, eds.), Princeton: Princeton University Press. [114,
115, 217, 218]

Kuhn, H. W. (1968), "Preface" to "Waldegrave's Comments: Excerpt
from Montmort's Letter to Nicholas Bernoulli", pp. 3–6 in *Precursors in Mathematical Economics: An Anthology* (Series of
Reprints of Scarce Works on Political Economy, 19) (W. J. Bau-
mol and S. M. Goldfeld, eds.), London: London School of Economics and Political Science. [30, 51]

Lewis, D. K. (1969), *Convention*. Cambridge: Harvard University Press.
[84]

Littlewood, J. E. (1953), *A Mathematician's Miscellany*. London:
Methuen. [71, 85]

Lucas, W. F. (1969), "The Proof That a Game May Not Have a Solution", *Transactions of the American Mathematical Society* **137**,
219–229. [279]

Lucas, W. F. (1992), "Von Neumann–Morgenstern Stable Sets", pp. 543–
590 in *Handbook of Game Theory with Economic Applications*,
Volume 1 (R. J. Aumann and S. Hart, eds.), Amsterdam: North-
Holland. [298]

Luce, R. D., and H. Raiffa (1957), *Games and Decisions*. New York:
John Wiley and Sons. [8, 15, 30, 64, 65, 160, 275, 293]

Madrigal, V., T. C. C. Tan, and S. R. da C. Werlang (1987), "Support Restrictions and Sequential Equilibria", *Journal of Economic Theory* **43**, 329–334. [254]

Maschler, M. (1976), "An Advantage of the Bargaining Set over the
Core", *Journal of Economic Theory* **13**, 184–192. [297]

Maschler, M. (1992), "The Bargaining Set, Kernel, and Nucleolus",
pp. 591–667 in *Handbook of Game Theory with Economic Appli-*

cations, Volume 1 (R. J. Aumann and S. Hart, eds.), Amsterdam: North-Holland. [298]

Maschler, M., and G. Owen (1989), "The Consistent Shapley Value for Hyperplane Games", *International Journal of Game Theory* **18**, 389–407. [298]

Maschler, M., and G. Owen (1992), "The Consistent Shapley Value for Games without Side Payments", pp. 5–12 in *Rational Interaction* (R. Selten, ed.), Berlin: Springer-Verlag. [298]

Maskin, E. S. (1985), "The Theory of Implementation in Nash Equilibrium: A Survey", pp. 173–204 in *Social Goals and Social Organization* (L. Hurwicz, D. Schmeidler, and H. Sonnenschein, eds.), Cambridge: Cambridge University Press. [195]

Maynard Smith, J. (1972), "Game Theory and the Evolution of Fighting", pp. 8–28 in *On Evolution* (J. Maynard Smith), Edinburgh: Edinburgh University Press. [51]

Maynard Smith, J. (1974), "The Theory of Games and the Evolution of Animal Conflicts", *Journal of Theoretical Biology* **47**, 209–221. [30, 51]

Maynard Smith, J. (1982), *Evolution and the Theory of Games*. Cambridge: Cambridge University Press. [51]

Maynard Smith, J., and G. R. Price (1973), "The Logic of Animal Conflict", *Nature* **246**, 15–18. [51]

Mertens, J.-F. (1995), "Two Examples of Strategic Equilibrium", *Games and Economic Behavior* **8**, 378–388. [254]

Mertens, J.-F., and S. Zamir (1985), "Formulation of Bayesian Analysis for Games with Incomplete Information", *International Journal of Game Theory* **14**, 1–29. [29]

Milgrom, P. R., and D. J. Roberts (1982), "Predation, Reputation, and Entry Deterrence", *Journal of Economic Theory* **27**, 280–312. [115, 254]

Milgrom, P. R., and D. J. Roberts (1990), "Rationalizability, Learning, and Equilibrium in Games with Strategic Complementarities", *Econometrica* **58**, 1255–1277. [65]

Milgrom, P. R., and N. Stokey (1982), "Information, Trade and Common Knowledge", *Journal of Economic Theory* **26**, 17–27. [85]

Milnor, J. W., and L. S. Shapley (1978), "Values of Large Games II: Oceanic Games", *Mathematics of Operations Research* **3**, 290–307. [297]

Moore, J. (1992), "Implementation, Contracts, and Renegotiation in Environments with Complete Information", pp. 182–282 in *Advances in Economic Theory*, Volume I (J.-J. Laffont, ed.), Cambridge: Cambridge University Press. [196]

Moore, J., and R. Repullo (1988), "Subgame Perfect Implementation", *Econometrica* **56**, 1191–1220. [196]

Moulin, H. (1979), "Dominance Solvable Voting Schemes", *Econometrica* **47**, 1337–1351. [64, 65]

Moulin, H. (1984), "Dominance Solvability and Cournot Stability", *Mathematical Social Sciences* **7**, 83–102. [64]

Moulin, H. (1986), *Game Theory for the Social Sciences* (Second Edition). New York: New York University Press. [8, 52, 64, 115, 275]

Moulin, H. (1988), *Axioms of Cooperative Decision Making*. Cambridge: Cambridge University Press. [275, 297, 298]

Muller, E., and M. A. Satterthwaite (1977), "The Equivalence of Strong Positive Association and Strategy-Proofness", *Journal of Economic Theory* **14**, 412–418. [189]

Muthoo, A. (1991), "A Note on Bargaining over a Finite Number of Feasible Agreements", *Economic Theory* **1**, 290–292. [131]

Myerson, R. B. (1977), "Graphs and Cooperation in Games", *Mathematics of Operations Research* **2**, 225–229. [297]

Myerson, R. B. (1978), "Refinements of the Nash Equilibrium Concept", *International Journal of Game Theory* **7**, 73–80. [254]

Myerson, R. B. (1980), "Conference Structures and Fair Allocation Rules", *International Journal of Game Theory* **9**, 169–182. [297]

Myerson, R. B. (1991), *Game Theory*. Cambridge, Massachusetts: Harvard University Press. [8, 275, 293]

Nash, J. F. (1950a), "Equilibrium Points in N-Person Games", *Proceedings of the National Academy of Sciences of the United States of America* **36**, 48–49. [29, 30, 51]

Nash, J. F. (1950b), "The Bargaining Problem", *Econometrica* **18**, 155–162. [312]

Nash, J. F. (1951), "Non-Cooperative Games", *Annals of Mathematics* **54**, 286–295. [30, 51]

Neyman, A. (1985), "Bounded Complexity Justifies Cooperation in the Finitely Repeated Prisoners' Dilemma", *Economic Letters* **19**, 227–229. [175]

Nikaidô, H., and K. Isoda (1955), "Note on Noncooperative Convex Games", *Pacific Journal of Mathematics* **5**, 807–815. [30]

Osborne, M. J. (1990), "Signaling, Forward Induction, and Stability in Finitely Repeated Games", *Journal of Economic Theory* **50**, 22–36. [115]

Osborne, M. J., and A. Rubinstein (1990), *Bargaining and Markets*. San Diego: Academic Press. [129, 131, 236]

Owen, G. (1982), *Game Theory* (Second Edition). New York: Academic Press. [275]

Pearce, D. G. (1984), "Rationalizable Strategic Behavior and the Problem of Perfection", *Econometrica* **52**, 1029–1050. [64]

Pearce, D. G. (1992), "Repeated Games: Cooperation and Rationality", pp. 132–174 in *Advances in Economic Theory*, Volume I (J.-J. Laffont, ed.), Cambridge: Cambridge University Press. [161]

Peleg, B. (1963a), "Bargaining Sets of Cooperative Games without Side Payments", *Israel Journal of Mathematics* **1**, 197–200. [298]

Peleg, B. (1963b), "Existence Theorem for the Bargaining Set $M_1^{(i)}$", *Bulletin of the American Mathematical Society* **69**, 109–110. [297]

Peleg, B. (1967), "Existence Theorem for the Bargaining Set $M_1^{(i)}$", pp. 53–56 in *Essays in Mathematical Economics* (M. Shubik, ed.), Princeton: Princeton University Press. [297]

Peleg, B. (1968), "On Weights of Constant-Sum Majority Games", *SIAM Journal on Applied Mathematics* **16**, 527–532. [297]

Peleg, B. (1992), "Axiomatizations of the Core", pp. 397–412 in *Handbook of Game Theory with Economic Applications*, Volume 1 (R. J. Aumann and S. Hart, eds.), Amsterdam: North-Holland. [275]

Piccione, M. (1992), "Finite Automata Equilibria with Discounting", *Journal of Economic Theory* **56**, 180–193. [175]

Piccione, M., and A. Rubinstein (1993), "Finite Automata Play a Repeated Extensive Game", *Journal of Economic Theory* **61**, 160–168. [175]

Postlewaite, A., and R. W. Rosenthal (1974), "Disadvantageous Syndicates", *Journal of Economic Theory* **9**, 324–326. [275]

Radner, R. (1980), "Collusive Behavior in Noncooperative Epsilon-Equilibria of Oligopolies with Long but Finite Lives", *Journal of Economic Theory* **22**, 136–154. [115]

Radner, R. (1986), "Can Bounded Rationality Resolve the Prisoners' Dilemma?", pp. 387–399 in *Contributions to Mathematical Economics* (W. Hildenbrand and A. Mas-Colell, eds.), Amsterdam: North-Holland. [115]

Raiffa, H. (1992), "Game Theory at the University of Michigan, 1948–1952", pp. 165–175 in *Toward a History of Game Theory* (E. R. Weintraub, ed.), Durham: Duke University Press. [30]

Rapoport, A. (1987), "Prisoner's Dilemma", pp. 973–976 in *The New Palgrave*, Volume 3 (J. Eatwell, M. Milgate, and P. Newman, eds.), London: Macmillan. [135]

Reny, P. J. (1992), "Backward Induction, Normal Form Perfection and Explicable Equilibria", *Econometrica* **60**, 627–649. [115]

Reny, P. J. (1993), "Common Belief and the Theory of Games with Perfect Information", *Journal of Economic Theory* **59**, 257–274. [115]

Repullo, R. (1987), "A Simple Proof of Maskin's Theorem on Nash Implementation", *Social Choice and Welfare* **4**, 39–41. [195]

Robinson, J. (1951), "An Iterative Method of Solving a Game", *Annals of Mathematics* **54**, 296–301. [52]

Rockafellar, R. T. (1970), *Convex Analysis*. Princeton: Princeton University Press. [263, 274]

Roemer, J. E. (1988), "Axiomatic Bargaining Theory on Economic Environments", *Journal of Economic Theory* **45**, 1–31. [312]

Rosenthal, R. W. (1979), "Sequences of Games with Varying Opponents", *Econometrica* **47**, 1353–1366. [51]

Rosenthal, R. W. (1981), "Games of Perfect Information, Predatory Pricing and the Chain-Store Paradox", *Journal of Economic Theory* **25**, 92–100. [115]

Roth, A. E. (1977), "Individual Rationality and Nash's Solution to the Bargaining Problem", *Mathematics of Operations Research* **2**, 64–65. [309, 312]

Roth, A. E. (1979), *Axiomatic Models of Bargaining*. Berlin: Springer-Verlag. [312]

Roth, A. E., and R. E. Verrecchia (1979), "The Shapley Value As Applied to Cost Allocation: A Reinterpretation", *Journal of Accounting Research* **17**, 295–303. [297]

Rubinstein, A. (1979), "Equilibrium in Supergames with the Overtaking Criterion", *Journal of Economic Theory* **21**, 1–9. [160]

Rubinstein, A. (1982), "Perfect Equilibrium in a Bargaining Model", *Econometrica* **50**, 97–109. [131]

Rubinstein, A. (1986), "Finite Automata Play the Repeated Prisoner's Dilemma", *Journal of Economic Theory* **39**, 83–96. [175]

Rubinstein, A. (1989), "The Electronic Mail Game: Strategic Behavior Under 'Almost Common Knowledge'", *American Economic Review* **79**, 385–391. [85]

Rubinstein, A. (1991), "Comments on the Interpretation of Game Theory", *Econometrica* **59**, 909–924. [51, 115]

Rubinstein, A. (1994), "Equilibrium in Supergames", pp. 17–27 in *Essays in Game Theory* (N. Megiddo, ed.), New York: Springer-Verlag. [160]

Rubinstein, A. (1995), "On the Interpretation of Two Theoretical Models of Bargaining", pp. 120–130 in *Barriers to Conflict Resolution* (K. J. Arrow, R. H. Mnookin, L. Ross, A. Tversky, and R. B. Wilson, eds.), New York: Norton. [131]

Rubinstein, A., Z. Safra, and W. Thomson (1992), "On the Interpretation of the Nash Bargaining Solution and Its Extension to Non-Expected Utility Preferences", *Econometrica* **60**, 1171–1186. [312]

Rubinstein, A., and A. Wolinsky (1994), "Rationalizable Conjectural Equilibrium: Between Nash and Rationalizability", *Games and Economic Behavior* **6**, 299–311. [64–65]

Samet, D. (1990), "Ignoring Ignorance and Agreeing to Disagree", *Journal of Economic Theory* **52**, 190–207. [85]

Satterthwaite, M. A. (1975), "Strategy-Proofness and Arrow's Conditions: Existence and Correspondence Theorems for Voting Procedures and Social Welfare Functions", *Journal of Economic Theory* **10**, 187–217. [195]

Savage, L. J. (1972), *The Foundations of Statistics* (Second Revised Edition). New York: Dover. [5]

Scarf, H. E. (1967), "The Core of an *N* Person Game", *Econometrica* **35**, 50–69. [269, 275]

Schelling, T. C. (1960), *The Strategy of Conflict*. Cambridge, Mass.: Harvard University Press. [8]

Schmeidler, D. (1969), "The Nucleolus of a Characteristic Function Game", *SIAM Journal on Applied Mathematics* **17**, 1163–1170. [297]

Schmeidler, D., and H. Sonnenschein (1978), "Two Proofs of the Gibbard–Satterthwaite Theorem on the Possibility of a Strategy-Proof Social Choice Function", pp. 227–234 in *Decision Theory and Social Ethics* (H. W. Gottinger and W. Leinfellner, eds.), Dordrecht: D. Reidel. [195]

Selten, R. (1965), "Spieltheoretische Behandlung eines Oligopolmodells mit Nachfrageträgheit", *Zeitschrift für die gesamte Staatswissenschaft* **121**, 301–324 and 667–689. [115]

Selten, R. (1975), "Reexamination of the Perfectness Concept for Equilibrium Points in Extensive Games", *International Journal of Game Theory* **4**, 25–55. [254]

Selten, R. (1978), "The Chain Store Paradox", *Theory and Decision* **9**, 127–159. [115]

Sen, A. (1986), "Social Choice Theory", pp. 1073–1181 in *Handbook of Mathematical Economics*, Volume 3 (K. J. Arrow and M. D. Intriligator, eds.), Amsterdam: North-Holland. [182]

Shaked, A. (1994), "Opting Out: Bazaars versus 'Hi Tech' Markets", *Investigaciones Económicas* **18**, 421–432. [129]

Shaked, A., and J. Sutton (1984a), "Involuntary Unemployment as a Perfect Equilibrium in a Bargaining Model", *Econometrica* **52**, 1351–1364. [131]

Shaked, A., and J. Sutton (1984b), "The Semi-Walrasian Economy", Discussion Paper 84/98 (Theoretical Economics), International Centre for Economics and Related Disciplines, London School of Economics. [131]

Shapley, L. S. (1953), "A Value for *n*-Person Games", pp. 307–317 in *Contributions to the Theory of Games*, Volume II (Annals of Mathematics Studies, 28) (H. W. Kuhn and A. W. Tucker, eds.), Princeton: Princeton University Press. [297]

Shapley, L. S. (1959), "The Solutions of a Symmetric Market Game", pp. 145–162 in *Contributions to the Theory of Games*, Volume IV (Annals of Mathematics Studies, 40) (A. W. Tucker and R. D. Luce, eds.), Princeton: Princeton University Press. [274, 275]

Shapley, L. S. (1964), "Some Topics in Two-Person Games", pp. 1–28 in *Advances in Game Theory* (Annals of Mathematics Studies, 52) (M. Dresher, L. S. Shapley, and A. W. Tucker, eds.), Princeton: Princeton University Press. [52]

Shapley, L. S. (1967), "On Balanced Sets and Cores", *Naval Research Logistics Quarterly* **14**, 453–460. [274]

Shapley, L. S. (1969), "Utility Comparison and the Theory of Games", pp. 251–263 in *La Décision*, Paris: Éditions du Centre National de la Recherche Scientifique. (Reprinted on pp. 307–319 of *The Shapley Value* (Alvin E. Roth, ed.), Cambridge: Cambridge University Press, 1988.) [298]

Shapley, L. S. (1971/72), "Cores of Convex Games", *International Journal of Game Theory* **1**, 11–26. (See also p. 199.) [275, 297]

Shapley, L. S. (1973), "On Balanced Games without Side Payments", pp. 261–290 in *Mathematical Programming* (T. C. Hu and S. M. Robinson, eds.), New York: Academic Press. [269, 275]

Shapley, L. S., and M. Shubik (1953), "Solutions of *N*-Person Games with Ordinal Utilities", *Econometrica* **21**, 348–349. (Abstract of a paper presented at a conference.) [275]

Shapley, L. S., and M. Shubik (1967), "Ownership and the Production Function", *Quarterly Journal of Economics* **81**, 88–111. [275]

Shapley, L. S., and M. Shubik (1969a), "On Market Games", *Journal of Economic Theory* **1**, 9–25. [274]

Shapley, L. S., and M. Shubik (1969b), "On the Core of an Economic System with Externalities", *American Economic Review* **59**, 678–684. [275]

Shubik, M. (1959a), "Edgeworth Market Games", pp. 267–278 in *Contributions to the Theory of Games*, Volume IV (Annals of Mathematics Studies, 40) (A. W. Tucker and R. D. Luce, eds.), Princeton: Princeton University Press. [274, 275]

Shubik, M. (1959b), *Strategy and Market Structure*. New York: Wiley. [160]

Shubik, M. (1962), "Incentives, Decentralized Control, the Assignment of Joint Costs and Internal Pricing", *Management Science* **8**, 325–343. [297]

Shubik, M. (1982), *Game Theory in the Social Sciences*. Cambridge, Mass.: MIT Press. [8, 275]

Sorin, S. (1990), "Supergames", pp. 43–63 in *Game Theory and Applications* (T. Ichiishi, A. Neyman, and Y. Tauman, eds.), San Diego: Academic Press. [161]

Sorin, S. (1992), "Repeated Games with Complete Information", pp. 71–107 in *Handbook of Game Theory with Economic Applications*, Volume 1 (R. J. Aumann and S. Hart, eds.), Amsterdam: North-Holland. [161]

Spence, A. M. (1974), *Market Signaling*. Cambridge, Mass.: Harvard University Press. [237]

Spohn, W. (1982), "How to Make Sense of Game Theory", pp. 239–270 in *Philosophy of Economics* (W. Stegmüller, W. Balzer, and W. Spohn, eds.), Berlin: Springer-Verlag. [64, 84]

Ståhl, I. (1972), *Bargaining Theory*. Stockholm: Economics Research Institute at the Stockholm School of Economics. [131]

Thompson, F. B. (1952), "Equivalence of Games in Extensive Form", Research Memorandum RM-759, U.S. Air Force Project Rand, Rand Corporation, Santa Monica, California. (Reprinted on pp. 36–45 of *Classics in Game Theory* (Harold W. Kuhn, ed.), Princeton: Princeton University Press, 1997.) [209, 217]

Tversky, A., and D. Kahneman (1986), "Rational Choice and the Framing of Decisions", *Journal of Business* **59**, S251–S278. [209]

Van Damme, E. (1983), *Refinements of the Nash Equilibrium Concept* (Lecture Notes in Economics and Mathematical Systems, Vol. 219). Berlin: Springer-Verlag. [64]

Van Damme, E. (1989), "Stable Equilibria and Forward Induction", *Journal of Economic Theory* **48**, 476–496. [115]

Van Damme, E. (1991), *Stability and Perfection of Nash Equilibria* (Second Edition). Berlin: Springer-Verlag. [8, 52]

Van Damme, E. (1992), "Refinements of Nash Equilibrium", pp. 32–75 in *Advances in Economic Theory*, Volume I (J.-J. Laffont, ed.), Cambridge: Cambridge University Press. [254]

Van Damme, E., R. Selten, and E. Winter (1990), "Alternating Bid Bargaining with a Smallest Money Unit", *Games and Economic Behavior* **2**, 188–201. [131]

Varian, H. R. (1992), *Microeconomic Analysis* (Third Edition). New York: Norton. [275]

Vickrey, W. (1961), "Counterspeculation, Auctions, and Competitive Sealed Tenders", *Journal of Finance* **16**, 8–37. [30]

Vives, X. (1990), "Nash Equilibrium with Strategic Complementarities", *Journal of Mathematical Economics* **19**, 305–321. [65]

von Neumann, J. (1928), "Zur Theorie der Gesellschaftsspiele", *Mathematische Annalen* **100**, 295–320. (Translated as "On the Theory of Games of Strategy", pp. 13–42 in *Contributions to the Theory of Games*, Volume IV (Annals of Mathematics Studies, 40) (A. W. Tucker and R. D. Luce, eds.), Princeton University Press, Princeton, 1959.) [29, 30, 51]

von Neumann, J., and O. Morgenstern (1944), *Theory of Games and Economic Behavior*. New York: John Wiley and Sons. [5, 8, 9, 30, 114, 274, 279, 297]

Zamir, S. (1992), "Repeated Games of Incomplete Information: Zero-Sum", pp. 109–154 in *Handbook of Game Theory with Economic Applications*, Volume 1 (R. J. Aumann and S. Hart, eds.), Amsterdam: North-Holland. [161]

Zemel, E. (1989), "Small Talk and Cooperation: A Note on Bounded Rationality", *Journal of Economic Theory* **49**, 1–9. [175]

Zeuthen, F. (1930), *Problems of Monopoly and Economic Warfare*. London: George Routledge and Sons. [312]

Index

Page numbers in boldface indicate pages on which objects are defined; as in the text, the symbols ▶, ■, ◇, and ? indicate definitions, results, examples, and exercises respectively.